E-Agriculture and Rural Development:

Global Innovations and Future Prospects

Blessing M. Maumbe
West Virginia University, USA

Charalampos Z. Patrikakis
T.E.I. of Piraeus

Information Science
REFERENCE

Managing Director: Lindsay Johnston
Editorial Director: Joel Gamon
Book Production Manager: Jennifer Romanchak
Publishing Systems Analyst: Adrienne Freeland
Development Editor: Hannah Abelbeck
Assistant Acquisitions Editor: Kayla Wolfe
Typesetter: Christina Henning
Cover Design: Nick Newcomer

Published in the United States of America by
Information Science Reference (an imprint of IGI Global)
701 E. Chocolate Avenue
Hershey PA 17033
Tel: 717-533-8845
Fax: 717-533-8661
E-mail: cust@igi-global.com
Web site: http://www.igi-global.com

Library of Congress Cataloging-in-Publication Data

E-agriculture and rural development: global innovations and future prospects / Blessing M. Maumbe and Charalampos Z. Patrikakis, editors.
p. cm.
Includes bibliographical references and index.
Summary: "This book is a comprehensive collection of research on the emerging trends and advances in the global application of information and communication technology use in agriculture and rural development"--Provided by publisher.
ISBN 978-1-4666-2655-3 (hbk.) -- ISBN 978-1-4666-2868-7 (ebook) -- ISBN 978-1-4666-2717-8 (print & perpetual access) 1. Agricultural innovations--Developing countries. 2. Internet in agriculture--Developing countries. 3. Communication in agriculture--Developing countries. 4. Communication in rural development--Developing countries. I. Maumbe, Blessing M., 1966- II. Patrikakis, Charalampos Z.
S494.5.I5E23 2013
338.1'620285--dc23
2012029112

British Cataloguing in Publication Data
A Cataloguing in Publication record for this book is available from the British Library.

Editorial Advisory Board

Table of Contents

Detailed Table of Contents

despite some projects meeting their goals. This study examines how the environments in which such ICT-based MIS are deployed affect their performance. It specifically uses two ICT-based market information service projects, the DrumNet and Kenya Agricultural Commodity Exchange (KACE) projects, to assess how the socio-economic, physical, and institutional environments in such projects are deployed affect the performance of such projects. The study finds that a number of environmental factors related to socio-economic, physical, market, and legal environment affect the performance of ICT-based projects. Some of these factors exacerbate transactions costs thus undermining the performance and even sustainability of ICT-based MIS projects. It discusses policy implications of these findings.

Development partner efforts and private sector initiatives on ICT applications in agriculture have brought new opportunities for farmers and traders to reduce transaction costs and increase incomes. The applications are primarily used for linking actors in the agricultural value chain, accessing real time information on prices, buyers and sellers, transport and haulage, and other relevant information services in the agricultural value chain. Limited evidence from Ghana and elsewhere show that cell phone applications have resulted in increased incomes but the impacts and sustainability of other ICT applications have proven elusive. The role of ICT in overcoming the key constraints in the agricultural value chain and for making evidence-based decisions will be greatly enhanced if farmers, aggregators, and other stakeholders in the value chain pay attention to their business scope and schedule planning, executing, monitoring and control, procurement, risk planning, and stakeholder communications in a "project management" context. When this is done, ICT applications will facilitate supply chain management through sharing of timely and pertinent information on producers, buyers and other services, thereby helping to promote industry competitiveness. The major challenges to the widespread use and sustainability of ICT remain access to the appropriate ICT tool, poor road and storage infrastructure (particularly in the farming communities) and illiteracy on the part of the majority of smallholder farmers.

Geographic Information Systems (GIS) and Remote Sensing (RS) techniques are increasingly being used in agricultural management to more efficiently support emergency agro-climatic information needed in this century. Although several studies have been carried out on application of GIS and remote sensing for agricultural and rural development in other parts of the world, little research has been done in sub-Saharan Africa in terms of developing geospatial data infrastructure to enhance agricultural practices especially agricultural risk management in this age of climate change. This chapter therefore examines technological aspects of geographic information sharing as a useful platform for sharing information among agricultural agencies in Nigeria and to formulate relevant prescriptions for the future. The study shows how an integrated GIS modeling system can allow agricultural producer as well as policy makers to know the impact of variation in climate from one place/region to another for better management, productivity, and profitability. The study also develops agricultural GDI prototype for agricultural emergency management. This chapter concludes by suggesting that developing country (i.e. African countries) should embark on agricultural policy reform to enhance investment in ICT infrastructure in agricultural production.

A number of telecentres have been established in places like shops, schools, community centre, police stations and clinics. The population of Nigeria, according to the national population commission (NPC) figures stands at over 140, 000,000, and 60% of this number is made up of unemployed youths. With the institutions of learning in Nigeria churning out graduates of various levels and degrees on a yearly basis, a rising trend has seen these graduates coming out of the nation's universities and polytechnics to join those who graduated ahead of them but without any means of livelihood for years. This chapter examines the role played by unemployment on the making of the Nigerian Gross Domestic Product (GDP) for a period of nine years (2000 - 2008). The objectives of the study are to examine the effects of unemployment on the Nigerian GDP for the selected years, to observe the kind of association that existing between the unemployment and the makings of the Nigerian GDP. Data was collected and analyzed using the regression analysis. Findings showed that unemployment has an enormous effect (over 65%) on the making of the Nigerian GDP, and there exists an inverse relationship between the model (unemployment) and the GDP - increase in the model leads to decrease on the GDP and vice versa. The role of ICT on unemployment and GDP is reviewed. In addition ICT as a tool of combating unemployment corruption is discussed. Recommendations are proffered based on the study that unemployment can be combated through the public sector reforms and the use of ICT.

The transformational role of ICT in the social and economic spheres of man engendered development of the communication technology in Nigeria. Based on the telecom deregulation policy of the Nigerian Government in 2001, the country witnessed an upsurge of private investment in ICT development in the country and its consequential ICT revolution in Nigeria. An empirical survey of diffusion and adoption pattern of ICT in the Nigerian agricultural system showed that mobile phone constitutes the most widely adopted ICT device by the agricultural researchers, extension agents, and rural farmers. In addition to this was computer and internet adoption by the researchers and radio transistors by the rural farmers. All eight agricultural research organisations and four extension agencies surveyed had electronic mail addresses and telephone lines for information exchanges. Six of the agricultural organisations - CRIN, FRIN, NIHORT, IAR&T, NIOMR, and NAERLS, and one of the extension agencies – LSADA, had organisational websites for hoisting generated agricultural information. Based on rural farmers' access to radio, and to some extent television, the research and extension organisations aired agricultural programmes on these electronic media for the farmers. ICT usage in the Nigerian agricultural system is

services, thereby facilitating an environment for more remunerative agricultural practices to meet the new challenges. This chapter provides an in-depth analysis of the existing rural agrarian digital divide in Tamilnadu, India. The chapter describes the information needs, ICT access, and new opportunities for farmers in rural Tamilnadu. The chapter also examines the status of E-Agriculture in Tamilnadu and the ICT interventions that are available to Tamilnadu farmers. The study outlines the emerging opportunities for rural farmer' participation in the knowledge society and offers some guidance on how the farmers can take advantage of ICT and utilize the information management system to derive maximum benefit out of the technology.

The role of Information Communication Technology in enhancing food security is well established and has been recognized at worldwide forums like World Summit of Information Society (WSIS) 2004-2005. Information Technology includes use of computer, internet, geographical information system, and mobile phone, besides traditional media. Every sphere having to do with agriculture has scope for information technology to make significant contribution, but in a few areas, its impact can be instantaneous. The present era, also known as the information era, requires creation of information expressway interlinking farmers, markets, corporate, agricultural scientists, and buyers of agricultural products. This may help producers to realize maximum share from the price paid by end consumer. In the present agricultural marketing system, middlemen share a significant portion from price paid by end consumer, and what the farmer gets is residue. Thus, the two major areas where information technology intervention can create progress are creation of information expressway for farmers; and strengthening farmers' linkages to supply chain. These two aspects have been discussed with the help of initiatives like AGMARKNET, NDDB Procurement Process, e-Choupal, and EID's Parry's Corner. The effort is to facilitate learning with real life examples as contrasted against pure and abstract discussion based learning.

"E-National Pest Reporting and Alert System" in pulse crops is a unique ICT based decision support system, which is very effective and easy to operate through a centralized server system at National Centre for Integrated Pest Management, New Delhi, connected with internet and mobile phones. This system has developed to cater to the needs of rural farmers of India, who grow pulse crops. The useful information is collected, stored, processed, and interpreted, and the appropriate advisories are sent to the registered farmers through centralized server system. They apply suitable corrective measures as per advisories at right time, and thereby, heavy loss caused by various pests can be checked/minimized below economic threshold level. Based on the past experiences and larger response of the stakeholders, Department of Agriculture and Cooperation, Govt. of India, suggested that this program should be extended and implemented in all pulse growing states. This system is quiet useful bottom to top level officials/policy makers, involved in E-Pest Surveillance programme.

Preface

The use of information and communication technologies (ICT) in agriculture and rural development has surged in the past decade. Growing attention has focused on the contributions of these technologies to agriculture and rural development. Evidence linking ICT to an increase in Gross Domestic Product (GDP) has spurred a global effort to study the socio-economic benefits, challenges, and policy implications of ICT in agriculture in both developed and developing countries. ICT have transformed agriculture value chains--input procurement, production, processing, storage, distribution, and marketing to the final consumer--and thus global agriculture (e-Agriculture) and the rural development landscape. Yet, despite the proliferation of ICT, their application in agriculture and rural development, enhancing competitiveness of agriculture value chains, modernizing rural service delivery, raising living standards by alleviating poverty remains unclear in many countries. The fragmented and undocumented evidence of successful applications of ICT in agriculture highlights the need to understand the socio-economic benefits that could accrue to governments, rural agribusinesses, and society. This book brings together the results of years of hard work from various scientists from around the world and it makes valuable contribution to new knowledge and our understanding of e-Agriculture and rural development globally.

This publication describes key developments and emerging challenges in ICT use in agriculture and rural development throughout the world. The book covers case studies; conceptual and implementation frameworks; agriculture market information systems; value-added agriculture services, e-agriculture policies, institutions and regulations that facilitate the use of ICT in agriculture and rural development. Collectively, the topics in this book provide readers with a broad perspective and a deeper understanding of global trends in e-Agriculture and rural development. The book documents empirical evidence of leading ICT-led agricultural projects, new innovations, support institutions, and policies driving e-agriculture success. E-agriculture and rural development provides readers with a unique opportunity to learn about the global expansion of e-Agriculture and rural development by drawing upon case studies, theoretical frameworks, best practices, and potential solutions to ICT challenges facing agricultural communities in both developed and developing nations. The authors use context specific case studies, qualitative and quantitative analyses, illustrations, and tables to describe key findings and suggest policy recommendations. This book appeals to a wide range of readers interested in e-Agriculture and rural development, agribusiness management, agriculture value chain management, global agri-food marketing, agriculture management information systems and decision support, agribusiness innovations, e-government service delivery, and sustainable agriculture development.

The diffusion and adoption of ICT in rural areas improves the efficiency of agriculture production and marketing through timely decision making that result in better performance across the agriculture value chain. In the 1990s, the use of computers and Internet to enhance agriculture value chains was concentrated mostly in developed countries. However, the diffusion of modern ICT such as mobile phones has created new opportunities for developing countries to participate in lucrative global markets. Evidence linking ICT to economic growth and improved access to agricultural markets signals the numerous opportunities for future agriculture expansion and rural development. ICT is an enabler for information sharing and knowledge dissemination, and it has the potential to revitalize rural economies through enhanced access to global markets and resources. The availability of affordable ICT or low-cost access to information infrastructure is a necessary prerequisite for the successful utilization of ICT in rural areas where the majority poor people live. E-agriculture or ICT adoption in agriculture has prompted the development of modern agriculture value chains, and use of new value-added agricultural services. Further, harnessing ICT effectively creates new economic opportunities by improving agricultural productivity, market linkages, and rural employment. Thus, successful integration of ICT in agriculture promotes economic vitality of agricultural communities and is a key driving force for agribusiness competitiveness and sustainability.

Given that the global population is estimated at over 6.5 billion and expected to reach nearly 9 billion in 2050, food production will need to increase by 70 percent in order to feed these 9 billion people (FAO). The effective appropriation of ICT in rural areas is central to meeting the goal of rising projected food demand and alleviating the problem of undernourishment affecting an estimated 1.02 billion people worldwide. Given that mobile networks are now available to 90 percent of the world population and 80 percent of the population living in the rural areas, e-agriculture can be a tool for sustainable food production systems and global marketing strategies. As developed countries adopt the next generation wireless platforms (4G) while developing nations increase their proportion of mobile subscriptions from 53 percent to 73 percent at the end of 2010, e-agriculture will be instrumental in the delivery of sustainability solutions for farmers, agribusinesses, and agricultural communities. The fact that an estimated 80 percent of the population in developing countries is not online implies that commercial opportunities for expanding e-agriculture are yet to be fully exploited. This exploitation of e-agriculture benefits will be possible for as long as key investments are directed towards the development and maintenance of the necessary ICT infrastructure required to support agriculture and rural development.

E-agriculture is happening at a time of profound changes in agriculture. The widespread use of ICT in rural areas coincides with other revolutions in agriculture such as bio-fuels and biotechnology development. In addition, there are heightened global concerns about climate change as ecosystems around the world are being subjected to vagaries of climatic extremes than previously experienced. These include floods, drought, desertification, melting of ice from mountaintops, and rising sea levels, among other changes. As a result, raising the awareness of the role of ICT and its significant contributions in food security, poverty alleviation, and rural development is competing with demands for greener energy technologies, clean water, housing, health, and other basic needs. Although linkages between ICT use in agriculture and food security are improving, connections are still being made on its role in minimizing the effects of climate change. Researchers are investigating linkages between ICT and sustainability, especially its role in reducing carbon emissions. Similarly, the growing realization of the need for effective e-waste management is indicative of efforts to produce ICT that can not only facilitate food production and access to markets, but are environmentally sustainable. As vulnerabilities for agricultural communities increase, especially in developing countries, there is growing focus on ICT use to enhance agricultural productivity, access local and regional markets, and provide quality information services. ICT use in agriculture has gradually become a potent tool designed to improve efficiency of agricultural markets, climatic change management, food security, and rural poverty alleviation, especially in developing countries.

Global experiences in e-agriculture and rural development are wide and diverse. In this book, successful case studies are documented and key lessons drawn for other countries and government policy makers. The book covers three continents, [sub-Saharan] Africa, Europe, and Asia, where some of the innovative uses of ICT in agriculture and rural development have been experienced over the past decade. In sub-Saharan Africa, successful case studies of e-agriculture and rural development are drawn from Ghana, Ethiopia, Nigeria, Kenya, and South Africa. In Europe, e-agriculture and rural development cases are drawn from Greece and Ireland and North America in general. The cases that are reported from Asia are drawn from India, a country that has made huge strides in e-agriculture and rural development. These cases from Indian provide lessons for other emerging economies in the sub-region that are ready to take major leaps in investment in e-agriculture projects and programs. Collectively, the case studies identify patterns of ICT use in agriculture, policy development, social, economic, and technological challenges, and future prospects for an increasingly innovative global economy.

Three key issues emerge from the publication of this book. First, e-agriculture is still in its infancy, and relatively unknown to most people including some government policy makers and agribusiness experts. Second, e-agriculture is a growing global phenomenon and as such both industrialized and developing nations can derive clear opportunities by implementing e-agriculture to enhance the quality of life in rural areas and society in general. Although the transformations in e-agriculture and rural development are still unfolding, long-term prospects suggest the dawning of a new development paradigm that has the potential to redefine the way agriculture and agribusiness is conducted in the twenty-first century. The rest of the book is divided into four sections and seventeen different chapters. The first chapter provides an overall introduction to global e-agriculture and rural development. This is followed by chapters describing case studies or projects from countries in three regions, sub-Saharan Africa, Asia, and Europe and North America, respectively. The following section provides a summary of the chapters in this book.

Chapter 1 provides a general introduction to e-agriculture and rural development around the world. It describes the evolution of e-agriculture and the e-value creation attributed to the use of information and communication technologies (ICT) in agriculture supply chains and rural development. The chapter sets the scene for the next several chapters that document different unique experiences in ICT use in agriculture and rural development globally. This introductory chapter explores briefly the meaning of e-agriculture and it raises fundamental questions around a broader understanding of e-agriculture. Next it highlights global trends in ICT diffusion and examines e-value creation and key characteristics of e-agriculture development. The chapter identifies implementation challenges, emerging lessons, and potential future directions for e-agriculture. This chapter provides an aerial view of the global e-agriculture landscape and it emphasizes the growing significance of e-agriculture in rural development including implications for food security and poverty alleviation. The multiplier effects from e-agriculture projects underscore its strategic importance in socio-economic development, agricultural transformation, and the revitalization of the rural economy.

Chapter 2 describes how the environment in which ICT-based market information services (MIS) affect their performance based on two ICT-based market information service projects, the DrumNet and Kenya Agricultural Commodity Exchange (KACE) projects. The chapter indicates that the need to provide agricultural information to farmers has led to emergence of numerous electronic-based MIS projects in developing country. The experiences from past and present projects that use ICT to link farmers to markets reveal a mixture of success and failure. The chapter reports that a number of environmental factors related to socio-economic, physical, market, and legal environment affect the performance and sustainability of ICT-based projects.

Chapter 3 examines the use of ICT to overcome constraints in the agriculture value chain in Ghana. The chapter indicates that ICT applications in agriculture have brought new opportunities for farmers and traders to reduce transaction costs and increase incomes. The authors highlight that these applications are primarily used for linking actors in the agricultural value chain, accessing real time information on prices, buyers and sellers, transport and haulage, and other relevant information services in the agricultural value chain. Cell-phone applications have resulted in increased incomes, but the impact and sustainability of other ICT applications have been elusive. The chapter observes that ICT applications facilitate supply chain management through sharing of timely and pertinent information on producers, buyers, and other services, thereby helping to promote industry competitiveness. Key challenges to the widespread use and sustainability of ICT remain access to the appropriate ICT tool, poor road and storage infrastructure, and illiteracy among majority of smallholder farmers.

Chapter 4 deals with geospatial data infrastructure for sharing agro-climatic information to agriculture productivity in Nigeria. The chapter argues that geographic information system (GIS) and remote sensing (RS) are increasingly being used in agricultural management since the tools provide more efficiently support for emergency agro-climatic information needed in the 21st century. The authors bemoan the fact that several studies have been carried out on application of GIS and remote sensing for agricultural and rural development in other part of the world, but little research has been done in sub-Saharan Africa in terms of developing geospatial data infrastructure to enhance agricultural practices and especially agricultural risk management in this age of climate change. The chapter examines technological aspects of geographic information sharing as a useful platform for sharing information among agricultural agencies in Nigeria and to formulate useful prescriptions for the future. The study show how an integrated GIS modeling system can allow agricultural producer as well as policy makers to know the impact of variation in climate spastically from one place/region to other for better management, productivity and profitability. The study also develops agricultural GDI prototype for agricultural emergency management. This chapter concludes by suggesting that developing country (i.e. African countries) should embark on agricultural policy reform to enhance investment ICT infrastructure in agricultural production.

Chapter 5 deals with ICT and unemployment in Nigeria. The chapter considers ICT to be veritable tools to tackle the rising unemployment in Nigeria. The chapter argues that increase in mobile phones has led to job creation with telecentres being set up in places like shops, schools, community centre, police stations and clinics. Given that Nigeria has a population of about I40 million according and 60 percent of this number is made up of unemployed youths, resolving unemployment is critical. This chapter examines the effects of unemployment on the Nigerian GDP for the period 2000-2008. Results showed that unemployment has an enormous effect on the Nigerian GDP and there exist an inverse relationship between the model (unemployment) and the GDP. The chapter discusses the role of ICT on unemployment and GDP and proposes using ICT as a tool to combating unemployment corruption.

Chapter 6 presents a discussion on ICT diffusion and adoption in Nigeria's agricultural sector. After the telecom deregulation policy of the Nigerian Government in 2001, the country witnessed an upsurge of private investment in ICT development in the country and the consequential ICT revolution in Nigeria. Empirical survey on diffusion and adoption pattern of ICT in the Nigerian agricultural system showed that mobile phone constitutes the most widely adopted ICT device by the agricultural researchers, extension agents and rural farmers. In addition to this, evidence showed computer and Internet adoption by the researchers, and transistor radio by the rural farmers. All the eight agricultural research organizations and the four extension agencies surveyed had electronic mail addresses and telephone lines for information exchanges. Six of the agricultural organizations - CRIN, FRIN, NIHORT, IAR&T, NIOMR and NAERLS, and one of the extension agencies – LSADA, had organizational websites for hosting generated agricultural information. Radio and television were the frequently used media to air farmer programs by agriculture and extension system. Future challenges lie in exploration and widespread use of modern ICT in agricultural system, electricity supply, and human capital development for effective utilization.

Chapter 7 examines the impact of ICT on livestock production by rural farmers in Nigeria. Questionnaire, interview, and personal observation methods were employed to elicit information on the impact of ICT on livestock production on rural areas of Nigeria. The chapter reveals the importance of personal characteristics of the respondents in ICT use. The findings also reveal that rural farmers need to be encouraged by providing them with relevant information and communication technology gadgets in order to enhance effective access to information on veterinary and extension services to improve productivity. The chapter recommends the establishment of Internet facilities in rural communities as a priority for the State and Federal Governments in order to encourage digital literacy.

Chapter 8 deals with marketing efforts by the government of South Africa (GSA) to provide e-services to its citizens. The Batho Pele Gateway Project is the access portal to government information and a key driver of the e-government transformation in South Africa. Although the GSA is determined to make e-government service delivery successful, the average citizen does not fully comprehend the key transformations driving government engagement with civil society and businesses. Therefore, the Cape Gateway Project (CGP) is confronted with the immense task to develop and implement a strategic marketing program for e-government by the Provincial Government of the Western Cape (PGWC) which is not only a portal, but also a call centre and walk-in centre. Various marketing tactics have been deployed, but their efficacy in elevating awareness levels and converting that to actual sustained use of e-government service delivery remains unknown. The chapter proposes an e-government marketing framework and uses CGP experience to examine communication and branding strategies for e-government in a South African context. Key challenges facing e-government market communication strategy are highlighted.

Chapter 9 deals with information security and the risks poised to South Africa's bid to build a broad-based information society. The integration of information security in e-government is no longer an option, but an imperative given the resulting "information overload" and the need to filter "good" from "bad" information. Unless South Africa integrates information security in its e-government development policy and practices, the acclaimed benefits of e-government will not be realized. The moral hazard problems arising from bad information behavior such as human manipulation, withholding information, unauthorized access, and violation of individual privacy and confidentiality heightens the need to combat info-security risks and vulnerabilities. South Africa's readiness to deal with the information security risks has come under scrutiny in the past decade. The information security infrastructure in South Africa is also not clearly understood. The chapter highlights South Africa's information security landscape and describes how institutional and agency coordination could help improve information security in e-government.

Chapter 10 describes the development of Knowledge Management (KM) centers for extension advisory services and communication in rural areas in Ethiopia. The KM centers provide an opportunity for agricultural development organizations to achieve higher efficiency and increased output. With the objective of attaining opportunities in KM at district level, Woreda Knowledge Centers (WKC) were established and used in ten districts in Ethiopia from 2005 to 2010. WKC is a tele-center used to gather, share, classify, access, and use knowledge at district (woreda) level. The chapter explains WKC usefulness, establishment steps, challenges, and opportunities. In ten districts, of the 500 survey respondents, 79%

and 71% agree on an increase in knowledge delivery and availability respectively. Temporal comparison showed that WKC increased staff capacity to document, access, share, and use knowledge leading to improved extension communication. The chapter recommends that WKC must be established by ministry of agriculture and its' strategic partners to enhance KM at district level for agricultural development.

Chapter 11 presents a situation analysis of e-agriculture in India. The chapter provides an in-depth analysis of the existing rural agrarian digital divide in Tamilnadu, India. The UN International Telecommunication Union (ITU) estimates that one billion people worldwide – most of whom depend in some way on agriculture for their livelihoods – still lack connection to any kind to ICT. This chapter assesses the information needs and access to ICT tools by rural farmers. In addition, the chapter provides an overview of ICT-led opportunities for farmers in rural Tamilnadu, and the status of E-Agriculture in Tamilnadu. The chapter also describes the ICT interventions that are available for Tamilnadu farmers and highlights the emerging opportunities for rural farmers to take part in the knowledge society. The chapter offers some guidance on how the farmers can take advantage of ICT and utilize the information management system to derive maximum benefit out of the technology.

Chapter 12 discusses ICT and its potential to transform Indian Agriculture. The chapter indicates that the role of ICT in enhancing food security has been well established and has got recognized at worldwide forums such as the World Summit of Information Society [WSIS] 2004-2005. ICT tools include use of computer, Internet, geographical information system (GIS), mobile phone, and traditional media such as television and radio. The author argues that the information age requires ICT to create an expressway of information interlinking farmers, markets, corporate, agricultural scientists and experts, and buyers of their products. ICT has a responsibility to ensure that producers get maximum share from the price paid by end consumer to purchase the produce. Thus, the two major areas where ICT can create opportunities happen to be, first Creating Information Expressway for Farmers; and second Strengthening Farmers' Linkages to Supply Chain. These two aspects have been discussed with the help of initiatives taken by Government as well as Corporate Sector. The chapter discusses models such as AGMARKNET, NDDB Procurement Process, e-Choupal and EID's Parry's Corner. A sound knowledge of these ICT experiences is expected to deepen understanding of the role and importance of information technology.

Chapter 13 deals with the "E-National Pest Reporting and Alert System" in pulses crops in India. This is a unique ICT based decision support system, which is very effective and easy to operate through a centralized server system at National Centre for Integrated Pest Management, New Delhi. The decision support system is connected with Internet and mobile phones. This system was developed to cater for the needs of rural farmers of India who grow pulse crops. This decision support system collects, stores, processes, and interprets useful information which is then disseminated as appropriate advisory services for registered farmers through a centralized server system. They apply suitable corrective measures as per advisories at the right time and thereby heavy loss caused by various pests can be minimized or kept below economic threshold level. Based on the past experiences and larger response of the stakeholders, Department of Agriculture and Cooperation, Government of India, suggested that this program should be extended and implemented in all pulse growing states. This system is quiet useful for bottom to up level officials or policy makers that are involved in E-Pest Surveillance program.

Chapter 14 describes the role of ICT in agribusinesses in Ireland. The chapter focuses on the opportunities for extranet services to provide information to farmers. This chapter critiques the level of adoption of ICT among the farming community in Ireland and compares this with adoption by the general population. The chapter reports that computer access and Internet use among Irish households have been increasing rapidly in recent years, but farm households have been lagging behind in adoption of these technologies. Broadband access has been particularly problematic in rural areas, where farm households are located. The chapter also evaluates the development of a novel extranet service by a large agribusiness firm. This service allows dairy farmer clients to access all information relevant to their accounts with this agribusiness through a secure website. A structured survey of clients was carried out and respondents were stratified on the basis of their usage of this extranet service. The readers might find this chapter useful as it provides key insights on ICT adoption in a developed country. The chapter identifies three key factors that constrain adoption and use of ICT among Irish farmers: low levels of computer skills; lack of awareness of the potential of ICTs to contribute to the farm business, and thirdly, access to Broadband in rural areas. Resolving the fundamental problems that constrain broadband penetration and adoption of ICT will enhance opportunities for extranet services for rural farmers and agribusinesses.

Chapter 15 deals with e-learning in forest worker's training for sustainable forest management. This chapter is unique in that it addresses a sub-sector that is closely aligned with agriculture and whose successful uptake of ICT will generate positive spillover effects for sustainable agriculture production. Forest workers' training is recognized as a necessary precondition for sustainable forest management. As sustainable forest management sets higher levels of standards, the adoption of new ICT in forest workers' training and extension systems is a necessity in the Information Age, which will help the current and future workforce to better fit in the increasingly digitalized and demanding world of forest work. The chapter noted that limitations on the use of ICTs do exist and should be seriously taken into consideration. Digital divide is evident both in developed and developing countries, making the implementation of ICT applications difficult, especially in the case of African and Asian countries. Furthermore, the development of e-learning systems for forest professionals must adhere to very specific design standards and educational theories. The chapter suggests research that focuses on the needs of the end-users at regional or national level. The chapter emphasizes the significance of a stable framework of political and financial support; combination of long-term planning, clear target setting and cost effectiveness is critical for the successful promotion of sustainable forest management through e-learning.

Chapter 16 brings in the issue of soft computing modelling of wild fire risk indices. The chapter is based on risk profile of Peloponnesus region in Greece. Forest fire risk estimation is a major issue and the necessity for more efficient allocation of fire fighting resources becomes more and more urgent, especially in the areas located around the Mediterranean basin (Portugal, Spain, France, Italy, Greece). The chapter explains the need to reduce the risk of devastating forest fires that have multiple negative ecological and financial consequences. This preliminary research effort is based on the implementation of an intelligent rule based fuzzy inference system evaluating wild fire risk in the forest departments of Greece. The system uses Soft Computing techniques and it was built in the *Matlab* integrated environment. The research is based on wild fires in Greece during the period 1983-1997 with data coming from the general forest management service. The estimation of the risk indices was done by using fuzzy Triangular membership functions and Einstein fuzzy conjunction T-Norms. Moreover the system produces a profile of the forest departments located in the geographic area of "*Peloponnesus*"- a region located in the southern part of Greece that has a vast number of annual forest fire breakouts. Meteorological, topographic and historical (total burned area and intervention time) features were considered for the determination of the risk indices. The system has shown a good performance which can be improved further if more data is gathered and used. Its main advantage is that it offers an innovative and reliable model that can be employed in any part of the globe as a basis for natural disasters' risk estimation.

Computer applications in wild fire risk estimation have positive implications for agriculture given potential ecological damage that could arise from poor wild fire risk estimation. The chapter highlights the linkages between ICT applications in agriculture and other key sub-sectors.

Chapter 17 discusses the unprecedented growth in information and communication technologies (ICT) through the widespread use of personal computers, Internet, and mobile phones. The chapter examines trends in ICT use in agriculture, identifies key success factors for ICT utilization in agriculture, and investigates the implications of ICT-enabled value chains for the agribusiness industry. The chapter describes the strategic role of ICT in the development of both e-commerce and mobile commerce in agriculture in North America and globally. The chapter identifies the leading areas of ICT use in agriculture and agribusinesses as input procurement, production, marketing, food traceability, and financial service delivery. Producers are increasingly seeking ways to add value to their businesses by integrating ICT in the value chain. Similarly, consumers are becoming more knowledgeable about how they could use ICT to articulate their preferences. The chapter discusses key success factors for ICT applications affecting both the internal and external environment of agribusiness firms. The chapter concludes by drawing implications for ICT use in agriculture and agribusiness value chains.

This book is culmination of a two year-long collaborative effort between Dr. Charalampos, Z. Patrikakis, and me, and indeed our two institutions, the *Informatics Laboratory of the University of Athens, Greece and West Virginia University, USA*. In addition, some of the research works published were a result of collaborative efforts between Cape Peninsula University of Technology and Provincial Government of the West Cape, in Cape Town South Africa. This book marks the beginning of a new chapter on global research collaboration in field that is receiving growing attention from researchers, non-governmental organizations (NGOs), agribusiness experts, policy makers, and other practitioners. Although e-agriculture is not yet considered a field of study, a number of students in developing countries are actively working on various topics and projects that address challenges in this emerging field. We hope that several books, journal articles, and international conferences on e-agriculture will result in an effort to seek new knowledge and understanding of this field. I believe that the publication of this book is a clear demonstration that the effort of many scholars and researchers working in this field is paying dividends by providing new ideas and better understanding of e-agriculture and rural development. I hope that readers will find this global publication useful and exciting, and that it will inspire students and other researchers to do more work in this field in the future.

Blessing Mukabeta Maumbe
West Virginia University, USA

Section 1
Introduction

Chapter 1
Global e-Agriculture and Rural Development:
E-Value Creation, Implementation Challenges, and Future Directions

Blessing Mukabeta Maumbe
West Virginia University, USA

ABSTRACT

E-agriculture is a newly emerging field that is receiving increasing attention from governments, agribusiness industry, and the agriculture community worldwide. Over the past decade, the world has witnessed major growth and expansion of e-agriculture projects that integrate information and communication technologies in agriculture value chains and rural development. The rise of e-agriculture has led to increased focus on the need to fully understand its impact on rural development, food security, and poverty alleviation by policy makers, researchers, and farmers. Global institutions such as the Food and Agriculture Organization (FAO), World Summit for Information Society (WSIS), and the World Bank have identified the development of e-agriculture as a key priority. This chapter sets the platform for a key publication designed to advance the global debate on e-agriculture development and policy. The chapter outlines key insights into the conceptualization of e-agriculture, innovative project initiatives, e-value creation, lessons learnt, and emerging challenges associated with e-agriculture implementation in various parts of the world. The main lesson is that e-agriculture is a growing field that is pivotal to the global agriculture development agenda in the 21st century.

INTRODUCTION

The past decade has witnessed a rapid growth in the use of information and communication technologies (ICT) in agriculture around the world (World Bank, 2008; Maumbe, 2010). The growing use of ICT in agriculture or "e-agriculture," has begun to play a pivotal role in global agriculture and rural development. The global significance of e-agriculture which stands for "electronic" or "digital" agriculture is being felt through its diverse influences on agricultural input procure-

DOI: 10.4018/978-1-4666-2655-3.ch001

ment, production, processing, distribution, market exchanges, agribusiness development, food security, and rural poverty alleviation. E-agriculture development is driving major transformations in agriculture today. The benefits from e-agriculture are derived from high value agricultural chains, productivity gains, global market access, e-agriculture services, efficient decision making, and enhanced communication with key industry stakeholders (Okello, 2010; Lio, 2006).

Although e-agriculture is now widely spread in the world, its development is not homogenous. The uneven nature of global e-agriculture development landscape presents challenges and opportunities. One of the challenges of e-agriculture development is its unintended consequence to create new forms of "digital divide" that further isolate other developing countries. Nonetheless, vast opportunities lie in the untapped potential in the future growth and expansion of e-agriculture. Currently, there are approximately 5.3 billion mobile cellular subscriptions and that number is expected to grow as more countries upgrade to third generation (3G) and fourth generation (4G) ICT with even faster broadband speeds (ITU, 2011). Recent evidence show increasing use of ICT especially mobile phones in rural areas (Aker, 2008). E-agriculture is of particular importance in developing nations given that agriculture is a key economic sector that contributes the most to employment, exports, and gross domestic product (GDP), it also affects the livelihoods of billions of people.

Despite the euphoria surrounding the development and growth of e-agriculture globally, it is still not well understood. Many questions are still being asked by academic scholars, policy makers, agribusiness industry experts and researchers about e-agriculture. Some of these questions include: What exactly is e-agriculture? What are its benefits to farmers, agribusinesses, governments, and society at large? How does e-agriculture differ from traditional agriculture practices that have existed for centuries? What are the policy and institutional issues surrounding the development of e-agriculture globally? How does e-agriculture enhance the performance of production agriculture, rural small and medium-sized agribusinesses, and the global food system in general? These are legitimate questions, and while answers may not be readily available, addressing them is critical to building a knowledge base and understanding of what e-agriculture really entails for society and the future generations.

As global interest about e-agriculture increases, it has become imperative to answer these questions surrounding e-agriculture development. This publication is one of the first major global efforts to examine key projects, agribusiness strategies, and government policies affecting e-agriculture and rural development. This book is motivated by the need to provide a holistic assessment of e-agriculture through a collaborative effort to analyze the scope and extent of e-agriculture worldwide. The findings are expected to not only enhance our understanding of e-agriculture, but highlight key successes and practical challenges associated with its implementation. The unanswered questions about e-agriculture and the socio-economic challenges brought about by its implementation over the past decade warrant a systematic inquiry in order to expand our knowledge and understanding of e-agriculture. The knowledge gained will inform farmers, policy makers, academics, agribusinesses, and society about a relatively new form of agriculture that has emerged and is spreading rapidly around the world.

E-agriculture is redefining the way agriculture production is practiced, food and agriculture value chains managed, product safety and quality evaluated, and how the basic food and fiber requirements of consumers are met in the twenty-first century. It should be noted that e-agriculture continues to evolve and has not yet reached full maturity. This is partly true because e-agriculture is primarily driven by modern ICT which are constantly changing and have become more versatile and sophisticated. However, as e-agriculture is diffused across various regions of the world, it is essential to document the dynamic changes taking place in agriculture and rural development. The need to highlight the key successes, emerging challenges, and future prospects will put e-agriculture into perspective.

This book on e-agriculture provides readers with an opportunity to better understand the positive impact of ICT on agriculture, rural development and the revitalization of lagging rural economies throughout the world. The book's chapters capture the major developments and challenges confronting e-agriculture globally. With chapter contributions drawn from various countries around the world, the book describes systematically the unique e-agriculture experiences, key challenges, and future prospects based on case studies from Sub-Saharan Africa, Europe, and Asia. It provides an assessment of the e-agriculture landscape by making a compelling case of its importance and potential future directions based on the various case studies analyzed. Although the book does not include cases from all sub-regions in the respective continents, the findings indicate that despite numerous challenges, the diffusion and adoption of e-agriculture has transformed agricultural communities and rural development globally. The benefits from ICT use include improved agriculture production, enhanced access to global markets, reduced transaction costs, and facilitation of real-time communication and collaboration among supply chain actors in geographically dispersed regions. As more farmers and agribusinesses adopt e-agriculture, it is likely to become a cornerstone of global agriculture and rural development.

The rest of this chapter is organized as follows: The next section describes the meaning of and raises key questions about e-agriculture. This is followed by a brief overview of the evolution of e-agriculture and the key global ICT trends that underpin its development. The next section highlights e-value creation associated with e-agriculture. Next, e-agriculture characteristics, implementation challenges, and key lessons are highlighted leading to concluding remarks on the future directions and implications of e-agriculture.

What is E-Agriculture?

It is not surprising that not much literature is available on e-agriculture. The concept of e-agriculture is relatively new to many people although social scientists and other scholars have grappled with the meaning of e-agriculture and its policy implications. The concept has different meanings to different people because of the way e-agriculture is practiced around the world. The application of ICT to agriculture, or the use of ICT by agricultural communities, is not a standard phenomenon. The experiences with ICT use in agriculture differ simply because agriculture production systems vary from region to region, and also country to country including within country variations. In that regard, e-Agriculture is defined as *the application of modern ICT to agriculture input or ingredient procurement, production, storage, distribution, processing, and marketing with the goal of transforming the people's lives.* This definition assumes that ICT such as computers and mobile phones are deployed in various agriculture value chains in order to enhance efficiency, improve market coordination, lower information search costs, facilitate market exchanges, enhance transparency and fairness, and stimulate better performance along the entire supply chain. E-agriculture can also be defined as *the application of ICT to enhance the efficiency and effectiveness of agricultural service delivery.* This second definition assumes that ICT are deployed to enhance accuracy, timeliness, relevance, and value-addition in agricultural e-service delivery. The past decade has seen ICT being applied to farm management advisory services, agricultural tax management, agricultural extension, input and product marketing, and financial service delivery, among others, to benefit people living in rural areas who traditionally did not have access to such formal services. The third definition says e-Agriculture *refers to the adoption and diffusion of ICT to enhance agricultural and rural development.* Behind this definition is the assumption that the use of ICT is primarily driven by a desire to modernize and upgrade rural infrastructure, and uplift rural livelihoods. E-agriculture implementation is accompanied by the development of modern telecommunication infrastructure including telecenters, village Internet kiosks, e-hubs, and Internet cafes. Investments in movable assets such as mobile motorcycles with satellite-based Internet access, and brokers that facilitate transfer of rural remittances are an integral part of e-agriculture development. The multiplier effect of e-agriculture has resulted in the expansion of rural agribusiness employment opportunities and the revitalization of the rural economy worldwide.

Summary of the Key Questions About E-Agriculture Addressed in this Book

Some key questions that remain unanswered or that need further clarification are identified and classified as follows:

- What are the major factors driving the development of e-agriculture?
- What are the successful examples of e-agriculture implementation around the world?
- What are the rural agribusiness strategies that are driving e-agriculture deployment?
- What can we learn about e-agriculture adoption and diffusion from the past decade?
- What are the gender issues that are significantly affecting e-agriculture in rural areas?
- What are the key challenges of e-agriculture in various countries in the world?
- What are the institutions, policies, and procedures that are guiding the development of e-agriculture in various regions of the world?
- What are the future prospects of e-agriculture?

A serious attempt to answer these key questions helps consolidate our understanding of e-agriculture issues. As a caution, careful attention is required to balance the "old" and "new" ICT being used in e-agriculture. As universal access to ICT becomes a human rights issue, the ability to identify and analyze barriers and challenges that impede the use of modern ICT in agriculture and rural development becomes essential. In that regard, ICT use in agriculture include computers, Internet, geographic information systems (GIS), geographic positioning systems (GPS), lap-tops, mobile phones, personal digital assistants (PDAs) and traditional media such as radio and television. Farmers have also adopted social networking tools such as *Blogs, Facebook, LinkedIn,* and *Twitter* as strategy to communicate with key stakeholders such as suppliers, potential customers, and market their products globally.

EVOLUTION OF E-AGRICULTURE AND GLOBAL ICT TRENDS

Understanding the "driving forces" behind the quiet "e-agriculture revolution" is necessary for an accurate evaluation of the successes, failures, and future potential for e-agriculture. E-agriculture is a result of historical developments in information technology (IT). The IT trajectory can be traced from static websites of the dot com era, Internet-led e-commerce boom, interactive e-government online services, to the mobile technology revolution (See Figure 1). The recent socio-economic transformations from e-government to mobile government service delivery have led to an emergence of e-agriculture and mobile agriculture

Figure 1. E-agriculture development pathway, 2000-2010

respectively. As the use of ICT in the government and private sectors (i.e., agribusiness industry) intensified, e-agriculture emerged as an alternative for rural farmers and society to conduct agribusinesses market transactions and obtain government services. Figure 1 shows a trajectory of ICT applications in agribusiness and government sectors leading to the birth of e-agriculture. The initial IT of the 1990s comprised static websites and dial-up networks that were mostly operated by dot.com firms. After mixed results from the dot.com firms, in early 2000, e-commerce use among agribusinesses increased fueled by the adoption of high speed Internet and the use of more interactive websites. However, many e-commerce ventures experienced some difficulties in late 2000 (Moss et al. 2003).

Internet use started spreading to remote rural areas with the rise of mobile telephony at the turn of the century. As most governments embarked on the transformation of public service delivery with more rural services migrating online, e-agriculture emerged and is set to become the next major ICT wave after e-commerce use in agribusiness. Therefore, the period 2000 to 2010 is generally considered as the first decade of e-agriculture. In this decade, ICT use spread beyond formal businesses as more ordinary people including farmers, fishermen, rural traders, and civil society started using these modern tools in their business transactions and everyday lives.

According to the International Telecommunication Union (ITU, 2011), mobile cellular telephone subscriptions are growing at an exponential rate. As shown in Figure 2, it is estimated that the number of mobile cellular subscriptions per 100 inhabitants more than doubled from about 30 in 2005 to 70 in 2010. The number of mobile cellular subscriptions per 100 inhabitants in 2010 is estimated to be seven-fold that of fixed broadband subscriptions per 100 inhabitants in the same year. The number of fixed telephone lines per 100 inhabitants has been declining since 2005, as a result of the increase of mobile telephone use. The number of global Internet users is growing at more than twice the rate of mobile broadband users although the decline in the use of fixed telephone lines suggests a substitution of mobile broadband for fixed telephony. The rising trend in mobile ICT uptake is reflected in its growing use in rural agricultural market integration thus raising its significance in e-agriculture development.

E-Agriculture Services and E-Value Creation

The use of diverse media channels such as computers, mobile telephones, and the Internet has led to numerous potential applications and value creation opportunities in agriculture and rural development. In the context of e-agriculture, ICT value creation is derived from both monetary (i.e., cost savings and revenue growth) and non-monetary benefits. Cetin et al., (2004), differentiated customer impact level of ICT into informational, transactional and relational. Understanding e-value creation in e-agriculture requires a sound knowledge of its scope and extent of its potential impact. Value creation in electronically-enabled digital world arises from four main sources- efficiencies, complementarities, lock-in, and novelty (Amit and Zott, 2001). Bryceson, 2006 argues that efficiencies arise

*Figure 2. Global ICT developments, 2000-2010**

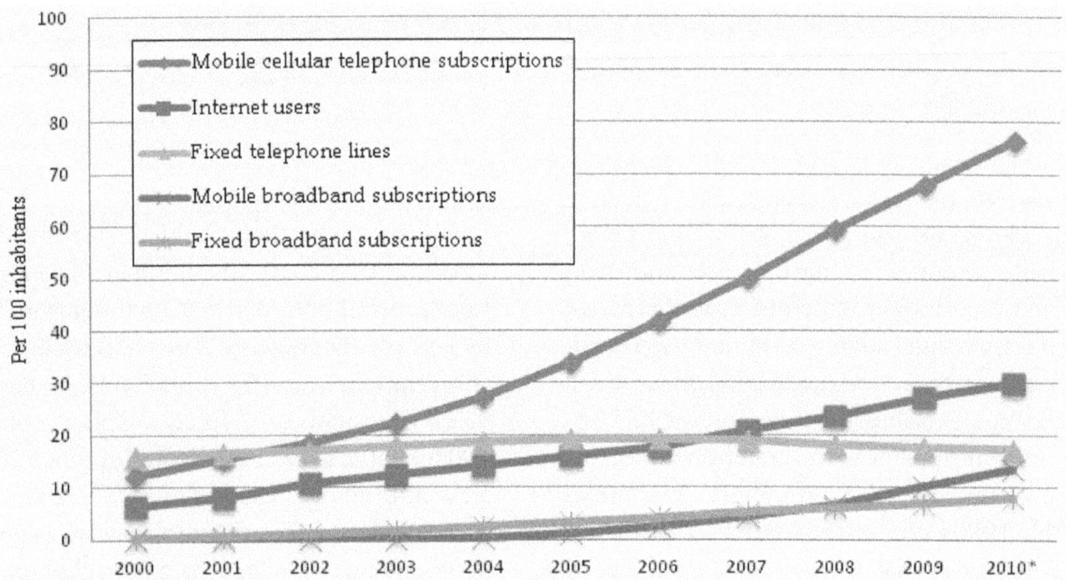

*Estimates
Source: Adpated from ITU World Telecommunication, 2011

from a reduction in information asymmetries in market exchanges, complementarities arise from product bundling, lock-in is attributed to the ability to generate repeat transactions while novelty deals with limitless possibilities or opportunities to innovate transactions. Maumbe, et al., (2007), identify internal, external and technical prime movers of e-value creation. The authors argue that e-value creation depends on the extent to which a government portal addresses specific user requirements that underpin public confidence. The potential impact of e-agriculture value creation can be measured at the individual, household, national, regional and global level. E-agriculture value-creation can be divided into four basic areas of ICT impact assessment: e-development in the rural sector, agriculture value chains, e-services in agriculture, and e-agriculture management information systems (See Figure 3). Each of these four dimensions of ICT application can be viewed as a major source of e-value creation for the global agricultural community. First, value creation in e-agriculture development requires *investments in key rural infrastructure* (e.g., tele-centers, e-kiosks, e-hubs, mobile phone towers), and *ICT policies and regulations* that create new opportunities for raising rural standards of living. The second dimension of e-value creation is related to *upgrading agriculture value chains*, and it arises from the deployment of ICT across the different stages of the food and agribusiness value chain.

The third dimension of e-value creation arises from *innovative e-service activities* associated with extension, financial management, risk management, and marketing (e.g., e-traceability in the global food supply chain). The fourth dimension of e-value-creation deals with functions related to the *development of e-databases and e-agriculture agro-portals (*one-stop shops for agricultural market information for farmers and other rural clients). A summary of the potential areas of ICT value-addition in agriculture, the e-value creation sources, and the expected benefits from e-agriculture is provided in Table 1. E-value creation sources include input procurement, agriculture financial services, farm production

Figure 3. E-agriculture development services and e-value creation framework, 2011

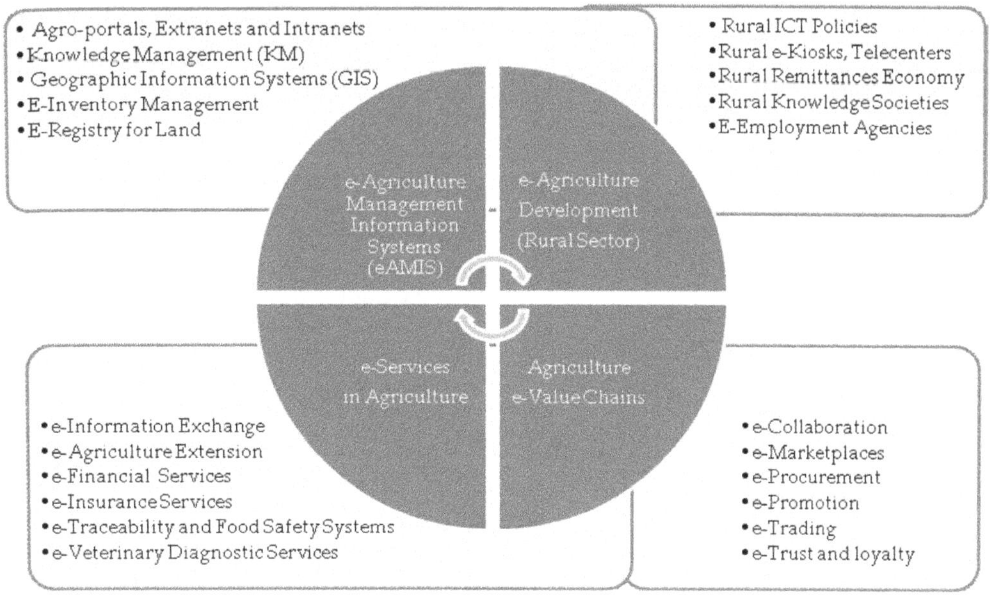

management, food marketing and retailing, supply chain management (SCM), risk management and rural development. The benefits derived from value creation in e-agriculture are linked to better performances in inventory management, efficiencies in crop protection, access to agricultural credit, flexibility in decision making, faster order completion, customization, information symmetry, price discovery, online marketing, and increased rural agribusiness employment. E-value creation from e-agriculture can also be obtained from: (1) better execution of, and cooperation in, agriculture contracts, (2) social bonding and social capital formation, (3) reduction in transaction costs, (4) trust-based importer and exporter relationships, (5) joint decision making, (6) poverty alleviation, (7) rural disaster and emergency response, (8) minimization of opportunism, market fairness and transparency, (9) increased commitment, trust and loyalty, and (10) gains in sustainable competitive advantage. Clearly, numerous direct and indirect benefits can be obtained from e-agriculture if properly deployed. More specific e-agriculture value creation is associated with the convenience from e-financial services such as electronic fund transfers (EFTs) or rural remittances to agricultural communities, e-trading on commodity future exchanges, and the use of electronic payment systems such as pay-pal to support agriculture-based global market transactions. Farmers and agribusinesses are shifting toward more convenient ICT-based services such as e-extension, e-health, and e-insurance.

In terms of electronic exchanges and strategic relationships, e-agriculture is manifested through government to consumer (G2C), government to business (G2B), and government to government (G2G) transactions. E-agriculture development also encompasses business to business (B2B) and business to consumer (B2C) transactions conducted by rural agribusinesses. Unlike the initial adoption of e-commerce that was driven by a few large agribusinesses, experiences from various parts of the world indicate that e-agriculture has

Table 1. Trends in e-value creation linked to e-agriculture and rural development, 2011

Applications in Agriculture	E-value Creation Sources/Processes	e-Agriculture Benefits and Potential Development Impacts
Input Procurement	E-procurement E-signatures and certifications	Contract management efficiencies, e-quotations, etc. Reduction of moral hazard, contractual power, fairness, etc.
Production Management	E-record keeping E-irrigation management GIS and precision farming GPS and farm mapping	Efficient inventory management, farm budgeting, etc. Water use efficiency, energy conservation, etc. Variable rate input applications-fertilizer, seed, etc. Location-based services and spatial information
Extension Services	E-advisory services E-consultation E-training- business webinars	E-learning, e-newsletters, knowledge development, etc. Long distance veterinary diagnosis, knowledge transfer, etc. Virtual meetings and joint decision making
Financial Services	E-banking in agriculture Mobile banking in agriculture E-tax management Electronic fund transfers (EFT)	Access to financial markets, e-billing, e-payment, etc. Financial security, mobile money solutions, etc. Credit rating, e-filing, reduction in tax evasion, etc. Rural remittances- poverty alleviation, education, etc.
Supply Chain Management (Distribution)	E-logistic management E-fleet management E-scheduling, RFID, EDI, JIT E-packaging	E-tracking and traceability- food safety and recalls Fuel use efficiencies, reduction of GHGs, etc. Product delivery efficiency –timeliness, convenience, etc. Waste management and recycling
Agricultural Marketing (Retailing)	E-customer relations services E-retailing, EDI, JIT, etc. E-trading, e-catalogues, etc. E-auctions- tea, livestock, etc. Agro-portals, e-databases, etc.	Customer relationship management image E-trust and e-loyalty, multifaceted strategic partnerships, etc. Customization and personalization Symmetric information One-stop shop services- low search costs
Risk Management	Weather monitoring Wildfire hazard monitoring Animal disease surveillance Crop pest management	Early warning systems-tsunami, tornadoes, etc. Emergence and disaster response Animal identification preservation systems-branding Crop protection- yield increase
Rural Development	E-collaboration E-development projects E-policies and e-strategies E-legislation and e-petitions	Rural communities of practice- project forums E-kiosks, e-hubs, price information, planting techniques, etc E-agriculture job creation, multiplier effects, etc. E-democracy in rural institutions, governance, etc.

Source: Adapted from Maumbe, B.M. 2010.

Notes: *EDI- Electronic data interchange, GHGs- Greenhouse gases, JIT-Just-in-time, RFID-Radio frequency identification*

been embraced by both large and small agribusinesses and society in general. E-agriculture is both a transformative rural development process and a strategy to facilitate interaction between government agencies, agribusiness firms, and final consumers. Further, the rapid growth in mobile technology especially in developing countries has given rise to "mobile agriculture," an integral part of "mobile government" development. Mobile agriculture simply refers to the application of mobile technology in agriculture. Similarly, "mobile government" refers to the application of mobile technology in government service delivery. The emergence of mobile agriculture is linked to the exponential growth in mobile telephony over the past five years and continues to gain considerable momentum. Mobile agriculture is characterized by mobile government to consumer (mG2C), mobile government to business (mG2B), mobile government to government (mG2G), mobile business to consumer (mB2C), and mobile business to business (mB2B) transactions.

Consequently, mobile commerce use in agribusiness industry is increasing. The development of "mobile agriculture value chains" is a result of a convergence of high speed mobile broadband, interactive websites, mobile Internet, blue-tooth, and use of 3G and 4G mobile technologies. This convergence of mobile ICT underpins the integration of social networking channels into food and agricultural marketing, customer relationship management, and other forms of global supply chain collaboration. The ICT transformations over the past decade demonstrate that both e-agriculture and mobile agriculture are products of this technological revolution. Given that common ICT and social media tools are behind the development and growth of e-agriculture and mobile agriculture, it is rather difficult to separate one from the other. Therefore, it is argued that mobile agriculture is subset of e-agriculture development. Clearly, the convergence of modern ICT has brought about new advances in agriculture information sharing, knowledge development, scientific discoveries, and value-added e-services to promote sustainable agriculture and rural development.

E-agriculture is at the center of agribusiness sustainability which focuses on markets, prices, technologies, consumer preferences, institutions, governance, policies, regulations, natural environment and industrial organization (Conforte, 2011). ICT are key enablers of sustainable development that influence markets, prices, and institutional governance through real-time global exchange of data, information and knowledge resulting in increased agricultural competitiveness (Rao, 2007). Modern ICT facilitate value chain upgrading, eliminate marketing bottlenecks and increase farmer bargaining position (Trienekens, 2011). The availability of ICT in rural area has neutralized the problem of deficient infrastructure that has impeded agriculture development for decades and exacerbated information paucity and thin markets. The use of modern, affordable, and "distance neutral" ICT has enabled rural farmers to overcome market barriers and access lucrative global markets. The ability to access virtual information sources in real time allows rural farmers and agribusinesses to participate in emerging national and international markets. Advances in ICT facilitate coordination of agriculture value chains by small producers despite the presence of asymmetric power relations attributed to increasing bargaining power of large retailers and industries (Trienekens, 2011).

KEY CHARACTERISTICS OF E-AGRICULTURE DEVELOPMENT

As the evidence from e-agriculture development over the past decade has been analyzed, a number of characteristics of e-agriculture have emerged. These characteristics are not uniform, but instead they vary from place to place depending on the prevailing socio-economic context. Essentially, these characteristics demonstrate the uniqueness of e-agriculture as a new development paradigm in agriculture and its potential impact on rural development. These key characteristics of e-agriculture are highlighted and briefly explained below.

1. E-agriculture has transformed the way the governments, agribusinesses, farmers and society *collect, analyze, and share agriculture information* for their daily decision making purposes. With new ICT use at their disposal, agricultural communities have joined the rest of the world on the path to become information or knowledge-based societies.
2. E-agriculture development has unleashed a *race for investment in ICT infrastructure and human capital* in rural areas including telecenters, village information kiosks, multifunctional agro-portals and information experts. A new breed of ICT entrepreneurs has emerged in geographically remote rural areas and rural small towns creating intellectual capital and new innovations that spur the development and growth of e-agriculture.

3. E-agriculture has led to *greater efficiencies in the operation of rural markets* by lowering transaction costs, eliminating information asymmetries, improving market coordination, and making thinly traded rural markets more transparent and profitable.
4. E-agriculture has resulted in the development of *trust-based relationships* between farmers, suppliers, and customers. Trust-based relationships have mushroomed between importers and exporters of food and agricultural products around the globe.
5. E-agriculture has the potential to *worsen the digital divide* by excluding poor communities who may not afford the ICT tools that are required to participate actively in emerging e-marketplaces and related value-added e-service delivery channels.
6. E-agriculture has facilitated the development and growth of *new global networks for agricultural information sharing and knowledge societies*. Many organizations and agencies are interested in the dissemination of quality information and the spread of e-agriculture innovations including FAO's e-agriculture.org, IDRC, Technical Center for Agricultural and Rural Development (CTA), CIRAD, IST-Africa, WSIS, e-India, and many others.
7. E-agriculture has created *new approaches for responding to sustainability and reduction of greenhouse gas (GHG) emissions.* For instance, mobile phones are increasingly viewed as positive agents in the reducing GHG emissions and the risks of climate change.
8. E-agriculture has transformed the *delivery of health services to rural farmers* as mobile phones and computers, Voice over Internet Protocol (VoiP) or Skype, and other ICT channels are readily deployed in global communications and long distance diagnosis of farmer's health status through the establishment of rural e-health programs.
9. E-agriculture has led to the development of *value added services such e-extension for farm management consultations, electronic fund transfers (EFTs), urban to rural remittances, mobile agriculture banking* for rural farmers to help with input purchases, school fees. Weather monitoring has been used to aid farm planning and scheduling of daily labor activities.
10. E-agriculture has led to *e-value creation along the agribusiness value chain* with the transformations affecting all the 4 Ps of marketing -pricing, promotion, place, and product. Farmers are more knowledgeable about input and product prices prevailing in local and regional markets, use GIS to locate markets, and are able to advertise the quantity and quality of their products using computer websites, short message services (SMS), and social networking tools such as Facebook, Twitter, and LinkedIn.
11. E-agriculture is *preceded by liberalization of the telecommunication sector or a reduction of monopoly participation by the state* in most developing countries enabling the entry of new players that have increased competition in the provision of broad-band, Internet, and mobile services.
12. E-agriculture has created opportunities for *ICT-driven solutions* to numerous agricultural problems in rural areas including inadequate infrastructure, lack of access to market information, poor record keeping and management, inaccessible roads, and poor living standards.
13. E-agriculture has led to the *development of ICT institutions and policies* in several countries. A number of countries that are serious about adopting e-agriculture have established dedicated government Ministries of ICT and others have crafted long-term strategies in ICT adoption and implementation in the rural sector.

14. E-agriculture is associated with *privacy, confidentiality and security* challenges. Given that the governance of the Internet remains contentious, coupled with the passing of information rights laws, many in society are concerned about the risks that are linked to ICT use when transacting online or sharing personal information with third parties.
15. The global nature of e-agriculture has led to the development of *virtual communities of farmers, customers, and agribusiness coalitions* thereby expanding opportunities for *global strategic collaboration and gains in competitive advantages through access to global markets.*
16. E-agriculture is associated with concerns from some poor rural communities that their *indigenous knowledge is inadequately integrated in the provision of* e-agriculture services.
17. E-agriculture is characterized by *vociferous antagonistic groups* of "traditionalists" who are stuck in the old ways and "resistant to changes" that are brought about by e-agriculture development around the world. Others are driven by the belief that *basic needs such as food, water, shelter, education and health should be given priority* instead of ICT investments.
18. E-agriculture has triggered a debate on the need for *best practices* in its provision around the world. As global markets become more sophisticated through traceability systems, harmonization of certification and standardization, global best practices and customization are increasingly being sought in e-agriculture and rural development.
19. As the world moves towards the "innovation economy," e-agriculture is being looked upon as both a *"driver" and a "new frontier" in the development of new product and service innovations.*
20. With an increasing use of mobile phones and PDAs in the provision of digital services, e-agriculture is positioned as the *"backbone" to the "mobile agriculture revolution."*

Without doubt, e-agriculture has affected not only rural agricultural communities, but policy makers, agribusiness industry experts, academics, and other stakeholders. E-agriculture is a transformative and is affected by other revolutions as well. This implies that e-agriculture is intractably linked to other on-going transformations such as mobile technology, biotechnology, and the biofuel revolutions. The rapid changes in modern ICT, the growth in knowledge societies, challenges posed by climate change, sustainability, and fierce competition in agribusiness suggests that agriculture is at the intersection of forces of change that will require e-agriculture to respond to competing global demands and overcome the emerging challenges.

Implementation Challenges of E-Agriculture and Rural Development

E-agriculture is unfolding at a time when global agribusinesses are facing numerous challenges from various fundamental forces (Olson and Boehlje, 2010). Further, the risk of technological change depends on the size and role of the technology, the speed of technical change and the socio-economic context (Maumbe et al. 2008). Similarly, the challenges facing e-agriculture implementation vary based on the pace and pathway of its development, socio-economic context, national ICT policies, and the structure of agriculture in the respective countries. The major global challenges associated with e-agriculture implementation are summarized as follows:

- Poor e-agriculture infrastructure development and smallholder ICT illiteracy.

- Understanding the unique ICT needs and socio-economic context of rural communities.
- Accessibility and inclusivity problems due to the diffusion of inappropriate ICT.
- Marginalization of rural women with respect to ICT use in agriculture.
- Privacy, confidentiality, and e-security concerns in agriculture financial services.
- Weak national ICT policy development and strategies for e-agriculture.
- Poor e-agriculture content development and integration of indigenous knowledge.
- Fragmented institutions and legislative framework for e-agriculture.
- Low levels of e-agriculture best practices, customization, and personalization.
- Global e-agriculture value chains not well coordinated or integrated.
- Underutilization of e-agriculture and lack of awareness of existing e-agriculture services.
- E-agriculture is a dynamic field with a learning curve.
- Lack of research investment in e-agriculture in developing countries.
- Capacity building initiatives for integration of ICT into agriculture and the environment.
- Dealing with spill-over effects into e-forestry, e-wildlife, and e-fisheries management.
- Decline with public expenditure on agriculture in developing countries.

Successful implementation of e-agriculture requires among others, a reliable telecommunication infrastructure, diffusion and adoption of appropriate ICT, relevant content, understanding the potential of ICT contribution to agribusinesses, improving the ICT skills of smallholder farmers, and a concerted global effort to address these challenges. Knowledge is becoming an increasingly significant factor of agriculture production. A realization of the strategic importance of ICT in facilitating knowledge transfer and dissemination is essential to restore the vitality of the rural economy via e-agriculture. The global nature of e-agriculture raises additional issues of the quality of e-content, standard operating practices, and harmonization of processes and procedures. The distribution of economic benefits across large and small rural farmers/agribusinesses will have to be resolved in order to minimize problems of inequality or digital divide. As the availability of more sophisticated ICT increases, fueled by the demand for customization, unintended problems of affordability could arise for the majority who are economically disenfranchised, but a key constituent expected to reap tangible benefit from e-agriculture development.

KEY LESSONS FROM GLOBAL E-AGRICULTURE DEVELOPMENT

The development and growth of e-agriculture helps us draw some key lessons and insights. The lessons are shaped by the global aspect of e-agriculture, positive contribution to socio-economic development, improvements in the performance of food and agricultural markets, transformations in agricultural information management systems, and development of knowledge societies in rural areas. These lessons and insights are covered in more detail in the next chapters and can be broadly summarized as follows:

- Growing contribution to global agriculture and rural development.
- Emerging new agriculture and economic development frontier.
- Alternative tool for socio-economic development, food security, and poverty alleviation.
- Global food and agriculture e-marketplace and information exchange.
- Key driver for new agricultural innovations and agribusiness competitiveness.
- Multi-segment food and agriculture value chain coordination and collaboration.

- Tools for advanced agriculture information management and knowledge development.
- Transformation of rural and agricultural communities into knowledge societies.

Considerable attention has focused on the use of ICT by small farmers, fishermen and other actors in the agriculture value chain (Jensen, 2010). As e-agriculture diffusion expands in scope and scale across the various regions of the world, its impact will include uplifting rural livelihoods, promoting food security, alleviating hunger and poverty, and stimulating agriculture knowledge development. The degree and extent of such impacts should be quantified through empirical impact assessment studies; and it will vary based on the diffusion rates and adoption levels of e-agriculture in various rural communities. To better understand the current and potential impacts of e-agriculture, it is necessary to conduct both cross-sectional and longitudinal impact studies based on e-agriculture projects. As the record of successful execution of e-agriculture strategies and projects increases, and our knowledge of the benefits and risks associated with e-agriculture implementation deepens, its overall contribution will be more accurately assessed especially given that it is still evolving in many parts of the world.

CONCLUSION

The Future of E-Agriculture and Rural Development

The rapid growth in the use of ICT over the past decade has produced a quiet revolution in agriculture. With Internet use projected to increase, and the number of mobile subscribers expected to reach 6.6 billion by 2016, e-agriculture is likely to take center stage in the 21 century. From what we now know so far about global e-agriculture, we can make some cautious predictions about its likely future contribution. Although there are no guarantees that what is highlighted in this section will be influenced by e-agriculture alone, it is fair to state that e-agriculture is expected to play a pivotal role in the modernization of agriculture, agribusiness innovation and competitiveness, agriculture knowledge development, agriculture sustainability, rural e-service delivery, and e-agriculture-based rural employment. Changes are expected in the agriculture curriculum and new job opportunities will arise as e-agriculture experts are sought in government, academics, and agribusiness industry. The following is a generalized summary of future areas where e-agriculture is likely to play a critical role in the successful modernization of agriculture and rural development:

- Agriculture production and marketing transformations in the innovation economy.
- Upgrading global agriculture value chains and stimulating global commerce.
- Global linkages, reducing climate change, and increasing agriculture sustainability.
- Development of centers of excellence in e-agriculture and global information networks.
- Development of e-extension and e-health services for rural farmers.
- Crafting a multidisciplinary e-agriculture curriculum.
- Development and implementation of ICT policy reforms, regulations, and best practices.

Experiences documented in this book indicate that e-agriculture is driven by innovations from the agribusiness industry, government sector, and farming community. The three pillars for e-agriculture are 1) e-government service delivery, 2) convergence of e-commerce and mobile commerce in agribusiness value chains, and 3) development of global information society still undergoing metamorphosis into a knowledge society characterized by discerning consumers, changing preferences, and a heavy reliance on ICT in daily decision making. Rural agribusinesses have responded to changes of how modern consumers in the so-called "knowledge societies" acquire and use information, purchase products and related services, and interact with one another. With the number of Internet users doubling between 2005 and 2010 and some nations even declaring access to the Internet a legal right (ITU, 2011), the future of e-agriculture will depend on access to reliable Internet and broadband services.

For as long as e-agriculture solves input procurement, production, storage, processing, distribution, and marketing problems without unnecessarily worsening the digital divide, its development is likely to increase with positive implications on agriculture productivity, agribusiness competitiveness, and the quality of life in rural areas. It is unlikely that e-agriculture will become the dominant form of agriculture anytime soon because of (i) underlying resistance from those who may be unwilling to change and (ii) traditional agriculture practices embedded in social and cultural norms. However, what is likely to happen in the short to medium term is growing awareness among the global population of the tangible and intangible benefits or value-addition from e-agriculture investments and projects. Based on the evidence in this book, it is predicted that in the short to medium term, farmers, agribusinesses, and individual consumers will become more specialized in, and knowledgeable about, e-agriculture to the extent that its application in agriculture and rural development will become more widespread than is currently the case. As e-agriculture reaches a critical mass of development, it will hold the future to increasing global agriculture productivity, agribusiness competitiveness, and rural development. Finally, rural small businesses stand to benefit from "first mover advantages" in e-agriculture and could gain sustainable competitive advantages from early adoption of e-agriculture.

REFERENCES

Aker, J. C. (2008). *Does digital provide or divide? The impact of cell phones in grain markets in Niger.* Center for Global Development, Working Paper 154.

Amit, R., & Zott, C. (2001). Value-creation in e-business. *Strategic Management Journal, 22*, 493–520. doi:10.1002/smj.187

Bryceson, K. P. (2006). *E-issues in agribusiness: The what, why and how*. USA: University of Queensland, CAB International. doi:10.1079/9781845930714.0000

Cetin, B., Akmar, A., & Ozsaym, D. (2004). The uses of information and communication technologies as critical success factor for marketing Turkish agri-food companies. *Food Reviews International, 20*(3), 221–228. doi:10.1081/FRI-200029420

Conforte, D. (2011). The nature of agribusiness management research. *International Food and Agribusiness Management Review, 14*(2), 143–147.

International Telecommunication Union. (2011). *The world in 2010 ICT facts and figures: The rise of 3G.*

Jensen, R. T. (2010). Information, efficiency, and welfare in agricultural markets. *Agricultural Economics, 41*, 203–216. doi:10.1111/j.1574-0862.2010.00501.x

Lio, M., & Liu, M. (2006). ICT and agricultural productivity: Evidence from cross-country data. *Agricultural Economics*, *34*, 221–228. doi:10.1111/j.1574-0864.2006.00120.x

Maumbe, B. M. (Ed.). (2010). *E-agriculture and e-government for global policy development: Implications and future directions*. Hershey, PA: IGI Global.

Maumbe, B. M., Owei, V., & Alexander, H. (2008). Questioning the pace and pathway for e-government development in Africa: Case of South Africa's Cape Gateway Project. *Government Information Quarterly*, *25*, 757–777. doi:10.1016/j.giq.2007.08.007

Maumbe, B. M., Taylor, W. J., Erwin, G., & Wesso, H. (2007). E-value creation in a government web portal in South Africa. In Tutnall, A. (Ed.), *Encyclopedia of portal technologies and applications*. Melbourne, Australia: Victoria University. doi:10.4018/978-1-59140-989-2.ch065

Moss, C. B., Schmitz, T. G., Kagan, A., & Schmitz, A. (2003). Institutional economics and the emergence of e-commerce in agribusiness. *Journal of Agribusiness*, *21*(1), 83–101.

Okello, J. J., Ofwona-Adera, E., Mbatia, O. L. E., & Okello, R. M. (2010). Using ICT to integrate smallholder farmers into agriculture value chain: The case of DrumNet Project in Kenya. *International Journal of ICT Research and Development in Africa*, *19*(1), 23–37. doi:10.4018/jictrda.2010010102

Olson, K., & Boehlje, M. (2010). Theme overview: Fundamental forces affecting agribusiness industries. *Choices (New York, N.Y.)*, *25*(4).

Rao, N. H. (2007). A framework for implementing information and communication technologies in agricultural development in India. *Technological Forecasting and Social Change*, *74*, 491–518. doi:10.1016/j.techfore.2006.02.002

Trienekens, J. H. (2011). Agriculture value chains in developing countries A framework for analysis. *International Food and Agribusiness Management Review*, *14*(2), 51–82.

World Bank. (2008). *Innovative uses of information technology to link farmers to markets in India and West Africa* (p. 121). World Development Report, World Bank.

Section 2
E-Agriculture and Rural Development in Sub-Saharan Africa

Chapter 2
ICT-Based Market Information Services (MIS) Projects, Deployment Environment, and Performance:
Experiences from KACE and DrumNet Projects in Kenya

Julius J. Okello
University of Nairobi, Kenya

ABSTRACT

The need to provide agricultural information to farmers has led to emergence of numerous electronic-based MIS projects in developing countries. These projects aim at promoting farmer linkage to better markets. However, experiences from past and present projects show mix cases of success and failure, despite some projects meeting their goals. This study examines how the environments in which such ICT-based MIS are deployed affect their performance. It specifically uses two ICT-based market information service projects, the DrumNet and Kenya Agricultural Commodity Exchange (KACE) projects, to assess how the socio-economic, physical, and institutional environments in such projects are deployed affect the performance of such projects. The study finds that a number of environmental factors related to socio-economic, physical, market, and legal environment affect the performance of ICT-based projects. Some of these factors exacerbate transactions costs thus undermining the performance and even sustainability of ICT-based MIS projects. It discusses policy implications of these findings.

1. INTRODUCTION

Smallholder farmers comprise the majority of the farming community in developing countries. However, most smallholder farmers face poor functioning input and output markets. The markets for agricultural inputs fail for most smallholder farmers in developing countries due to many factors (Kydd & Doward, 2004; Key & Runsten, 1999; Poulton, et al, 2006). First, smallholder farmers face high transactions costs in the input markets. Markets for agricultural inputs, especially

DOI: 10.4018/978-1-4666-2655-3.ch002

agricultural information, fail for smallholder rural farmers due, among other things, to illiteracy, distance to information sources, or absence of the type of information the farmer needs to produce the desired crop (Key & Runsten, 1999). Yet access to market information is essential especially where buyer(s) have quality and safety specifications that must be met by farmers. Poor access to information also promotes opportunistic behavior and results in traditional information asymmetry problems namely, moral hazard and adverse selection. Thus poor access to agricultural information results in poorly functioning input markets which in turn constraints the participation of smallholder farmers in such markets. In particular, poorly functioning input markets impede smallholder farmer investment in and use of yield-enhancing agricultural technologies. Low investment in agricultural production is the leading cause of low equilibrium trap in agriculture (Kydd & Doward, 2004). Low equilibrium trap is a steady state situation in which low levels of investment sustains imperfect markets and poverty.

Lack of agricultural information also constrains the functioning of output markets. It especially results in thin markets characterized by trade in low volumes, personalized exchange characterized by visual inspection of commodities being traded and the tendency to sell produce at the farm gate rather than in better-paying output markets (Fafchamps & Hill, 2005). Information asymmetry leads to lack of defined quality parameters and trade in non-standard scales that further fuel opportunistic behavior by traders. Prices earned in markets that operate under information asymmetry therefore tend to be low and do not compensate farmers for services/functions such as selection/sorting, grading, and packaging. Consequently produce tends to be sold in undifferentiated form (Poulton et al, 2006). These kinds of markets are common in many rural setting of developing countries. Such markets discourage smallholder commercialization and are seen as the major reason why many smallholder farmers remain trapped in subsistence agriculture and the low equilibrium poverty trap (Barrett, 2008)

The need to smallholder farmers' participation in input and output markets has led to search for new models of providing market information to such farmers. Recent attention has especially shifted to the use of ICT-based technologies. Hence projects that use information and communication technologies (ICT) to provide input and output market information to farmers have proliferated in developing countries (De Silva, 2010; Okello et al, 2010a). A recent study found 39 such agricultural projects with ICT components in Kenya alone (Munyua, 2007). In Africa, examples of ICT-based projects include the Kenya Agricultural Commodity Exchange and DrumNet in Kenya, Busoga Rural Open Source and Development Initiative and FoodNet in Uganda, Malawi Agricultural Commodity Exchange in Malawi, Manobi in Senegal, TradeNet in Ghana, Kilosa Rural Services and Electronic Communication in Tanzania, and Program for Promoting Rural Revenues in Madagascar. The major goal of these ICT-based projects is to facilitate farmer access to market information and help them transition from subsistence to commercial agriculture.

Recent studies suggest that these ICT-based market information service (MIS) project do facilitate smallholder farmers' access to market information and hence linkage to input and commodity markets (Aker, 2008; Chigona et al, 2009; De Silva, 2010). Nonetheless ICT-based projects operate under considerable challenges (Mukhebi, undated). Indeed such projects tend to collapse when donors withdraw funding. This paper examines how the environment within which ICT-based MIS projects operate affects their performance. It specifically: i) highlights the environmental factors/challenges that affect the success of ICT-based MIS projects in linking farmers to markets and ii) discusses the strategies that have been used by these projects to respond to some of these challenges.

The paper focuses on two projects in Kenya that targeted the farmers with input and output information to facilitate their linkage to better-paying markets. The projects, known as DrumNet and the Kenya Agricultural Commodity Exchange (KACE), used new and old generation ICT-tools (i.e., radio, mobile phones and a computer-based platform) to provide farmers with agricultural information. The rest of this paper is organized as follows: Section 2 presents the study context and introduces the two projects while Section 3 discusses the conceptual framework. Section 4 discusses the challenges the projects have faced, how they affected the performance of the projects and the strategies they used in response. Section 5 concludes and provides policy implications of the findings.

2. STUDY CONTEXT

The DrumNet and KACE projects operated in overlapping areas but differed in terms of geographic coverage. DrumNet consisted of two projects implemented in two localized regions of Kenya namely, Western province and Central provinces, focusing on sunflower and French bean farmers, respectively. KACE has, on the other hand, nationwide coverage although some of its activities are concentrated only in western part of Kenya. We provide brief details of the operations of the two projects below.

2.1. The DrumNet Project

DrumNet project launched its operations in Kirinyaga district (central Kenya) in 2003, later extending its activities to Meru district. The main goal of the project was to shorten the value chain of the target commodities and hence improve farmer's revenues. DrumNet therefore aimed at replacing the numerous intermediaries in the value chains by intermediating the transactions. The project used mobile phone-based platform to provide market information to the various project partners. The partners included farmers organized into small-holder farmer groups averaging 30 members per group, a financial organization (Equity Bank), a buyer and an agro-input dealer. In 2006, (under phase 2), DrumNet moved to western Kenya where it launched its operation in Kakamega district. However, within the 3 years that DrumNet was in western Kenya, it constantly migrated to a new district, almost yearly. From Kakamega, the project migrated to Bungoma, Busia and eventually Migori and Homabay districts. The farmers in western Kenya grew sunflower for sale to Bidco Oil Ltd while farmers in central Kenya grew French beans for sale to Kenya Horticultural Exporters (KHE) Ltd, one of the leading fresh produce exporters in Kenya.

In both central and western regions market-relevant information was sent via an SMS from DrumNet central computer-based platform located in Nairobi. Okello et al (2010 a) discuss the DrumNet model in detail and indicates that the information provided by the project ranged from production to market information. Small-holder farmers participating in the project signed contracts with DrumNet that entailed paying a commission of 6% and12% in western and central Kenya, respectively, for a bundle of services including agricultural information. The other services provided by DrumNet were, in-kind credit and coordination, produce collection and transportation.

The fees for provision of market information and other services were recovered upon the sale of beans (in central Kenya) or sunflower (western Kenya). Hence the project used an interlinked credit scheme to recover the costs of providing agricultural information to farmers. The link between farmers and DrumNet was through a member selected by the group known as transaction agent (TA). The TA acted as the information hub. Information from DrumNet was passed to the TA via mobile phone short messaging system (SMS) and the TA was expected to relay it to members. The TA later sent as SMS to DrumNet indicating that the information has reached the group.

2.2 The KACE Project

KACE was first launched in 1996 as a commodity exchange with auction floor located in Nairobi. The company's goal was to provide a forum that brings buyers and sellers/farmers together thus elimination the myriad intermediaries along commodity supply chains and hence reducing the transaction costs. KACE specifically had 3 major goals namely i) serve as market for buyers and sellers, ii) to provide information on demand, supply, prices and trends and iii) to establish a system of commodity trading. The commodities traded included staple and non-staple food crops, livestock and livestock products, and cash crops. In 1998, KACE launched a website through which subscribed sellers and buyers could place bids for commodities. In addition, the company initiated a mobile phone-based information provision program with nationwide coverage. Through this service, sellers (farmers) and buyers could obtain commodity prices by sending a mobile phone-based text message using short messaging system (SMS) to KACE database at a cost of Kshs 5.00 for prices of a commodity in different markets.

KACE also runs an FM radio program called "soko hewani" (meaning supermarket-on-air). The program allows buyers and sellers to offer bids for commodities by calling the KACE presenter/moderator or sending an SMS during a live radio broadcast. The bids are collected and processed by KACE staff at the radio station and immediately broadcasted. The program however has limited coverage. It covers western province and its adjacent areas namely, parts of Rift Valley, Nyanza and eastern Uganda. Same commodity range as those in the online service is covered under the FM radio broadcast auction program. In addition, however, the program also sells agricultural inputs including land.

KACE has a team of enumerators that collect market information from major and regional markets each day. The enumerators collect prices from wholesalers at different times of the morning. The data collected is sent (usually by email) to KACE head office in Nairobi where it processed and the average prices for major markets computed and then uploaded onto the KACE website and computer-based platform. Prices for each day are available for download from KACE website or via SMS at about noon each day.

KACE also runs market information points (MIPs) in some major towns in western Kenya. The MIPs operate as franchises. Each MIP has computer with internet connection and hence can be used by farmers/sellers and buyers to download information. They also have an information board with bids placed by buyers and sellers as well as commodity prices. The other add-on services provides by MIPs are moisture testing and storage.

3. CONCEPTUAL AND EMPIRICAL METHODS

3.1. The Conceptual Framework

Market information problems can be analyzed using the transaction cost theory. The theory posits that exchange between any two actors in a market entail costs. These costs arise from the need to i) search and screen transaction partners, ii) negotiate the terms of exchange, iii) monitor the terms of exchange, iv) enforce the terms of agreement, and v) renegotiate the terms of exchange should changes in the market/physical environment demand.

Transaction costs are exacerbated by lack of information between buyers and sellers, a situation called information asymmetry. Asymmetry of information encourages opportunistic behavior by the more informed transaction partners and is itself exacerbated by uncertainty, asset specificity, and frequency of transactions (Williamson, 1985). Opportunistic behavior is one of the major reasons why markets for agricultural inputs fail for smallholder farmers and why such farmers tend to have difficulty being linked to markets (Kydd & Doward, 2004; Omamo, 1998).

One of the strategies for governing transactions between exchange partners is contracting (Martinetz, 2005). Contracting also helps resolve farmer-specific (idiosyncratic) market failure such as lack or poor access to information necessary when the missing information is provided under the contract (Bagetoft & Olesen, 2004; Key & Runsten, 1999). In the context of the relationship between a farmer and an MIS project, contracting can take of the form of periodic subscription. Two types of subscription can be identified in this case. The first is where the client (farmer) pays up the fees for using the market information services. This type of subscription however requires clients that are financially endowed and hence able to afford the usually large annual fee. It therefore excludes smallholder farmers. In the second type of subscription, the services can be recovered from farmers' sale of contracted crop through interlinked credit scheme. This arrangement works where the contracted crop is sold through a project that has a contractual arrangement with members, and is common in developing countries (Bagetoft & Olesen, 2004).

The success of interlinked credit scheme is affected by a number of factors. First, where there are several buyers, there tends to be a problem of side-selling (Jaffee, 1995). That is, farmers get input loan (i.e., information) from a project but sell the produce elsewhere hence avoiding loan repayment. Side-selling tends to be acute where many buyers compete for produce making known sanctions such as threat of suspension or contract termination ineffective. Second, some farmers can join a project mainly for the purpose of getting credit but having no intention of repaying the loan, a situation known as strategic default. Strategic default can lead to unraveling of the market (Robinson, 1998).

The above transaction-related problems can further be affected by, among others, the socio-economic, physical, and legal financial environment in which the MIS is deployed. For instance, the socio-economic environment may relate to the norms of behavior and work and the economic status of the clients (farmers) that affect honesty. The physical environment, in the context of agriculture may relate to problems that arise from lack or unreliability of rain, pest and disease outbreak and poor soil fertility conditions. Strategic defaulters often blame their inability to service their loan on these physical environmental factors (Hueth, 1999).

The literature offers some theoretical solutions for dealing with transaction -related problems such as the above (Bagetoft & Olesen, 2004; Besley, 1998; Hueth, 1999; Williamson, 1985). These include instituting a system of monitoring the behavior of the farmer to verify claims used to justify default (Hueth, 1999), using local individuals with good knowledge of local farmers to screen out the dishonest (i.e, the would-be strategic defaulters) (Besley, 1998), and structuring the payment system to provide incentive for the client to be truthful (or to comply with contractual requirements) (Williamson, 1985; Bagetoft & Olesen, 2004).

3.2 The Empirical Methods

This paper is based on detailed case-study interviews with various stakeholders of the two MIS projects. The interviews were conducted between September 2009 and February, 2010. The stakeholders interviewed included project staff based in the field and in the head office, farmers, produce buyers, the commercial bank involved, local key informants and opinion leaders, and local government agricultural staff. How many of these individuals were interviewed in each case depended on the evidence being pursued but overall six DrumNet staff (four field-based and two head office staff) and five KACE (two head office and three field-based) staff. Farmers interviewed comprised group leaders (six in Kirinyaga, five in Kakamega, four in Bungoma, seven in Busia and six in Migori). Interviews with field staff yielded information on field-based operations and challenges while interviews with head office staff was intended to reveal the strategies used by the projects to respond to challenges posed by the deployment environment. The key informants were local leaders, traders, input sellers. Several farmers were also interviewed in each area covered where issues raised by the leaders needed to be corroborated. The information was generated through personal interviews.

Evidence collected from these various sources were triangulated using Yin's (1989) case study methodology guide. The information was further supplemented with quantitative data collected from a random sample of 379 farmers stratified by participation in ICT-based projects including KACE and DrumNet. The data was collected in April and May 2010 from central (Kirinyaga) and western (Bungoma and Migori) regions of Kenya through personal interviews using pre-designed and pre-tested questionnaires. The data was entered and analyzed using SPSS. Table 1 presents the summary statistics of the respondents and presents tests of differences in means between farmers who participate in

In this paper, we define performance of the project by the extent to which it meets its major goals. The DrumNet project had a major goal of linking farmers to better paying markets and hence improving their household incomes. Consequently, we use participation in market (proxied by the share of sales) to measure its performance in linking farmers to market. We also use differences in income as a proxy for increase in household incomes. KACE, on the other hand, had a major goal of providing systems of efficient price discovery by facilitating direct contact between sellers (farmers) and buyers in order to reduce the transaction costs small farm households face. We therefore use the differences in transaction costs between users and non-users of KACE services to gauge its performance. Based on previous literature, two most prevalent types of transaction costs in agricultural exchange namely, search and transport costs (Omamo, 1998; Okello et al, 2010b) are used to assess the differences in transaction costs between users and non users of KACE service.

The t-values suggest that there are differences between participants and non participants of ICT-based projects with respect to farmer-specific, farm-level and capital endowment characteristics. Specifically, there are differences in household size and hence household adult equivalent. Interestingly, non-participating households are bigger and hence have a bigger adult equivalent.

Results further show significant difference between ICT-based project participants and their counterparts with regard to distances to the local market and nearest agricultural extension office. The participating households were further from the markets than their counterparts. Participants also have more crop enterprises than the non-participating households, suggesting that the participating households are more risk averse than their counterparts. There are also regional difference between participants and non-participants.

Table 1. Characteristics of ICT-based project participants and non participants

Variable definition	Participants (N=215)	Non participants (N=164)	Mean difference	t-values
Dependent variables				
Awareness of ICT-based project	0.65	0.33	0.32***	6.49
Project participation	0.74	0.58	0.16***	3.19
Independent variables				
Farmer/Household specific variables				
Age (years)	44.32	42.88	1.44	1.00
Education (years)	8.62	8.21	0.41	1.08
Household size (count)	5.41	6.00	-0.60***	-2.68
Gender (1,0)	0.52	0.49	0.03	0.60
Occupation (1=farmer, 0 otherwise)	0.92	0.87	0.05	1.45
Farm-specific variables				
Local market distance (km)	4.85	4.11	0.74*	1.71
Extension agent distance (km)	4.56	3.74	0.83**	2.18
Non-ICT agric project	0.64	0.22	0.42***	9.05
Capital endowment variables				
Land size (acres)	2.58	2.21	0.37	1.54
Farming experience (years)	18.63	17.22	1.40	1.16
Ln assets (ks	10.65	10.51	0.13	0.89
Ln income	8.43	7.41	1.02**	2.36
Ln other non-farm income	9.96	10.44	-0.48***	-3.03
Group membership (1,0)	0.68	0.57	0.11**	2.09
Regional dummies				
Kirinyaga	0.31	0.35	-0.05	-0.95
Bungoma	0.49	0.25	0.24***	4.96
Migori	0.20	0.40	-0.19***	-4.00

Note: Significance level: *10%, **5% and ***1%. Kshs = Kenya Shillings ; 1 USD = Kshs 75

4. PERFORMANCE OF DRUMNET AND KACE PROJECTS, CHALLENGES THEY FACED AND THEIR RESPONSE

4.1 Performance of DrumNet and KACE Projects

Table 2 provides the results of analysis of market participation, household incomes and transaction costs for KACE and DrumNet project members and non-members. The results show statistically significant difference (as shown by p-values) between members and non-members of ICT-based projects with regard to share of sales, income, and transaction costs. DrumNet members had significantly higher share of sales and agricultural income than their counterparts. The finding that DrumNet farmers benefitted financially from the project corroborates the findings of Okello et al (2010a). Table 2 further shows that farmers that use KACE services spent less on travel and search

Table 2. Mean share of sales, income and transaction costs for ICT-based project members and non members, 2010

Variable	DrumNet				KACE			
	Members (N=100)	Non members (N=279)	Mean difference	p-value	Members (N=37)	Non members (N=342)	Mean difference	p-value
Mean share of sales	0.59	0.51	0.07	0.070	0.49	0.54	-0.043	0.460
Mean agric income (Kshs)	51966.26	33400.50	18565.76	0.008	35682.16	38202.24	2520.07	0.805
Mean total HH income (Kshs) per season	120995.83	119162.69	1833.14	0.922	87926.89	73482.82	2863.08	0.016
Total travel costs (Kshs)	83.01	51.64	31.37	0.263	26.21	62.92	-36.71	0.049
Total search costs (Kshs)	129.52	73.27	56.24	0.148	40.00	92.17	52.17	0.036

Source: Survey results; HH = Household; Kshs = Kenya Shillings ; 1 USD = Kshs 75

costs than those who do not. Consequently users of KACE services had more total household income than their counterparts. These results suggest that members of the DrumNet project received higher incomes due to stronger linkage to market. KACE, on the other hand, improved household incomes by reducing the transaction costs farmers face in selling their commodities. It is therefore plausible to argue that both projects met their major goals.

Despite meeting their goals, both KACE and DrumNet projects have faced several challenges that have affected their operations. DrumNet for instance relocated its operations several times and changed the target crop and hence beneficiaries. It moved from working with French bean farmers to sunflower farmers and relocated from central Kenya to western Kenya. In western Kenya, DrumNet changed the location/district of operation three times. KACE, on the other hand, has also had to revise its portfolio of services over the years. It, for instance, dropped the auction floor service and severally suspended the online service. It also has had to suspend/cancel its operations in some areas. A case in point is the closure of KACE office in some areas of Siaya district notably Ugunja market. So what explains the migration tendencies by DrumNet and the revision and suspension of services by KACE? In the section below, the challenges that DrumNet and KACE encountered and the strategies they used to respond to them are discussed.

4.2 Challenges Posed by Operational Environment and Adaptation Strategies Used by KACE and DrumNet Projects

In this section we assess the various challenges associated with environment in which the DrumNet and KACE projects were implemented and how they affected the operations of these projects. We address these environmental factors under four categories namely, socioeconomic, market, physical, and legal.

4.2.1. Socioeconomic Factors

The DrumNet project targeted low income smallholder farm households in western and central Kenya. The major income sources for such households were the sale of surpluses from food crops and remittances from family members in towns (see Okello et al (2010a) for details on the characteristics of DrumNet members). Due to shortage of income, many farmers were not able to meet household needs. The low farm household incomes created a strong incentive for project farmers to default on the in-kind loans (inform of fees for MIS services) they received from DrumNet. The DrumNet project did not have a strong field monitoring strategy hence did not deter default on loans. Farmers developed ingenious ways of avoiding the repayment of MIS loans resulting in high default rate of up to 30%. In many of the DrumNet groups, some farmers signed up for loans with no intention of repaying, a phenomenon known as strategic default. Such farmers sold the crop outside the project to other intermediaries after receiving the MIS services from the project thus avoiding the repayment of the loan altogether. In the less severe cases, the defaulters sold only part of the crop to the project, and the rest outside. Consequently, the revenues they earned from the sale of the produce delivered to the project were not enough to cover the value of their loans. In both cases, the defaulting farmers blamed some environmental factors (e.g., bad weather, pest attack, and disease outbreak) most of which were difficult for DrumNet to verify due to thin monitoring staff in the field.

Project farmers were influenced to engage in strategic default (and especially side-selling) by the prices that other intermediaries offered. Such prices tended to be higher than the net price paid by the project (Okello et al, 2010b). The other intermediaries enticed project farmers by offering higher prices than the project prices because project prices were net of the MIS fee, transport costs and other group charges. Side-selling was widespread in both western and central Kenya, indicating that the nature of the crop (whether export oriented or not) did not influence the tendency by project farmers to cheat on the contract. It also suggests that societal norms that typically encourage honesty did not provide sufficient restraint against this form of cheating on contract in both central and western Kenya.

Strategic default led to collapse or dissolution of many DrumNet groups, especially following sudden exit by the defaulting members. Since the group guaranteed the loan, DrumNet typically sought to minimize its losses by recovering the defaulters' loans (and its commission which included the fees for MIS) from the remaining farmers. Thus the remaining farmers in the affected groups earned less for their produce, became discouraged by the lower earnings and, in some cases, decided to cheat during the next season. In all of the cases of massive loan default, the group pressure intended to pressurize members to repay was ineffective for two reasons: i) the self help groups are non-legal entities and hence could not sue a defaulting member and ii) the group did not inflict sufficient reputational damage on the defaulter to deter others.

Strategic default can be reduced through close monitoring of project farmers. However, DrumNet did not have a strong team of field staff to do so. The inability of the project farmer group to sue defaulting members therefore exacerbated the opportunistic behavior. So what did the project do? In central Kenya, it attempted to recover the unpaid loan through legal recourse but was unable. In western Kenya, the project simply moved to new locations after the collapse of the older groups. The Project opened new operations in Kakamega, Butere-Mumias, Bungoma, Busia, Suba, and Migori in just 3 years, in response to problems related to loan default. The disadvantage of this strategy was that project field staff spent most of their time teaching new members how the DrumNet model works and had no time to build longer lasting group structure to make the groups sustainable. The second strategy that DrumNet used was to drop some of the items that comprised the bundle of services it offered the farmers. In Busia, for example, the project provided fertilizer and seed loan for only one season while in Suba, both fertilizer and seed loans never were given to farmers as part of the bundle of services DrumNet provided. The third strategy used by DrumNet in response to default in loans was the reduction in the fees charged on farmers. The fee was reduced from 12% in central Kenya to 6% in western in order to make the costs of the services that included ICT-based MIS affordable to farmers.

The second major problem was dishonesty of the project field staff especially in central Kenya. Some of the project staff colluded with farmer leaders and agro-input dealers to cheat the unsuspecting members. The cheating involved inflation of the volume of beans collected from a group. The cheating staff then earned free money by computing the unit price received by farmers using the inflated volume. This resulted in DrumNet members receiving lower price under the project than the spot market price, which further encouraged some to side-sell their beans. Dishonesty among field staff was fueled by the weak monitoring of field activities.

DrumNet head office became aware of the dishonesty after farmers boycotted selling their beans to the project but responded by only sacking the suspected staff and not suing the sacked staff to recover the stolen money. This punishment however provided inadequate sanction and further fuelled the culture of dishonesty among the remaining field staff. One staff thus said:

"The stealing went on, even much more, after the sacking of the first staff. He stole a lot of money from the farmers and was simply sack. He left with so much money that he immediately set up a good business. It was as though the project told him to put his stolen money to use. We all need money. If all that will be done to me is to be sacked, then why should I sit here and watch my opportunity to get rich pass by?"

Unlike DrumNet, KACE did not provide its services through interlinked credit arrangement. Instead, users of its services paid directly through higher fee for a text message sent to KACE database or through upfront subscription. In the case of "soko hewani", (i.e., supermarket-on- air) service, the buyer and seller do not meet hence the payment is made to KACE which enables it to deduct the fee for its services before remitting the rest to seller. Consequently KACE never got into problems of recovery of payments for services it provides. Never the less, KACE too faced challenges related to the socioeconomic conditions of the farm households. First, small farm households in some KACE target areas were semi-subsistence with very small surpluses for sale. Second, farmers in some of KACE target regions grew low value crops. Third, the level of income of small farm households was very low leading to lack of effective demand for market information. KACE responded to this problem by simply closing its operations and moving elsewhere once donor funding targeting such areas ended.

Other socio-economic factors affected the performance of the project included literacy level, access to electricity, and availability of mobile phone network. Okello et al (2009) discuss the effect of low literacy levels and poor access to electricity for charging mobile phone batteries on the use of mobile phones. Most farmers interviewed had on average 8 years of schooling. The low level of education reduced the use of some of the functions of mobile phones, especially SMS whose menus are often in English making it had for semi-literate and illiterate farmers to navigate. Indeed KACE system which uses SMS had the lowest usage in western Kenya. DrumNet also experienced difficulties getting SMS-based responses from it members.

Lack of electricity for charging phones caused farmers to limit the use of phones to only certain days of the week. Most farmers charged their phones on market days and used it for the next 2-3 days. This was a major problem for DrumNet project which depended on mobile phones to pass essential agronomic and market information to its farmers at critical points during the crop production cycle and to link project partners to each other. It often forced DrumNet to call the transaction agents up for information whenever the problem was acute. Lack of mobile network coverage also seriously affected the ability to commutate and pass agronomic and market information to some project groups. In areas where the phone network signal was too weak or absent, a farmer was forced to move to other locations to send or receive DrumNet or KACE SMS. When out of these areas, the farmer became inaccessible to the DrumNet project and could not access KACE information. The inaccessibility reduced timely access to information and resulted in costly delays to the project.

4.3 Market Environment

DrumNet faced a serious problem relating to the end market in its Kirinyaga (i.e., central Kenya) project. The produce from the project (i.e., French beans) was sold in the UK supermarket where demands for compliance with Global Good Agricultural Practices (GlobalGAP[1]) had reached its climax at the time the DrumNet project was setting up its operations in central Kenya. The buyer, Kenya Horticultural Exporters (KHE) Ltd, demanded that DrumNet farmers fully comply with GlobalGAP standards. In addition, KHE introduced a traceability system that allowed it to control the sources of beans it received from the DrumNet groups. However, the farmers resisted these supply-side requirements due to high costs of upgrading the facilities to meet GlobalGAP standards. DrumNet members were also not able to strictly adhere to the traceability requirements. Consequently, KHE Ltd stopped buying from group leading to lack of reliable market. DrumNet advised its members to find their own buyers when KHE left thus dropping the market linkage component of the project. A few months later, it closed the French bean project and moved to western Kenya, where it switched to sunflowers. Again, KACE did not experience this problem because it focuses on domestic crops only. The farmers started selling to brokers but eventually shifted to production of beans for canning and bananas.

4.4 Physical Environment

The main physical factors that affected the DrumNet project was bad weather (i.e., lack or unreliability of rain) and pests. Bad weather was a major problem in western Kenya where farmers do not use irrigation. It resulted in poor germination of seeds and eventually in poor crop yields. In terms of pests, the key problem farmers faced was bird attack especially at the milk stage of the sunflower growth stages. Farmers incurred huge losses, in some cases, of up to 50% due to the birds. Farmers in central Kenya did not have as much difficulty with pests and diseases, probably because of the many years of experience in growing French beans. Bean fly was mentioned by some farmers as one of the pests that can lead to heavy losses in bean. However, majority of the farmers knew how to control the pest unlike the problem of birds in western Kenya which was new to most farmers. In order to tackle the problem of birds on sunflowers in western Kenya, DrumNet arranged for farmer training on strategies for controlling birds. The training recommended that farmers plant sunflowers close to one another (a form of plantation) to reduce impact of birds. Secondly, it recommended that farmers employ workers to "chase" birds from the fields. However, most farmers found it hard to implement both recommendations. Thus the losses due to birds were never controlled and became one of the major problems in the project.

It is common for transaction actors who engage in strategic default to blame some factors that are hard to verify as the reason for failure to meet expected targets. In the DrumNet project, the hard-to-verify factors were bad weather and birds. Indeed, farmers who willfully defaulted in loan repayment blamed bad weather and/or pests for the default. This was the case in some parts western Kenya e.g., Kakamega where farmers blamed drought and damage by bird on their inability to repay the loan and Budalangi where farmers blamed flooding. Leaders of these groups however indicated that many farmers sold some of their beans outside the project thus making full recovery of loan difficult. Indeed default on loan repayment continued even after the project intensified farmer education on bird control in areas most affected by birds, suggesting bird damage was not the primary cause of default.

4.5 Legal Environment

DrumNet project worked with smallholder farmers organized as groups. These groups were registered by the government under the ministry of culture and social services rather than the ministry of cooperatives. The registration under ministry of culture and social services meant that the farmer organizations (groups) were not legal entities and therefore could not be sued in case of breach of terms of the contract. Indeed, DrumNet tried to overcome this problem by signing separate contracts with individual farmers and then collectively with a group. Nevertheless, enforcing such individual contracts required costly monitoring which increases transaction costs. At the same time DrumNet had very few field staff. This strategy of signing contracts with individual farmers did not therefore control side-selling and loan default. Instead, it increased the time costs of resolving farmers' complaints arising from financial transactions. In central Kenya (i.e., Kirinyaga) for instance, the field finance officer attended to long queues of farmers with complaints about the deductions made on their sales revenues every day. Many farmers would demand computer print-outs to check how their payments are arrived at, which further increased the administrative costs. DrumNet tried to get around these costly administrative procedures by charging farmers who demand additional information such as computer printouts of their records a fee. However, this angered the farmers even more because the farmers felt that they were paying for the mistakes made by the project staff.

The second major legal challenge the project faced was with regard to business regulations relating to banking at the time. The project signed a contract with the farmers to pay them within 48 hours of the collection of their produce. The arrangement was that the project would send a mobile phone text message alerting the bank to process the pay. The text message was to act as an invoice. However, when the project was implemented, the bank refused to act based on text messages sent by the project. It argued that it needed a signed paper invoice in order to process the payments because banking laws did not recognize text messages as legal documents. This forced the project to prepare paper invoices which usually caused lengthy delays and eroded the trust the farmers had on the project.

5. CONCLUSION AND RECOMMENDATIONS

Several projects have emerged in Africa that use new generation ICT tools such as mobile phones to provide market information to farmers. Studies have shown that such projects have benefits to farmers in that they link such farmers to better markets. Others studies have suggested that ICT-based MIS facilitate trade and improve the efficiency of agricultural markets. However, there have not been studies that systematically assess how the environments in which such project operate affect how they perform. This paper examines how the socioeconomic, physical, market and legal environments within which ICT-based MIS projects operate affect their performance. The study focuses on the DrumNet and KACE projects implemented in central and western Kenya. It finds that the two ICT-based MIS projects, in general, succeeded in meeting their major goals of linking farmers to markets and improving household incomes. However, the sustainability of the MIS projects has been less successful. The paper highlights how poor economic status of the farmers encourages them to engage in opportunistic behavior that undermines the projects' ability to function well. The culture of dishonesty also affects the level of trust in the project and ultimately its performance. The study further discusses how factors related to the physical environment in which ICT-based MIS are deployed reinforces the tendency of project farmers to behave opportunistically thus further affecting the project's ability to recover its loans and performance. In addition, inflexible business regulations and the lack of enforceable legal framework work against the performance of ICT-based MIS projects.

The paper highlighted the strategies used by MIS projects to overcome some these challenges. KACE has over the years changed its services and has been able to stay in business. DrumNet on the other hand adopted the strategies that eventually drove it out operation. This suggests that although the ICT-based MIS project might encounter similar environmental challenges, their sustainability greatly depends on the strategies they use in responding to the challenges. The implication of these findings is that ICT-based MIS projects require good socio-economic environment, supporting legal framework and flexible strategies. In addition, success of such projects might require investment in crop insurance to overcome some of the challenges especially those related to the physical environment.

REFERENCES

Aker, J. C. (2008). *Does digital divide or provide? The impact of cell phones on grain markets in Niger. Job Market Paper*. Berkeley: University of California.

Bagetoft, P., & Olesen, H. B. (2004). *The design of production contracts*. Copenhagen, Denmark: Copenhagen Business School Press.

Barrett, C. (2008). Smallholder market participation: Concepts and evidence from eastern and southern Africa. *Food Policy*, *34*, 299–317. doi:10.1016/j.foodpol.2007.10.005

Besley, T. (1998). How do financial failures justify interventions in rural credit markets? In Eicher, C. K., & Staatz, J. M. (Eds.), *International agricultural development* (3rd ed., pp. 370–389).

Chigona, W., Beukes, J., Vally, D., & Tanner, M. (2009). Can mobile Internet help alleviate social exclusion in developing countries? *The Electronic Journal of Information Systems in Developing Countries*, *36*, 1–16.

De Silva, H. (2010). ICT policy for Sri Lanka. *International Journal of ICT and Research Development*, *1*(39).

Fafchamps, M., & Gabre Madhin, E. (2006). Agricultural markets in Benin and Malawi. *African Journal of Agricultural and Resource Economics*, *1*, 67–94.

Fafchamps, M., & Hill, R. V. (2005). Selling at the farm gate or traveling to the market. *American Journal of Agricultural Economics*, *87*, 717–734. doi:10.1111/j.1467-8276.2005.00758.x

Hueth, B. (1999). Incentive instruments in fruits and vegetables: Input control monitoring, measuring and price risk. *Review of Agricultural Economics*, *2*, 374–398.

Jaffee, S. M. (1995). The many faces of success: The development of Kenyan horticultural exports. In Jaffee, S., & Morton, J. (Eds.), *Marketing Africa's high-value foods* (pp. 319–374). Washington, DC: World Bank.

Key, N., & Runsten, D. (1999). Contract farming, smallholders, and rural development in Latin America: The organization of agro-processing firms and scale of out-grower production. *World Development, 27*, 381–401. doi:10.1016/S0305-750X(98)00144-2

Kydd, J., & Dorward, A. (2004). Implications of market and coordination failures for rural development in least developed countries. *Journal of International Development, 16*, 951–970. doi:10.1002/jid.1157

Martinetz, S. (2005). *Vertical coordination of marketing systems: Lessons learned from the poultry, egg and pork industries*. United States Department of Agriculture. ERS Agricultural Economic Report No. 708.

Mukhebi, A. (n.d.). *The challenges and opportunities of a young commodity exchange in emerging market economy: The experience of Kenya Agricultural Commodity Exchange*. Retrieved from http://www.google.co.ke/#hl=en&source=hp&biw=1245&bih=486&q=The+challenges+and+opportunities+of+a+young+commodity+exchange+in+emerging+market+economy&btnG=Google+Search&rlz=1R2SKPB_enKE362&aq=f&aqi=&aql=&oq=&gs_rfai=&fp=ceb31e2ddf411e47

Munyua, H. (2007). (Submitted to). ICTs and small-scale agriculture in Africa: A scoping study. [International Development Research Center.]. *Draft Report, 1*.

Okello, J. J., Al-Hassan, R., & Okello, R. M. (2010b). A framework for analyzing the role of ICT on agricultural commercialization and household food security. *International Journal of ICT and Research Development, 1*, 38–50. doi:10.4018/jictrda.2010010103

Okello, J. J., Ofwona-Adera, E., Mbatia, O. L. E., & Okello, R. M. (2010a). Using ICT to integrate smallholder farmers into agricultural value chain: The case of DrumNet project in Kenya. *International Journal of ICT and Research Development, 1*, 23–37. doi:10.4018/jictrda.2010010102

Okello, J. J., Okello, R. M., & Ofwona-Adera, E. (2009). Awareness and use of mobile phones by smallholder farmers in Kenya. In Maumbe, B. M. (Ed.), *E-agriculture and e-government for global policy development: Implications and future directions*. Hershey, PA: IGI Global. doi:10.4018/978-1-60566-820-8.ch001

Omamo, S. W. (1998). Farm-to-market transaction costs and specialization in smallscale agriculture: Explorations with non-separable household model. *The Journal of Development Studies, 35*, 152–163. doi:10.1080/00220389808422568

Poulton, C., Kydd, J., & Doward, A. (2006). Overcoming market constraints on pro-poor agricultural growth in sub-Saharan Africa. *Development Policy Review, 24*, 243–277. doi:10.1111/j.1467-7679.2006.00324.x

Robinson, S. M. (1998). Microfinance: The paradigm shift from delivery to sustainable financial intermediation. In Eicher, C. K., & Staatz, J. M. (Eds.), *International agricultural development* (3rd ed., pp. 390–415).

Williamson, O. E. (1985). *The economic institutions of capitalism: Firms, markets, relational contracting*. New York, NY: The Free Press. doi:10.1007/978-3-8349-9320-5_6

Yin, R. (1989). *Case study design: Design and methods*. Thousand Oaks, CA: Sage Publication.

ENDNOTES

1 GlobalGap is a voluntary standard but an essential requirement for supplying high-value markets in Europe.

Chapter 3
Using ICT to Overcome Constraints in the Agriculture Value Chain:
Emerging Trends in Ghana

Siegfried Kofi Debrah
International Fertilizer Development Center (IFDC), Ghana

Isaac Kwadwo Asare
International Fertilizer Development Center (IFDC), Ghana

ABSTRACT

Development partner efforts and private sector initiatives on ICT applications in agriculture have brought new opportunities for farmers and traders to reduce transaction costs and increase incomes. The applications are primarily used for linking actors in the agricultural value chain, accessing real time information on prices, buyers and sellers, transport and haulage, and other relevant information services in the agricultural value chain. Limited evidence from Ghana and elsewhere show that cell phone applications have resulted in increased incomes but the impacts and sustainability of other ICT applications have proven elusive. The role of ICT in overcoming the key constraints in the agricultural value chain and for making evidence-based decisions will be greatly enhanced if farmers, aggregators, and other stakeholders in the value chain pay attention to their business scope and schedule planning, executing, monitoring and control, procurement, risk planning, and stakeholder communications in a "project management" context. When this is done, ICT applications will facilitate supply chain management through sharing of timely and pertinent information on producers, buyers and other services, thereby helping to promote industry competitiveness. The major challenges to the widespread use and sustainability of ICT remain access to the appropriate ICT tool, poor road and storage infrastructure (particularly in the farming communities) and illiteracy on the part of the majority of smallholder farmers.

DOI: 10.4018/978-1-4666-2655-3.ch003

INTRODUCTION

Globalization is indeed providing new opportunities for smallholder farmers to increase their share of domestic, regional and international markets. By becoming participants in the global economy, smallholder farmers will be able to raise their incomes and achieve food security for their families and rural populations in the long run. But to become competitive in today's global marketplace, farmers need to be integrated into the production chain - from the farm to the final consumer's plate. However, lack of adequate and affordable inputs, inefficient farming practices, poor product quality, inefficient transport, lack of access to information on pricing for buyers and sellers and several other factors are hindering farmers' efforts to become competitive in the global marketplace. The high penetration of mobile telephony in developing countries however offers practical options to farmers and other actors in the value chain to access real-time market, price and other pertinent information, enabling them to make informed and better decisions. This has huge potential for addressing key constraints that have hitherto limited productivity and agricultural competitiveness.

With more than 15 million mobile phone subscribers in Ghana, the potential market for cell phone-based services is significant (Dowuona, 2011). The recent introduction of mobile phone-based money service and the "e-zwich"[1] in Ghana reduces transaction costs of financial services for users in rural areas where financial services providers seldom exist. In the agricultural sector ICT tools[2] are helping address constraints in agricultural value chains and are offering new opportunities to reduce transaction costs, increase access to markets, improve productivity, provide better and frequent access to critical market information and improve communication throughout the value chain. Its use in support of market information collection and dissemination in Ghana dates back to the late 1990s when the International Fertilizer Development Center (IFDC) created the African Fertilizer Market Information Network (AFAMIN) to link Ghana, Nigeria, Togo and Burkina Faso on a web-based platform to exchange price information on-line among participating countries. Today, due to efforts of development partners and private sector initiatives, cell phones are no longer only used as convenient tools to stay in touch with people but have evolved as "markets in the pocket" business communication tools in the agricultural sector. Notwithstanding these advances, the poor nature of the roads and unreliable transport services remain a huge challenge to competitiveness and productivity.

This chapter describes development partner efforts and private sector initiatives on ICT applications in Ghana. The ICT applications are primarily used for linking actors in the agricultural value chain, accessing real time information on prices, buyers and sellers, transport and haulage and other relevant information services in the agricultural value chain. Limited evidence from Ghana and elsewhere show that cell phone applications have resulted in increased incomes but the impacts and sustainability of other applications have proven elusive. Following the review of current ICT use in agriculture, the paper describes agricultural production from a "project management" perspective and explores the potential of ICT (e.g. the emerging *mFarms* ICT platform) to coordinate the initiation, planning, execution, monitoring and control and closure processes of the "agricultural production project." Special emphasis is put on the platform's use to improve the negotiation power of farmers, early identification of buyers and the proper timing and execution of critical activities that will result in greater and sustainable incomes for actors in the value chain through better information exchange, improved productivity and cost reductions.

ICT APPLICATIONS IN AGRICULTURE

ICT applications have become important in facilitating decision making by actors in the agricultural value chain. Most of the applications have been for market price information collection and dissemination and are based on cell phone use. Such services, according to Action for Enterprise (2010, p. 2), are provided through four categories of providers: i) Mobile Network Operators (MNOs); ii) Third-party service providers; iii) Buyers providing information to producers; and iv) Market Information Service provided by the public sector. The MNOs (e.g., Grameen Phone or GP), offer information to their subscribers through the cell phone in Bangladesh in the form of interactive voice recognition (IVR), short message service (SMS), wireless application protocol (WAP) browsing and web-based Internet. Third-party private company providers (e.g., Esoko, Manobi, Reuters Market Light and Google Trade) also use voice, SMS and WAP to offer an array of services including up-to-date price information, weather and trade alerts to subscribers. The "Buyers providing information to producers" category (e.g., e-Choupal, mFarms), provides information on commodity offer prices and purchase intentions, weather, input, transport and haulage information and crop advisory services to their producers. The public sector MIS uses civil servants to collect price information on selected commodities from selected markets where the information is used mainly for food security purposes.

ICT-enabled information systems have proven beneficial to producers and traders. Limited evidence from India, Niger, Uganda and Ghana show that the use of cell phone networks alone results in increases in profit margins of between 9 and 29 percent (Aker and Mbiti, 2010; Muto and Yamano, 2009; Debrah, 2007a). Also, the Ghana Agricultural Producers and Traders Organization (GAPTO), a major beneficiary of the ICT-enabled Market Information Systems and Traders' Organizations of West Africa (MISTOWA) project, reportedly concluded trade deals worth $60,000 with other producer and trade organizations in Burkina Faso, Mali and Nigeria in 2006. These deals involved purchasing tomatoes, onions and potatoes without middlemen, reducing the transaction costs substantially (World Development Report, 2008).

Emerging Trends in ICT Applications in Ghana's Agriculture

The increasing use of ICT for agricultural development in Ghana has been pioneered since the mid-2000s by the joint efforts of development partner support and private sector initiatives. The development partner support came from flagship projects funded by the United States Agency for International Development (USAID). The projects were the Market Information Systems for Trader Organizations in West Africa (MISTOWA) implemented by IFDC in 2004 and the Trade and Investment Program for a Competitive Export Economy (TIPCEE) implemented by the American consulting firm Chemonics in 2005. The private sector initiative came in 2004 from BusyLab, a private Ghanaian company which together with the development partners introduced the application of ICT that has since been used in the fruits, agri-inputs and the staple crops subsectors in Ghana.

The Fruits Sub-Sector

In 2006, the TIPCEE project assisted in the provision of technical knowledge in the production, management and marketing of fruit crops such as mango and citrus and of late, food crops such as maize. TIPCEE-assisted citrus and mango farmers received support in the areas of GIS mapping, training in pest and disease identification and management, pruning, proper harvesting and packaging techniques (Wolter, 2008). TIPCEE then engaged *Mobile Content* to develop a mobile application (menu-driven SMS/GPRS java-based application) in partnership with MTN, a Mobile Network Operator, to improve upon data collection from their farmers which were previously collected and sent weekly by vehicle to the TIPCEE office in Accra. This initiative enabled farmers in the citrus and mango growing areas to send real-time scout information on their farm operations by a menu-driven text message using their cell phones to a hub (web-based application/platform) hosted by TIPCEE. The information was sorted, analysed and made accessible for consultants and buyers to provide advice on scout information collected on-farm to farmers. In 2010, the West Africa Fair Fruit (WAFF) also developed mobile phone application to facilitate the collection of real time data from smallholder farmers for the purpose of logistics planning in order to meet buyer demands in Europe.

Agri-Inputs Sub-Sector

Since 2009 the Alliance for a Green Revolution (AGRA)-funded Ghana Agro Dealer Development (GADD) Project has developed a mobile Short Message Service (SMS) and internet application platform to link seed producers to agro-input retailers. For example, at the beginning of the 2010 minor cropping season when agri-input dealers were unsure of seed supply, the GADD project sent mass SMS to about 1,000 agro-input dealers to announce the availability of seeds from M&B Seed Company based in Ho, in the Volta Region of Ghana. More than 700 agro-input dealers from Eastern, Greater Accra and different parts of the Volta Region of Ghana received the SMS text message and concluded business with the seed company. Given the new demand base, the seed company has opened more distribution points in Northern Volta, Southern Volta, and Eastern region to serve agro-input dealers in those localities. The SMS service has not only shortened the search and discovery time but has opened avenues for farmers to access high quality improved seeds from agro-input dealers without having to travel too far away from their localities (IFDC, 2010 p.21).

Staple Food Crops Sub-Sector

The use of ICT in the staple foods sub-sector began in 2004 when IFDC partnered Busylab in the MISTOWA project to develop and deploy web and mobile phone-based software to address the problems of inadequate market information necessary for efficient crop production, post-harvest management and marketing (Debrah, 2007a). The platform, www.tradenet.biz (now rebranded Esoko, www.esoko.com) has since been used in 15 West African countries for producers and traders to share information on real-time prices, alerts, news and other market intelligence.

Since 2009 Esoko has enhanced its market price information services to include *mobile alerts* (for real-time market intelligence via cell phone-based SMS and web), *sites* (for on-line marketing), *push* (for direct marketing and delivery of extension services) and *scout* (for polling users by mobile phone instead of field visits). Esoko thus allows different actors in the value chain to exchange real-time market information. Farmers receive current demands, prices for crops, and the locations of seeds and fertilizers directly on their cell phones and businesses can track how their produce is used and market themselves to new customers.

When donor funding for the MISTOWA project which partnered with BusyLab to promote the Esoko platform ended earlier than planned, neither the impact of ICT nor its a sustainable business model had been found. Fortunately, a new USAID-funded Agribusiness and Trade Promotion (ATP) project is currently partnering with Esoko and it is hoped that the impacts on farmers' incomes and a sustainable business model will be found by the end of the project. It is also interesting to note that IFC, a member of the World Bank Group, and the Soros Economic

Development Fund have invested $1.25 million into Esoko to give smallholder farmers and businesses timely crop information that can be shared via text messaging enabling them to increase their incomes (Esoko, 2010).

AGRICULTURAL PRODUCTION VIEWED FROM THE PERSPECTIVE OF "PROJECT MANAGEMENT"

The relatively successful applications of ICT in the agricultural value chain are from the use of cell phones to collect and disseminate market information but its use for production and supply chain management in Ghana is limited. *mFarms*, an emerging ICT-based platform is designed to fill this void by helping actors in the agricultural value chain to be linked and communicate with each other, establish business relationships, manage the flow of goods and services and serve as a decision support tool. To illustrate the potential of ICT in coordinating the activities in the value chain, it is necessary to consider agricultural production from the perspective of project management[3].

To the extent that agricultural production (e.g., maize production) is an endeavor that has beginning and end dates and produces a unique product, it qualifies as a "project" according to the definition of the Project Management Institute (PMI). Therefore as a project it requires adherence to the five project management process groups of *initiation, planning, executing, monitoring and control and closure* and the application of the nine knowledge areas over its life cycle (PMI, 2008). The discussion will focus on the planning process where activities are identified and scheduled, followed by an analysis of the potential of ICT for facilitating, coordinating and monitoring of the planned activities.

The planning process results in the identification, sequencing, scheduling and resourcing of the activities necessary to complete the project. In our maize project example, the activities include land preparation, procurement of seeds, fertilizer and other inputs, planting, weeding, harvesting, threshing, transporting and selling. Once the activities are identified and sequenced, the time required for each activity to be completed is estimated and a network logical diagram showing the activities and their logical relationship to each other may be created. The first step in constructing the network is to identify every activity and each of its predecessors (activities to be completed before this activity can start), its successor (activities that follow) or those that can be accomplished independently or in parallel with the activity. Table 1 illustrates a logical sequence of activities in our maize example showing predecessor, successor and independent activities.

The total duration is estimated to be 255 days, composed of 112 days for planting through harvesting, and 143 days for other related activities. By planning and mapping the sequence of activities, the farmer becomes aware of the critical periods within which some activities must be completed without adverse impacts. For example, maize must be planted and fertilized within a one-month window from May 20th to June 20th in the Guinea Savanna zone of Ghana to avoid a 0.750 tons/ha yield loss for a day's delay in planting and 1.1 tons/ha yield loss for a day's delay in fertilizer application (FAO, 2010).

The Potential of ICT for Facilitating, Coordinating, and Monitoring of the Planned Activities in the Agricultural Value Chain

The success of any business endeavor usually depends on how well it is planned and executed. Unfortunately in the agricultural sub-sector, limited and insufficient planning is done by smallholder farmers leading to reduced productivity and profits. Agricultural production planning helps for example, in the early identification and mapping of stakeholders and their requirements (e.g. produce buyers, inputs sellers, financial institutions, transport companies etc). It also helps to identify and plan for risks and to develop strategies for

Table 1. An illustrative sequence of activities in maize production from land preparation to final sale

Phase/Activity	Alphabetic Code	Immediate predecessor activity(ies)	Activity duration (days)
Phase 1 (Land Preparation) 39			
Acquire land (own or rent)	a	-	10
Procure tractor services for land preparation	b	a	15
Prepare land	c	a, b	14
Phase 2 (Production) 106			
Purchase seed and fertilizer	d	a, b, c	14
Planting and replanting	e	d	30
1st Weeding	f	e	25
1st fertilizer application	g	f	12
2nd Weeding	h	e, f, g	15
2nd fertilizer application	i	h	10
Phase 3 (Harvest and Post Harvest) 110			
Harvesting	j	i	20
Primary Processing (de-husking, sorting, bagging)	k	j	25
Seek buyers	l	j, k	30
Seek Transport and cart produce to buyer	m	l	25
Purchase by final buyer	n	m	10
TOTAL NUMBER OF DAYS 255			

effective communications with stakeholders. Because the agricultural sub-sector involves several thousands of farmers, buyers and other service providers dispersed over space, the tendency today is to take advantage of the growing penetration of ICT to improve communications and business linkages along the value chain.

The mFarms Platform

mFarms is an integrated ICT platform designed to help stakeholders in the agricultural value chain communicate with each other efficiently, establish and maintain business relationships and manage the flow of goods and services among them. Developed by *ImageAD* a private Ghanaian company, it is currently being piloted by IFDC to implement the "Ghana Agro Dealer Development (GADD)" and "Linking Farmers to Markets (FtM)" projects funded by AGRA.

A pre-requisite of *mFarms* is the availability of reliable and trusted geo-referenced databases of key actors (stakeholders)[4] in the value chain. The databases would normally include stakeholder profiles and requirements, background, business information, geographic location (i.e., longitude and latitude of the locality), and cell phone contacts. Stakeholder profiling and mapping is on-going but as of December 2010, electronic databases were created for four industrial and large buyers, 16 aggregators, 29 community warehouses, 20,000 farmers and their organizations and 3,500 agricultural input dealers who were profiled and mapped. Other service-providing entities like transport and haulage companies and artisanal produce processors in the value chain were also profiled but yet to be geo-referenced. The data have so far been migrated to the *mFarms* platform where the geographic locations of the various actors can easily be visualized (Figure 1).

Figure 1 is an example of the geographic distribution of farmers, input dealers and warehouses in the West Mamprusi District in the Northern Region of Ghana. When other stakeholder locations are added, it is easier for users to quickly identify where and how far to go to purchase, sell, store, or access services, and *mFarms* can be used to communicate and conclude business deals more efficiently.

mFarms's Use for Linking and Communicating Among Value Chain Stakeholders

The platform permits users to be linked to exchange business information using the cell phone and the Internet. Any potential user can register by sending an SMS to a special telephone number "short code", to the platform. An identification number will be sent back to him to provide other details by dialing a short code to the Unstructured Supplementary Service Data (USSD) protocol linked to the mFarms remote server. The USSD application is a menu-based information service that can be accessed by the use any cell phone (unlike java applications which require the use of slightly sophisticated cell phones). USSD messages create a real-time connection during a USSD session and remains open, allowing a two-way exchange of a sequence of data, making the USSD more responsive than services that use SMS (Wikipedia, 2011).

The categories of targeted *mFarms* platform users under the FtM project are industrial and large produce buyers and aggregators at the downstream demand end of the value chain. They

Figure 1. District map (West Mamprusi) showing selected value chain actor locations

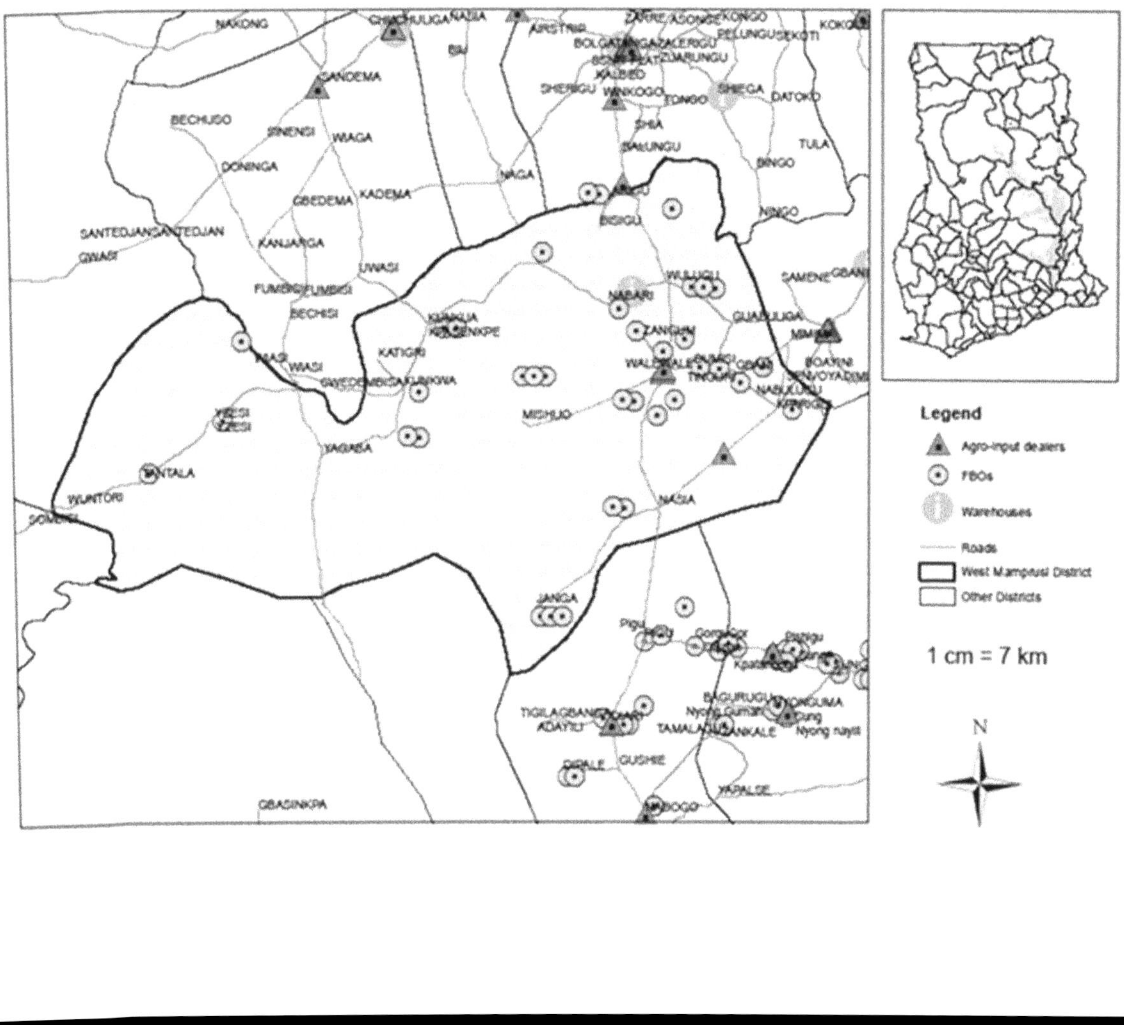

include agro-industries, the Ghana School Feeding Program markets, World Food Program markets, agribusiness centers, and buffer stock warehouses. The targeted aggregators are agents who aggregate produce from thousands of small-scale farmers for supply to the large buyers. At the upstream supply end, the platform targets farmers and their groups as well as input and other service providers. The targeted stakeholders are all linked vertically to each other and when they use the *mFarms* platform, they can manage[5] their own "network" or "groups" to whom or from whom business services are rendered or obtained (Figure 2).

In Figure 2, industrial buyers for example will manage their own network of different aggregators, the aggregators in turn will manage their own network of farmer groups while farmer groups manage their network of individual farmers within their catchment area. Individual farmers can also use the platform to receive crop advisories and alerts or make private business deals. The registered groups also have access to the pool of service providers that they can pull into their network or groups and similarly, service providers will have access to all actors on the platform to provide specialized services. For example, registered tractor service providers and other input dealers on the *mFarms* platform can directly communicate input or tractor service availability information through mass SMS messages to targeted farmers in their network to accelerate the delivery and as a result, cut down the service acquisition time. Other services such as purchase, transport and haulage could all be arranged through information sharing between provider and user in advance and cut the total time considerably.

Because each user category is free to pull stakeholders from other categories into their groups, they will be forced to provide the best services to retain them or face the risk of losing them to better performing competitors. It is this spirit of competition to capture a greater share of the

Figure 2. The mFarms platform: An ICT-based platform for business linkages among actors

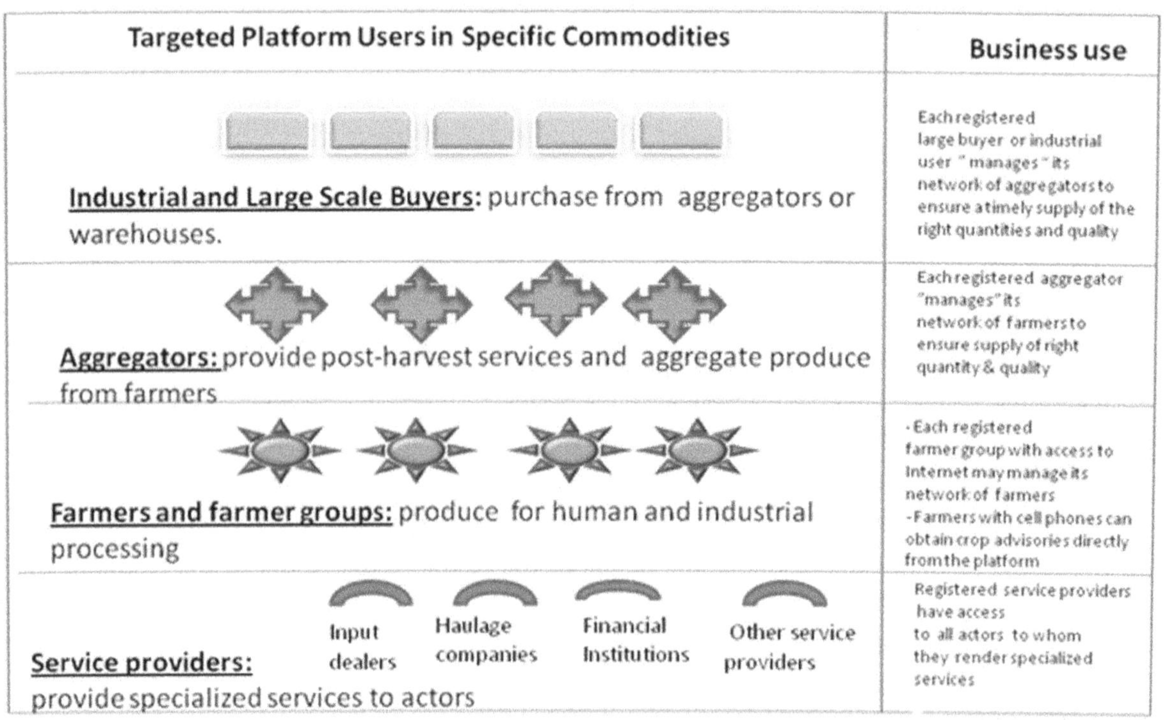

market that will ensure that quality services and competitive pricing are provided along the value chain. Since the user categories themselves are paying fees to use the platform it is likely to be more sustainable and scalable than the traditional donor-funded initiatives.

mFarms's Use for Pre-Season Production and Purchase Planning

Typical smallholder farmers do not usually produce what the market wants but try to sell what they produce. How can we blame them when they find themselves in an information vacuum - not knowing how much to produce, who would buy, for how much and when their produce would be bought. Similarly seed, fertilizer and other agro-chemical sellers are not sure of the type and what quantity of wares to stock. This is so because they do not know how much of what product would be needed and at what period. Aggregators may have fixed supply contracts to deliver (a given quantity and a given date) but are unsure how to get produce of uniform quality from smallholder farmers who are dispersed over space. The industrial buyer is unsure of how much raw material he can procure domestically, thus affecting import plans. The *mFarms* platform is useful for pre-season production and purchase planning as illustrated in Figure 3.

In Figure 3, suppose that prior to the 2010/2011 growing season, an aggregator registered on the *mFarms* platform secures the commitment of the World Food Program (WFP) to supply 2,000 tons of soybeans. The aggregator creates its own network of about 6,000 farmers who are willing to produce to meet the aggregator's requirements. The aggregator uses the medium of the *mFarms* platform to reach all the farmers to ascertain their soybean production plans for the season through a single targeted message by bulk SMS from the platform. Once the farmers transmit information to the aggregator on their intended production figures, the aggregator estimates their seed, fer-

Figure 3. Production and purchase planning with the mFarms platform

tilizer and other requirements and arranges their procurement by mass SMS to input dealers on the platform specifying the input type, quantity and where needed. This allows the input dealers to contact the farmers to solicit their business. The aggregator can also prompt farmers to plant on time by broadcasting through SMS the optimum planting period of the crop. Sometime after planting, the aggregator may send another mass SMS to its farmers to confirm the planted acreage and dates from which preliminary estimates of total production can be made. During the growing season, the aggregator can broadcast cultural techniques periodically to its farmers, coaching them on timely and appropriate practices leading to good yields. Before harvest, the aggregator ascertains when exactly the produce will be ready at each location and arranges transportation for pickup and delivery to the industrial buyer whose commitment was obtained prior to the season.

CONCLUSION

ICT has great potential for overcoming key constraints in the agricultural value chain, and there are limited quantitative and anecdotal evidence of the benefits this technology brings to users. Most of the applications have been made possible through donor funds but became unsustainable when donor support ceases. This is probably because sustainable business models demonstrating benefits to users beyond the costs of investments and maintenance are yet to be found. ICT applications that drive value to users (and therefore are likely to appeal to them), appear to be those that can reduce transaction costs, increase access to markets, improve productivity by providing real-time information on better farming practices, provide better and more frequent access to critical market information and improve communications throughout the value chain. The role of ICT in overcoming the key constraints in the agricultural value chain and for making evidence-based decisions will be greatly enhanced if farmers, aggregators and other stakeholders in the value chain pay attention to their business scope and schedule planning, executing, monitoring and control, procurement, risk planning and stakeholder communications in a "project management" context. When this is done, ICT applications will facilitate supply chain management through sharing of timely and pertinent information on producers, buyers and other services, thereby helping to promote industry competitiveness. The major challenges to the widespread use and sustainability of ICT remain access to the appropriate ICT tool, poor road and storage infrastructure (particularly in the farming communities) and illiteracy on the part of the majority of smallholder farmers.

ACKNOWLEDGMENT

The authors would like to acknowledge the practical and intellectual contributions to this paper of their IFDC colleagues, especially of Dr. Georges Dimithe, Project Leader of the MIR+ project, Mr. Patrice Annequin, MIS Project Leader and Dr. Andre de Jager, Agribusiness Program Leader. Finally, we would like to thank Judy Payne (e-Business and AFR ICT Advisor, USAID EGAT/I & E/ICT) and other anonymous reviewers for their insightful comments.

REFERENCES

Action for Enterprise. (2010). *Using ICT to provide agriculture market price in Africa*. Briefing paper supported by USAID's Fostering Agriculture Competitiveness Employing Information Communication Technologies (FACET) project. Washington, DC: United States Agency for International Development (USAID).

Aker, J., & Mbiti, I. (2010). *Mobile phones and economic development in Africa*. CGD working paper 211.

Debrah, S. K. (2007a). *Final report of the Market Information Systems for Trade Organizations for West Africa (MISTOWA) project*. Unpublished IFDC report submitted to USAID under the USAID Cooperative Agreement No. 641-A-00-04-00230-00, November 15, 2007.

Debrah, S. K. (2007b). *Compiled testimonies of beneficiaries of the MISTOWA project interventions*. Unpublished IFDC report submitted to USAID under the USAID Cooperative Agreement No. 641-A-00-04-00230-00, October 25, 2007.

Dowuona, S. N. N. (2011, February 4). Mobile telecom market share: Three gain, two lose. *MyJoyonline*. Retrieved from http://news.myjoyonline.com/business/201102/60519.asp

Esoko. (2010, October 14). Esoko: The market on your mobile. *Esokonetworks*. Retrieved from http://www.youtube.com/user/esokonetworks

FAO. (2010). *Crop calendar – A crop production information tool for decision making*. Rome, Italy: FAO. Retrieved from http://www.fao.org/agriculture/seed/cropcalendar

International Fertilizer Development Center. IFDC. (2010). *Biannual report of the Ghana Agro-Dealer Development Project (GADD)*, (p. 21). Unpublished report submitted by IFDC to the Alliance for a Green Revolution in Africa (AGRA), September 2010

Muto, M., & Tekashi, Y. (2009). *The impact of mobile phone coverage expansion on market participation: Panel data evidence from Uganda*. World Development Report, 2009.

Project Management Institute (PMI). (2008). *A guide to the Project Management Body of Knowledge (PMBOK Guide)* (4th ed.). Newtown Square, PA: Project Management Institute.

Wikipedia. (2011). *Unstructured supplementary service data*. Retrieved from http://en.wikipedia.org/wiki/Unstructured_Supplementary_Service_Data

Wolter, D. (2008). *Ghana: Agriculture is becoming business*. Paris, France: OECD. Retrieved from www.oecd.org/dev/publications/businessfordevelopment

World Development Report. (2008). *Innovative uses of information technology to link farmers to markets in India and West Africa* (p. 121). Washington, DC: The World Bank.

ENDNOTES

1. An electronic platform that enables loading and spending of electronic cash and settlement of inter-bank claims.
2. Information and Communication Technologies (ICT) include cell phone and Internet services, radio, and a wide range of digital devices and related tools such as cameras, GIS, and a wide range of hand-held computing devices.
3. The purpose here is not to describe the Project Management processes and the nine knowledge areas of *Integration, Scope, Time, Cost, Quality, Human Resources, Communications, Risk and Procurement* management (for further reading on project management, consult PMI, 2008).
4. Stakeholders are persons, institutions or organizations whose interests may be positively or negatively affected, and who may exert influence on the deliverables. The stakeholders here include agro-input dealers, farmers, buyers, transport services, warehouse owners, financial institutions etc.
5. "Manage" involves intra- or inter-group business communication usually through SMS or other tools in order to ensure that the right quantity and quality is produced and supplied at the right price to the aggregator or final buyer at the right time and at the right place.

Chapter 4
Geospatial Data Infrastructure (GDI) for Sharing Agro-Climatic Information to Improve Agricultural Productivity in Nigeria

A. Ayanlade
Obafemi Awolowo University, Nigeria

M. O. Olawole
Obafemi Awolowo University, Nigeria

O. S. Ayanlade
National Centre for Technology Management, Obafemi Awolowo University, Nigeria

O. Bolarin
University of Ilorin, Nigeria

ABSTRACT

Geographic Information Systems (GIS) and Remote Sensing (RS) techniques are increasingly being used in agricultural management to more efficiently support emergency agro-climatic information needed in this century. Although several studies have been carried out on application of GIS and remote sensing for agricultural and rural development in other parts of the world, little research has been done in sub-Saharan Africa in terms of developing geospatial data infrastructure to enhance agricultural practices especially agricultural risk management in this age of climate change. This chapter therefore examines technological aspects of geographic information sharing as a useful platform for sharing information among agricultural agencies in Nigeria and to formulate relevant prescriptions for the future. The study shows how an integrated GIS modeling system can allow agricultural producer as well as policy makers to know the impact of variation in climate from one place/region to another for better management, productivity, and profitability. The study also develops agricultural GDI prototype for agricultural emergency management. This chapter concludes by suggesting that developing country (i.e. African countries) should embark on agricultural policy reform to enhance investment in ICT infrastructure in agricultural production.

DOI: 10.4018/978-1-4666-2655-3.ch004

INTRODUCTION

Geospatial data plays an important role in decision-making and proper management before, during and after agricultural activities. Most importantly, geospatial data infrastructure is an indispensable means of information to response to the impacts and vulnerability of agricultural production to climate change/variability. A spatial data infrastructure is expected to facilitate and coordinate the exchange and sharing of static and dynamic agro-climatic information between all the agricultural agencies. Geographical Information Systems (GIS) and Remote sensing (RS) techniques are the main tools for this means of information sharing (Braun and Pantel, 2004).

GIS is a system specifically designed to work with data referenced by spatial or geographic coordinates. It is therefore both a database system with specific capabilities for spatially referenced data, as well as a platform for analytical operations to work with the data (Kufoniyi and Agbaje, 2005). GIS and RS are increasingly used in agricultural management since they provide tools for more efficient support of emergency agro-climatic information needed in this century. GIS and RS allow for an effective visualization of change in agricultural production which may result from change in climatic situation in a given region (ZhiGong *et al.*, 2009). Therefore, it is evident that geospatial information is now significantly relevant to the agricultural managers for better information sharing not only within the farmers but also essential for all agricultural agencies (Craig 1995; Croswell, 1989; Pandey, 2009; Igbal, 2010). For instance, Igbal (2010) notes that climate variability is a major factor which affecting agricultural practices in the arid region. He observes that application of traditional methods to assess drought impacts on crops production are time consuming and labored intensive. Thus, application of remote sensing and GIS is imperative to the development of agricultural practices in the developing countries.

Though, several studies have been carried out on application of GIS and remote sensing for agricultural and rural development, but little research has been done in the area of developing geospatial data infrastructure to enhance agricultural practices especially agricultural risk management in this age of climate change. Therefore, this chapter aims at examining the technological aspects of geographic information sharing as a useful platform for sharing information among agricultural agencies, taking example from Nigeria and to formulate useful predictions for the future. Major objectives of this chapter are to evaluate mode of data acquisition and information sharing among agricultural agencies; asses the importance of GIS in agro-climatic data capture and information sharing and give a case study of application of GIS for agro-climatic information sharing.

Information Sharing Among Some Agricultural Agencies

Over the past decades, agro-climatic information sharing has been based on manual means of data acquisition and dissemination. Also, majority of primary data employed in the description of agro-climatic conditions in many African countries come from inadequate and poorly equipped meteorological stations. For example in Nigeria, with a land area of 923,768 square km, there are less than 30 standard meteorological stations where observations in respect of the main weather elements are made on a daily basis. Under ideal situation, if the meteorological stations had been regularly spaced, there would be one within each unit area of up to 30,000 square kilometers. There are now available standard techniques for solving these problems (An *et al.*, 2003). Thus, application of modern day GIS and RS approaches in agricultural cropping, agricultural marketing, extension services, irrigation scheduling and management decision making are of significant importance (Pandey *et al.,* 2009). Commercialization and mechanization of agricultural production in many African countries can only be promoted through incorporation of ICT in all agricultural sectors.

Moreso, GDI provides a feasible alternative way to manual observation with a fast system of information collection and transmission. Nowadays, it is possible to manage an incredible large quantity of data such as climatic data, information from agricultural activities etc. These data provide most important inputs which form basic informative layers in GIS environment for agro-climatic information sharing. This means of information sharing provides improved knowledge of existing agricultural problems (such as climate change impact on crops) and allows better agricultural resources planning (Odekunle *et al.,* 2007). There is no doubt that this level of technological development will aid agricultural decision-making and partnerships at agencies, local, state and national levels (Bolarin & Ayanlade, 2010).

The Importance of GIS in Agro-Climatic Information Sharing

Geographical Information System (GIS) as a modern day means of information and communication technologies (ICT) offers a new way of sharing agro-climatic information among agricultural stakeholder such as farmers; agricultural ministries, rural and farm NGOs, agricultural extension agencies among others (Pandey *et al.,* 2007). In general, agro-climatic information sharing has two main goals: (1) it is a system of ICT that aims at enhancing the knowledge of rural farmers and agricultural stakeholders on the impact, vulnerability and adaptation to climate change and (2) it also aims at changing the behavior of rural people and agricultural stakeholders towards proper information sharing for better agricultural production.

Although, the world is rapidly moving towards ICT-based agricultural structure, but many African countries are still behind the circle (Adesina, 2005; Igbal, 2010). The reason for this is that most of African countries are wallowing in poverty and to invest in information sharing technology for agricultural development is not their first priority. The real fact here is that information technology itself would provide the means for countries to turn their disadvantages into advantages and change to ICT-based agricultural practices by putting in place the required infrastructure of information systems to aid agricultural production (Brodie, 1993; Igbal, 2010). Geospatial information field in many of the African countries is still immature with broad set of issues yet to be resolved: awareness and political support, execution of GDI policy and metadata management system in agricultural sector. This is because in the view of many agricultural practitioners, the greatest obstacles to information sharing seem to be behavioral, rather than technical.

Thus, the importance of GDI for sharing agro-climatic information cannot be overstated. It aids technology transfer; provides advisory work for all agricultural stakeholders; enhances agricultural resources development and management and facilitates rural employment. Technology transfer processes include both top-down and bottom-up approaches for agro-climatic information that delivers specific recommendations to farmers about the practices they should adopt when climate turns nasty. GDI provides advisory works in terms of persuasive and participatory. Agricultural extension agencies and agricultural NGOs that use participatory approaches to promote pre-determined packages of technology, will find this function of GDI appropriate in the field of work. Also, GDI aids in creating awareness in terms of teaching and enhancing farmers abilities to adapt to change in climate and aids Precision Agriculture (An *et al.,* 2003).

Application of GIS to Agro-Climatic Information Sharing: A Case of Middle Belt of Nigeria

In this section, we give example of application of GIS in agro-climatology study in the middle belt of Nigeria (See Figure 1). Information about climate change impact is rich and complex, as one of the top important issues in the agricultural production. Well-organized GDI and GIS model will help users of this information understand the influence of climate change on crop yield. Thus, analyzing relevant scientific information on this subject and researching a GDI approach to organize this information into a model becomes the main question in this research. The middle belt of Nigeria is made up of traditional small-holder farmers, who use simple techniques of production and the bush-fallow system of cultivation and accounts for around two-thirds of Nigeria's total agricultural production (Odekunle *et al.,* 2007). The region has a climate which is characterized by relatively high temperatures throughout the year. The average annual maximum temperature varies from 35°C and 31°C throughout the year while the average annual minimum temperature from 23°C and 20°C (Hauser, 2000; Odekunle *et al.,* 2007). Rainfall in the zone is largely seasonal and highly varies from year to year, with mean annual rainfall between 1000mm and 1200mm. Two distinct seasons are observed, dry and wet. The dry season extends over a period of about 6-7 months, from October to March/April while the wet season extends over a period of about 5-6 months, from May to September.

This study develops a GIS data infrastructure on inter-annual rainfall variability and crop yield in the middle belt of Nigeria and maps inter-annual changes in crop yields as a response to inter-annual rainfall variability in the study area. It also relates the spatiotemporal variability in rainfall with millet yields; and model future scenario of the impact of rainfall variability on crop yields. This is with a view to evaluate the spatio-temporal impacts of rainfall variability on crop yields in the area. Although data capture, manipulation, and management are important functions of GIS which are used to support agro-climatic data analysis and decision-making. The literature in the management information systems field is rich with descriptions of various decision support technologies to which GIS can be applied in agricultural sectors. For example, Aronoff, (1989) and Goodchild (1993) have provided useful frameworks for understanding the nature and role of agricultural decision support technologies of GIS. In particular, the framework proposed by Goodchild (1993) is quite useful for defining a practical framework for considering GIS in agricultural activies.

Figure 2 present the GIS framework developed to analyse agro-climatic data in the region. A GIS must not only be able to manage these types of data, but also manage and integrate spatial data (for example data which includes cartographic coordinates). For instance, drought is associated with a number of outcomes. GIS data infrastructure begins by mapping out direct physical consequences of a climatic event or change, such as, reduced crop yields (Igbal, 2010).

In this study, both climatic and agricultural data were used. The Geographic Information Systems (GIS) database was used to create interactive maps of the crops production areas, areas harvested and the yield. The objective of geospatial data analysis to this study is to transform data collected from field into useful visual information. Building a database is the key task in GIS modeling. The database in this research project contained feature datasets, domains, validation rules, and

Figure 1. Map of Nigeria showing middle belt where the study was conducted

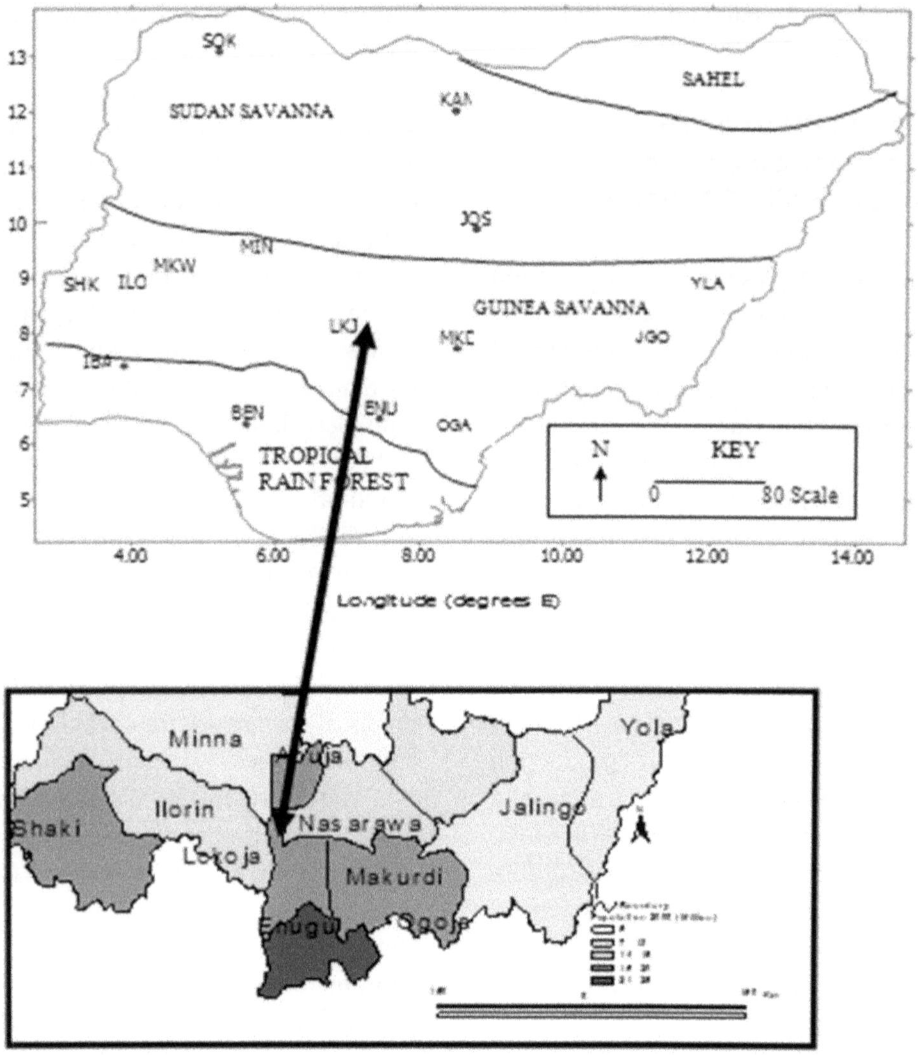

attribute datasets. The results of the analysis and the maps produced show that inter-annual variability of rainfall brings about the differences in the rate of millet yields in the study area. The result of the analysis shows that rainfall in the zone is on the increase but the amount differs. Figure 3 shows a good example of rainfall variability from 1970 to 2000 in the study area, with mean amount of about 1700mm per annum. The figure also shows abnormal low rainfall in 1975 and 1983 with amount less than 900mm. The zonal variability of rainfall, especially, is observed to bring about not only the differences in the types of crops cultivated but also the rate of yield of crops.

This study finds out that an important issue in climate variability is that the basic unit (rainfall) has a spatial distribution which does affect crop yield considerably. GIS data infrastructure is emerging as an important tool in mapping and modeling climate impact on crop production. These findings confirm the outcome of previous

Figure 2. GIS framework developed to analyze agro-climatic data

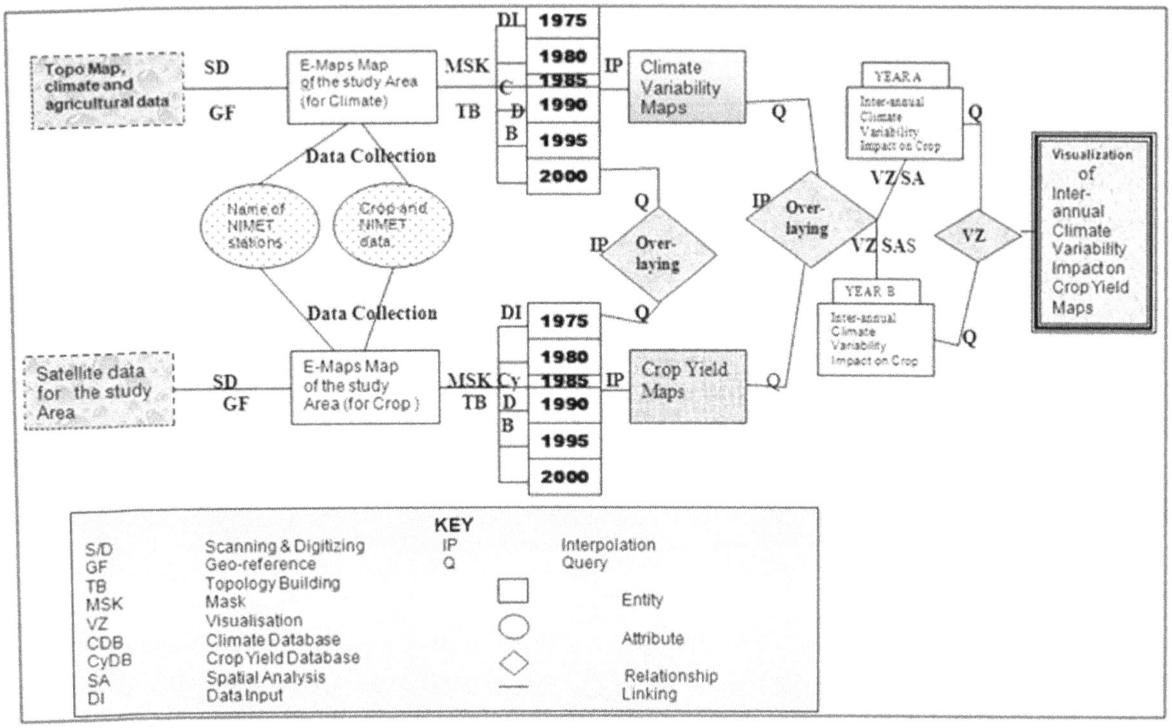

studies (Odekunle *et al.,* 2007; ZhiGong *et al.,* 2009; Igbal, 2010) which showed that GIS and remote sensing are useful tools in agricultural development. It is believed that integration of GIS data infrastructure in agricultural practices is helpful to assess the drought conditions at temporal basis using the meteorological and agricultural parameters.

The present study also finds out that the natural variability in rainfall is a major cause of variability in millet production in the study area. Therefore, the degree of vulnerability of any crop varies, depending on the nature of rainfall variability, the crop growth stage when it occurs and the area where the crop is grown.

Modeling with GIS tool offers a mechanism to integrate many scales of data developed for agricultural research. For example, observation shows that for millet yield, production increases as rainfall increases (See Figure 4). Generally in all stations, increase in length of growing season leads to increase in millet yield. Generally, maize is an important food crop in the study area where it is gradually replacing the traditional cereal crops such as sorghum and millet. Despite its high yield potential, millet production is however faced with numerous constraints, one of which is intermittent drought during the growing season.

This study contributes to knowledge of implementing agro-climatic information. The results from this study show how sharing of agro-climatic information, which is extremely useful, enables farmers to save resources, time and effort when trying to acquire information on climate change impacts on agricultural production. Moreover, the study shows how an integrated GIS modeling system can allow agricultural producers as well as policy makers to know the impact of variation in climate spatially from one place/region to the other for better management, productivity and

Figure 3. Map of the study area showing the rainfall variability pattern. The figure also shows abnormal low rainfall in 1975 and 1983 with amount less than 900mm.

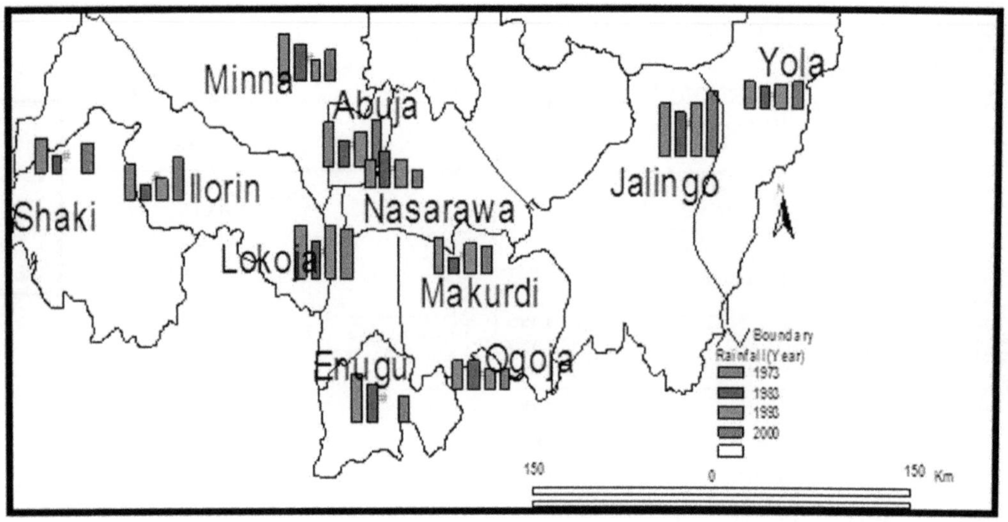

profitability. The study also develops agricultural GDI prototype for agricultural emergency management. Poor adoption of improved technologies in this study area due to combination of poverty, severe environmental conditions and high variability in annual rainfall makes it difficult to improve productivity. The outcome of the study can be used in agricultural decision making and enables wide variety of users to access and disseminate spatial information for better agricultural development in an easy and secure way. With the foregoing in mind, the key question now is:

Figure 4. Map of the study area showing average Millet Yield (1973-2000) and average annual rainfall variability. Millet yield and the degree of vulnerability of any crop varies, depending on the nature of rainfall variability.

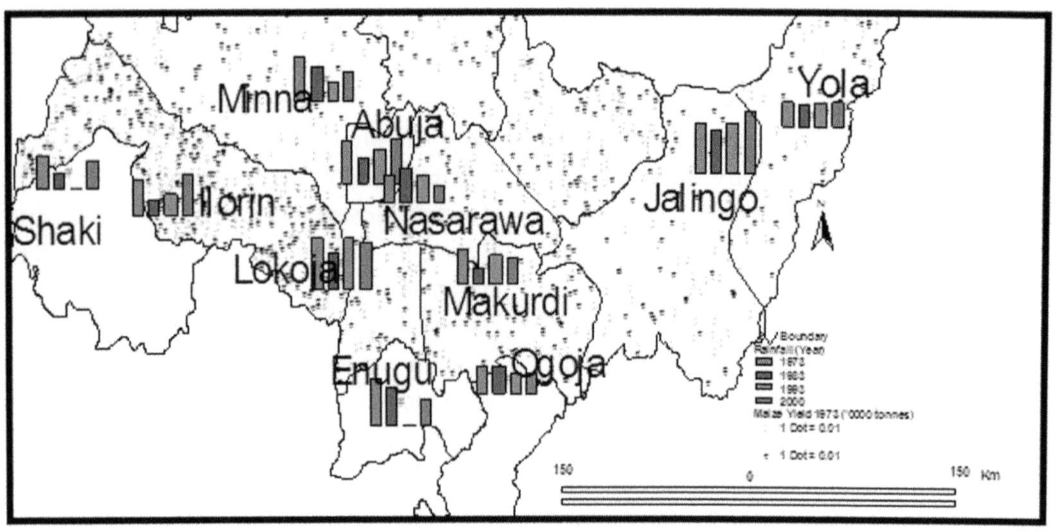

what can be done to achieve an increase in agro-climatic information sharing among the agricultural stakeholders in developing countries?

WHAT CAN BE DONE?

Government of developing nations should encourage agro-climatic information technology research to improve information sharing among agricultural sectors. Such research needs to address a means of improving information communication and technology to aid agro-climatic sharing in the present and future when the climatic conditions may be less favorable for agricultural production. It is obvious that the rate of growth in crop yields will be enhanced by extending agro-climatic information technology research. This will help both downstream to farmers and upstream to the use of tools derived from agro-climatic information technology and biotechnology to assist conventional breeding. Several studies have shown that participatory plant breeding and agro-climatic information technology play a key role for successful yield increases through genetic improvement in rain-fed environments (particularly in dry and remote areas).

Also, developing countries should embark on agricultural policy reform to enhance investment on ICT infrastructure for agricultural purposes. Emerging evidence shows that the right kind of ICT investments can boost agricultural productivities far more effectively than previously thought in many less advantaged countries (UNEP, 2001). We want to suggest that precision agriculture and improved agro-climatic information technology which has been used in the United States and other developed countries should be used in developing countries in Africa. Agro-climatic information technology along with research on integrated applied ICT to adapt conservation technologies for use in unfavorable rain-fed systems could have a large positive impact on agricultural production of any developing nation.

In addition, there is a need for governments in other developing countries to embrace agro-climatic information technology and GDI in agricultural sectors. Over the past 20 years, these ICT innovations have been practiced in many developed nations such as Northern Europe and North America. Introduction of these ICT innovations in other developing countries will aid the development of national database infrastructure on: the size of cultivated area; types of crops available in their countries; production and yield of the crops; time of plantation and harvest. This information will be available through internet servers for better and real-time decision making (Maumbe & Okello, 2010). For example, introduction of ICT in all phases of agricultural activities will be a powerful instrument for improving standards of living throughout Sub-Saharan Africa.

REFERENCES

Adesina, F. A. (2005) *Geoinformation and natural resources exploitation in Africa*. Economic Commission for Africa, Fourth Meeting of the Committee on Development Information (CODI IV) Addis Ababa, Ethiopia 23 – 28 April. [Online]. Retrieved from www.uneca.org/.../Geoinformation%20and%20natural20resources%20exploitation.doc

An, K., Xie, G., Leng, Y., & Xiao, Y. (2003). Design of farmland GIS for precision agriculture. *Chinese Geographical Science*, *13*, 20–24. doi:10.1007/s11769-003-0079-3

Aronoff, S. (1989). *Geographic information system: A management perspective* (pp. 24–37). Ottawa, Canada: WDL.

Bolarin, O., & Ayanlade, A. (2010). Utilization of satellite imaging and information communication technology in agricultural information dissemination in Nigeria. *Journal of Agricultural & Food Information*, *11*, 351–357. doi:10.1080/10496505.2010.520987

Braun, M., & Pantel, A. (2004). GIS for sustainable agriculture. *ArcNews Magazine*, pp. 26-39. Retrieved from http://www.esri.com/library/bestpractices/sustainable-agriculture.pdf

Brodie, M. L. (1993). The promise of distributed computing and the challenges of legacy information systems. In D. K. Hsiao, E. J. Neuhold, & R. Sacks-Davis(Eds.), *Interoperable database systems (DS-5)*, (pp. 1-31). North- Holland, The Netherlands: Elsevier.

Craig, W. J. (1995). Why we can't share data: Institutional inertia. In Onsrud, H. J., & Rushton, G. (Eds.), *Sharing geographic information*. New Brunswick, NJ: Center for Urban Policy Research, Rutgers University.

Croswell, P. L. (1989). Facing reality in GIS implementation: Lessons learned and obstacles to be overcome. In M. J. Salling & W. F. Gayk (Eds.), *Proceedings of the 1989 Annual Conference of the Urban and Regional Information Systems Association (URISA)*, Vol. IV (pp. 15-35). Washington, DC: URISA.

Goodchild, M. F. (1993). The state of GIS for environmental problem- solving. In Goodchild, M. F. (Ed.), *Environmental modelling with GIS* (pp. 8–15). New York, NY: Oxford University Press.

Hauser, S., Ndi, J. N., & Hulugalle, N. R. (2000). Yields of maize/cassava intercrops grown with hedgerows of three multipurpose trees on an acid Ultisol of Cameroon. *Agroforestry Systems*, *49*, 111–122. doi:10.1023/A:1006351025490

Igbal, F. (2010). *Remote sensing and GIS application for drought severity assessment: Using remote sensing and GIS to estimate drought severity* (pp. 45-56). VDM Verlag Dr. *Müller Inc.*, *ISBN-13*, 978–3639303261.

Kufoniyi, O., & Agbaje, G. I. (2005). *National geospatial data infrastructure development in Nigeria: The journey so far* (p. 3). A paper presented at Pharaohsto Geoinformatics FIG Working Week 2005 and GSDI – 8 Cairo, Egypt.

Maumbe, B. M., & Okello, J. (2010). Uses of information and communication technology (ICT) in agriculture and rural development in sub-Saharan Africa: Experiences from South Africa and Kenya. [IJICTRDA]. *International Journal of ICT Research and Development in Africa*, *1*(1), 1–22. doi:10.4018/jictrda.2010010101

Odekunle, T. O., Orimoogunje, I. O. O., & Ayanlade, A. (2007). Application of GIS to assess rainfall variability impacts on crop yield in Guinean Savanna part of Nigeria. *African Journal of Biotechnology*, *6*(18), 2100-2113. ISSN 1684–5315.

Pandey, A., Chowdary, V. M., & Mal, B. C. (2007). Identification of critical erosion prone areas in the small agricultural watershed using USLE, GIS and remote sensing. *Water Resources Management*, *21*, 729–746. doi:10.1007/s11269-006-9061-z

Pandey, V. K., Panda, S. N., Pandey, A., & Sudhakar, S. (2009). Evaluation of effective management plan for an agricultural watershed using AVSWAT model, remote sensing and GIS. *Environmental Geology*, *56*, 993–1008. doi:doi:10.1007/s00254-008-1201-8

UNEP. (2001). *Climate change 2001: Impacts, adaptation and vulnerability.* Summary for policy makers and technical summary of the Working Group II Report. UNEP Policy Series. Retrieved from fire.pppl.gov/ipcc_SWG2_2001.pdf

ZhiGong. P., Yu, L., Di, X., & Lei, W. (2009). Application of RS and GIS technique for water-saving potential estimation of regional crops. *Transactions of the Chinese Society of Agricultural Engineering*, *25*(7), 8-12. ISSN 1002-6819

Chapter 5
ICT and Unemployment in Nigeria:
Implication on Nigerian GDP

N.D. Oye
Federal University of Technology, Nigeria

Inuwa Ibrahim
Federal University of Technology, Nigeria

Muhammad Shakil Ahmad
COMSATS Institute of Information Technology, Faculty of Management, Pakistan

ABSTRACT

A number of telecentres have been established in places like shops, schools, community centre, police stations and clinics. The population of Nigeria, according to the national population commission (NPC) figures stands at over 140, 000,000, and 60% of this number is made up of unemployed youths. With the institutions of learning in Nigeria churning out graduates of various levels and degrees on a yearly basis, a rising trend has seen these graduates coming out of the nation's universities and polytechnics to join those who graduated ahead of them but without any means of livelihood for years. This chapter examines the role played by unemployment on the making of the Nigerian Gross Domestic Product (GDP) for a period of nine years (2000 - 2008). The objectives of the study are to examine the effects of unemployment on the Nigerian GDP for the selected years, to observe the kind of association that existing between the unemployment and the makings of the Nigerian GDP. Data was collected and analyzed using the regression analysis. Findings showed that unemployment has an enormous effect (over 65%) on the making of the Nigerian GDP, and there exists an inverse relationship between the model (unemployment) and the GDP - increase in the model leads to decrease on the GDP and vice versa. The role of ICT on unemployment and GDP is reviewed. In addition ICT as a tool of combating unemployment corruption is discussed. Recommendations are proffered based on the study that unemployment can be combated through the public sector reforms and the use of ICT.

DOI: 10.4018/978-1-4666-2655-3.ch005

INTRODUCTION

This chapter considers ICT to be veritable tools to tackle the rising unemployment in Nigeria. Being without a job is indeed an enforced idleness of wage earners and ICT can help generate youth employment. The increase in mobile phones has led to job creation. ICT is defined as any technology that facilitates communication and assists in capturing, processing and transmitting information electronically. Some commonly used ICTs in many developing countries include Radio, television and print media (Parliamentary Office of Science and Technology, 2006). Modern ICTs such as software, mobile phones and associated applications such as 'VOIP' (transmitting telephone calls over the internet) have become available to many countries worldwide in recent years. However, the most rapid growth is in mobile phone usage (Parliamentary Office of Science and Technology, 2006).

The Ministry of Labor and Productivity in Nigeria says the Federal Government is set to declare total war on unemployment and joblessness in the country with the use of ICT systems and operations which is aimed at training unemployed Nigerians. (Fakiyesi, 2010), opine that despite the exploitation of negative aspect of Information and Communication Technologies (ICT) by a few Nigerian youths, ICT has raised the economy of the nation, especially the Gross Domestic Product (GDP) to 1.62 per cent. Although he lamented that apart from the improvement on the economy, ICT has also enhanced falsehood, dissemination of harmful information as well as fraudulent activities in the transaction of businesses.

Being without a job is indeed an enforced idleness of wage earners who are able and enthusiastic to work but cannot find jobs. In societies in which most people can earn a living only by working for others, being unable to find a job is a serious problem. Because of its human costs in deprivation and a feeling of rejection and personal failure, the extent of unemployment is widely used as a measure of workers' welfare. The proportion of workers unemployed also shows how well a nation's human resources are used and serves as an index of economic movement (positive or negative).

Unemployment has call for a greater concern in the Nigeria economy. It has continued to be the major macroeconomic objectives of the government. Unemployment constitutes a series of serious developmental problems and is increasingly more serious all over Nigeria. The major policy of the government and the international agencies is targeted at reducing the rate of unemployment. Since the population explosion begun, the developing nations have been characterized by unemployment.

1. SMEs FOR JOB CREATION

It is widely recognized that SMEs are very important for economic growth and job creation in both developed and developing countries (Aris, 2006; Harzbo, Arnela, & Chun-yan, 2008; Mutula & Brakel, 2006). Researchers argue that SMEs play a major role in poverty alleviation in developing countries and also stimulate domestic and regional economic growth in national and regional economies (Berisha-Namani, 2009; Golding, Donaldson, Tennat, & Black, 2008). They help to diversify economic activity and are flexible to changing market demands.

The Library of Congress Country Studies and the CIA World Fact book (1991), reported that, measures taken under the Structural Adjustment Program (SAP) in Nigeria has resulted to instability in the unemployment rate - the national unemployment rate, estimated by the Office of Statistics as 4.3 percent of the labor force in 1985, increased to 5.3 percent in 1986 and 7.0 percent in 1987, before falling to 5.1 percent in 1988. Most of the unemployed were city dwellers, as indicated by urban jobless rates of 8.7 percent in 1985, 9.1 percent in 1986, 9.8 percent in 1987, and 7.3 percent in 1988. Underemployed farm labor, often referred to as disguised unemployed, continued to be supported by the family or village, and therefore rural unemployment figures were less accurate than those for urban unemployment. Among the openly unemployed rural population, almost two-thirds were secondary-school graduates. The largest proportions of the unemployed (consistently 35 to 50 percent) were secondary-school graduates. There was also a 40- percent unemployment rate among urban youth aged twenty to twenty-four, and a 31-percent rate among those aged fifteen to nineteen. Two-thirds of the urban unemployed were fifteen to twenty-four years old. Moreover, the educated unemployed tended to be young males with few dependents. There were relatively few secondary-school graduates and the lowered job expectations of primary-school graduates in the urban formal sector kept the urban unemployment rate for these groups to 3 to 6 percent in the 1980s.

(Alanana, 2003), opines that unemployment is potentially dangerous as it sends disturbing signal to all segments of the Nigerian Society. The rate of youth unemployment in Nigeria is high, even at the period of economic normalcy i.e. the oil boom of the 1970s (6.2%); 1980s (9.8%) and the 1990s (11.5%).

Economy watch, (2005) report that unemployment in Nigeria is one of the most critical problems the country is facing. The years of corruption, civil war, military rule, and mismanagement have hindered economic growth of the country. Nigeria is endowed with diverse and infinite resources, both human and material. However, years of negligence and adverse policies have led to the under-utilization of these resources. These resources have not been effectively utilized in order to yield maximum economic benefits. This is one of the primary causes of unemployment in Nigeria.

As per the report of the World Bank, (2005) the GDP at purchasing power parity of Nigeria was $170.7 billion. Unemployment in Nigeria is a major problem both economically and socially. Unemployment in Nigeria has resulted in more and more people who do not have purchasing power. Less consumption has led to lower production and economic growth has been hampered. Unemployment also has social consequences as it increases the rate of crime. The secondary-school graduates consist of the principal fraction of the unemployed accounting for nearly 35% to 50%. The rate of unemployment within the age group of 20 to 24 years is 40% and between 15 to 19 years it is 31%. This paper will examine unemployment within the scope of the Nigerian economy and its implication on the economic growth – the GDP to be precise for some selected years. How far has this unemployment issues have been hindering the positive movement of the Nigerian economic growth. To what extent has unemployment have an effect on the makings of the GDP over the years? The downstream effects of youth unemployment are fueling rapid alienation and social unrest across the Nigerian landscape, the immediate symptoms of which are evidence in the palpable rise in organized crime, armed insurgency, vandalism and drug trafficking.

2. ICT COMBATING UNEMPLOYMENT CORRUPTION

In Nigeria the bulk employment is from the public sectors. While the government declare embargo on employment, this is only meant for the children of the poor mass. The candidates of top rank officer can get job at any time in the place of their choice. The truth is that jobs are being reserved for special people.

A person is corrupt when he is dishonest in his intentions and actions. So corruption means dishonesty in thinking and action. Bribery using one's official position improperly and using money or material of others without their permission are some very common forms of corruption. Corruption is a manifestation of institutional weakness, poor ethical Standards, and insufficient enforce cement of rules and regulations.

When corrupt officials slowly drain the resources of a country, this will greatly affect the employment schedule of that country.

The use of ICT for faster communications, exchange of information, and improved recording and monitoring of information can greatly improve the operational efficiency of Government agencies. For example the issue of unemployment can be tackled easily. ICT facilitates citizen's right to information, augments citizen-government dialogue and encourages people's participation in the economic activities of the nation through employment. Therefore the use of ICT to combat unemployment corruption cannot be overemphasized. The use of ICT can improve the delivery of government services to the people, improve transparency, and accelerate economic growth and employment rate.

3. THE ROLE OF ICT ON UNEMPLOYMENT AND GDP

(Punch, 2009) states that, Lagos state has the plan to curb unemployment through the use of ICT. Again (Vanguard, 2009), wrote that the Lagos Chamber of Commerce and Industry (LCCI) has announce its readiness to help unemployed youths to become entrepreneurs. NICTSE, 2010, comment that Nigeria ICT Platform 2010 is to examine the contribution of ICT to various sectors of the economy, and how it has improved the county's GDP, what planning model can be formulated by the government in ensuring a sustained growth mechanism which can reduce abject poverty and increase business opportunities through the use of ICT among its citizenry. Some countries with high ICT penetration are fully utilizing their ICT potentials which have translated to wealth of nations.

(Vanguard, 2010), Wrote that, the rising rate of unemployment in Nigeria threatens the welfare of hundreds of thousands of individuals. However, industry experts believe that the only way out is to invest in Nigerians' ICT environment, this may effectively provide immediate strategies to compensate for the damages unemployment has created. The argument is that it is good tom invest in the country traditional physical infrastructure like roads, bridges, rail system, etc, however it is also very important to invest in the country's' ICT resources, or digital infrastructure. This will have positive impact on jobs and it will also be groundwork for sustained GDP and innovation.

4. ICT EMPLOYMENT GENERATION FOR THE YOUTHS

The worldwide expansion of mobile phone networks and the growth in the number of mobile phone subscribers has been phenomenal in recent years. According to a recent study, out of Nigeria's population of 140 million, 12.1 million own mobile phones and 64 million "are potential mobile phone users through mobile payphones at call centres." The increase in mobile phones has led to job creation as well.

According to a Nigerian Communications Commission, one million "indirect jobs" have been created by the mobile telephony sector in the past five years. "This figure indicates that mobile telephony could boost job creation and poverty alleviation if the conditions that would stimulate its spin-offs are introduced and nurtured in a consistent manner." About 10,000 "direct jobs," or people employed by the mobile phone operators in Nigeria, have been created by the industry. Telecentres are being set up through public and private initiatives in many developing countries in telephone shops, schools, libraries, community centres, police stations, and clinics. Sharing the expense of equipment, skills and access amongst an ever-increasing number of users also helps to cut costs and make these services viable in remote areas.

4.1 Research Objectives

- To see how ICT can promote youth employment.
- To observe the kind of association that exists between the unemployment and the makings of the Nigerian GDP.
- To determine the role of unemployment in the production of the GDP.

5. LITERATURE REVIEW

William Phillips, in his paper "*The Relationship between Unemployment and the Rate of Change of Money Wages in the United Kingdom 1861–1957*", describes how he observed an inverse relationship between money wage changes and unemployment in the British economy over the period examined (Phillps, 1958). Phillips in his work he made an explicit the link between unemployment and inflation: when unemployment was high, inflation was low, and vice-versa. (Samuelson, 2008) noted the same phenomena as he took Phillips' work and made explicit the link between inflation and unemployment: when inflation was high, unemployment was low, and vice-versa. In the 1920s an American economist in his work (Fisher, 1973) noted this kind of Phillips curve relationship. However, Phillips' original curve described the behavior of money wages A simple curve (Phillips curve) was developed to demonstrate the relationship between the rate of unemployment and inflation. The relationship on the curve shows that as unemployment rate decreases the inflation rate climbs. When the unemployment rate increases, the inflation rate drops. What does this mean? Well essentially, when unemployment decreases more people are employed and output is higher than normal. Higher output and employment leads to an increase in the price level because firms have to pay their workers more. Workers have to pay more because unemployment is low and it is easier for workers to find other jobs and it is difficult for firms to hire new workers; there aren't many people unemployed, i.e. looking for work.

Take a look at the math behind this Phillips Curve. Consider the following equation:

$$\pi_t = \mathrm{v} - \alpha(Ut)$$

where:

- π_t is inflation in year t
- v is a variable denoting exogenous economic shocks
- α is a constant
- U_t is the unemployment rate in year t

Unemployment, U_t, goes up the whole right side of the equation, $v - a(U_t)$ decreases because $-a(U_t)$ becomes bigger. That is, more can be subtracted from v so the whole thing becomes much smaller.

Or, conversely, as U_t gets smaller, i.e. unemployment goes down, the whole right side of the equation, $v - a(U_t)$, increases because $-a(U_t)$ becomes smaller. That is, less can be subtracted from v so the whole thing becomes much larger. This explains how inflation behaves:

- When U_t is HIGH, π_t is low.
- When U_t is LOW, π_t is high.

Not much attention was paid to the fact that Phillips curve was not stable over time. Despite that Phillip's (1958) and Lipsey's (1960) studies found that the relationship shifted over time as indicated by differences in the estimated coefficients for different periods. All the studies carried out in the 1960s tend to show a significant non-linear relation between wage inflation and unemployment. Phillips' work has generated many more empirical studies of the relation between the rate of change of money wages, inflation and unemployment. Other studies have incorporated other variables as explanatory variables for wage or price inflation. Some studies found unemployment to be insignificant in explaining wage inflation. Many more empirical studies have used price inflation instead of wage inflation. According to (Lavacic & Rebmann, 1982) "one can move from a relationship between the rate of change of money wages and unemployment to one between the rate of change of price level and unemployment by allowing for lung-run changes in the productivity of labour".

Many countries (In the 1970s) experienced high levels of both inflation and unemployment also known as *stagflation*. Theories based on the Phillips curve suggested that this could not happen, and the curve came under concerted attack from a group of economists headed by Milton Friedman argued that the demonstrable failure of the relationship demanded a return to non-interventionist, free market policies. The idea that there was one simple, predictable, and persistent relationship between inflation and unemployment was, at least, questioned. The Phillips Curve doesn't quite fit the facts perfectly. Remember this curve very accurately depicted the data available in the 1960s, but it actually fell apart and had to be revised thereafter.

6. NEW CLASSICAL VERSION OF THE PHILLIPS CURVE

New theories, such as rational expectations and the NAIRU (non-accelerating inflation rate of unemployment) arose to explain how stagflation could occur. The latter theory, also known as the "natural rate of unemployment", distinguished between the "short-term" Phillips curve and the "long-term" one. The short-term Phillips Curve looked like a normal Phillips Curve, but shifted in the long run as expectations changed. In the long run, only a single rate of unemployment (the NAIRU or "natural" rate) was consistent with a stable inflation rate. The long-run Phillips Curve was thus vertical, so there was no trade-off between inflation and unemployment. Edmund Phelps won the Nobel Prize in Economics in 2006 for this.

The Phillips curve equation can be derived from the (short-run) Lucas aggregate supply function. The Lucas approach is very different from that the traditional view. Instead of starting with empirical data, he started with a classical economic model following very simple economic principles.

Start with the aggregate supply function:

$$Y = Y_n + a(P - P_e)$$

where:

- $\boldsymbol{Y}$ is the log value of the actual output
- $\boldsymbol{Y_n}$ is the log value of the "natural" level of output
- $\boldsymbol{a}$ is a positive constant
- $\boldsymbol{P}$ is the log value of the actual price level
- $\boldsymbol{P_e}$ is the log value of the expected price level.

Lucas assumes that Y_n has a unique value. The equation indicates that when expectations of future inflation (or, more correctly, the future price level) are *totally accurate*, the last term drops out, so that actual output equals the so-called "natural" level of real GDP. This means that in the Lucas aggregate supply curve, the *only* reason why actual real GDP should deviate from potential—and the actual unemployment rate should deviate from the "natural" rate—is because of *incorrect expectations* of what is going to happen with prices in the future. This differs from other views of the Phillips curve, in which the failure to attain the "natural" level of output can be due to the imperfection or incompleteness of markets, the stickiness of prices, and the like.

In the non-Lucas view, incorrect expectations can contribute to aggregate demand failure, but they are not the only cause. To the "new Classical" followers of Lucas, markets are presumed to be perfect and always attain equilibrium (given inflationary expectations).

Most economists no longer use the Phillips curve in its original form because it was shown to be too simplistic. This can be seen in a cursory analysis of US inflation and unemployment data 1953-92. There is no single curve that will fit the data, but there are three rough aggregations—1955-71, 1974-84, and 1985-92—each of which shows a general, downwards slope, but at three very different levels with the shifts occurring abruptly. The data for 1953-54 and 1972-73 do not group easily, and a more formal analysis posits up to five groups/curves over the period.

According to two influential papers that incorporate a New Keynesian Phillips curve are Clarida,(Blanchard, Oliver, Gali, & Jordi, 2007). An equation like the expectations-augmented Phillips curve also appears in many recent New Keynesian dynamic stochastic general equilibrium models. In these macroeconomic models with sticky prices, there is a positive relation between the rate of inflation and the level of demand, and therefore a negative relation between the rate of inflation and the rate of unemployment. This relationship is often called the "*New Keynesian Phillips curve*." Like the expectations-augmented Phillips curve, the New Keynesian Phillips curve implies that increased inflation can lower unemployment temporarily, but cannot lower it permanently.

7. DATA AND METHOD OF ANALYSIS

The major focus of this paper is to examine the effects and the relationship existing between the unemployment and the GDP (at current prices). As such the GDP data corresponding to that of the unemployment was collected for the period of nine years (2000 - 2008) for the study as in Table 1.

In other to get the better, objectives and findings of this paper the econometrics (simple linear regression) method of analysis would be used on the data collected.

7.1 Findings

The results of the analysis presented below shows the effects, contribution and the relationship linking the unemployment and the making of the GDP for the years under study (See Table 2).

Thus, the result is graphically represented in Exhibit 1.

The results have shown that 65.5% variation of the GDP is explained by the model (unemployment). Similarly, as the unemployment tends to zero the GDP tends to $240,089,308.8. Increase

Table 1. Trends in GDP and unemployment, 2000-2008

S/N	YEAR	Y GDP ($)	X Unemployment
1	2000	110500000	34534590.16
2	2001	117000000	35457975.28
3	2002	105900000	33744170.04
4	2003	113500000	37486876.84
5	2004	114800000	38430877.24
6	2005	125700000	36056156.64
7	2006	175500000	4060102.718
8	2007	191400000	7831807.512
9	2008	294800000	8588118

Source: National Bureau of Statistics, Population Reference Bureau & UNAIDS, CBN, http://www.answers.com/topic/nigeria, CIA World Fact book.

Table 2. Regression result showing relationship between unemployment and GDP

Coefficient of determination	Coefficients	
R^2	α	β
0.655	240,089,308.8	-3.43665

in the unemployment will lead to decrease in the GDP figures by $3.436646181205 and vice versa. This has shown an inverse relationship between the model and the GDP for the indicated period (2000-2008) in Table 1.

8. CONCLUSION

Unemployment has shown an enormous effect (over 65%) on the making of the Nigerian GDP (that is depending on whether unemployment increase or decrease) for the years under study (2000 - 2008). Increase on the unemployment rate has drastically diminished the making of the Nigerian GDP over the years. Unemployment has a pivotal role to play as increase in its figures will lead to economic drawback and vice-versa. There is indeed an inverse relationship between the unemployment and the GDP for the period under study.

This study has revealed that external debt, exchange rate and capacity utilization among others are significant explanatory variables of unemployment in Nigeria. Mobile phones in Nigeria have transformed the country, economically, socially, and democratically as well.

9. RECOMMENDATIONS

Unemployment can be combated through the Public Sector Reforms. The government should enhance a proper man management system. The development of any country very much depends on the caliber and organization of the human resources. So, the government of Nigeria should set on ground a public sector reform that is strong and dependable. The government should encourage and support ICT employment generation through youth entrepreneurship.

The government should encourage Telecentres to be set up in places like shops, schools,

Exhibit 1.

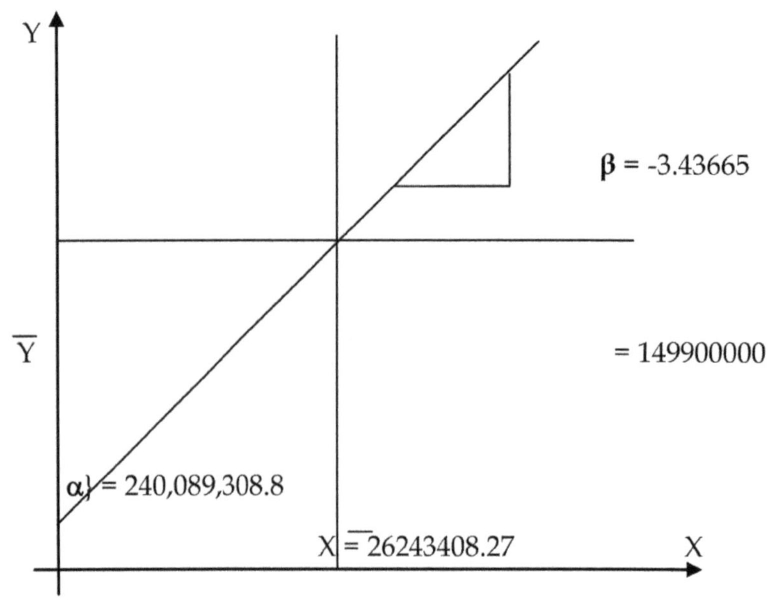

Notes: **Alpha (α):** Constant of the GDP as the model (unemployment) tends to zero. **Beta (β):** Slope of the model. Change on the GDP as a result of the corresponding changes on the unemployment.

community centers, police stations and clinics. People should be ready to engage themselves in one work or the other as over dependence on the government work will not be the right idea. To curb unemployment problems in Nigeria, the government must provide power (electricity) and security as this will give an enable grounds for the foreign investors to invest in the economy.

REFERENCES

Alanana, O. O. (2003). Youth unemployment in Nigeria: Some implication for the third millennium. *Global Journal of Socila Sciences*, *2*(1), 21–26.

Aris, N. M. (2006). *Building blocks for economic growth*. Paper presented at the National statistics Conference.

Berisha-Namani, M. (2009). *The role of information technology in small and medium sized enterprises in Kosova*. Paper presented at the Fulbright Academy 2009 Conference Small Places can Change the World.

Blanchard, O., & Gali, J. (2007). Real wages rigidity and the new Keynesian model. *Journal of Money, Credit and Banking*, *39*(1), 35–65. doi:10.1111/j.1538-4616.2007.00015.x

Fakiyesi, T. (2010). ICT raise Nigerian GDP. *ITRealms Link*.

Fisher, I. (1973). I discovered the Phillips curve: A statistical relationship between unemployment and price changes. *The Journal of Political Economy*, *81*(2), 496–502. doi:10.1086/260048

Golding, P., Donaldson, O., Tennat, V., & Black, K. (2008). *An analysis of factors affecting the adoption of ICT by SMEs in rural and urban Jamaica* [Electronic Version]. Retrieved from http://is2.lse.ac.uk/asp/aspecis/20080109.pdf

Harzbo, S., Arnela, C., & Chun-Yan, H. (2008). ICT adoption model of Chinese SMEs. *International Journal of Business Research*, *44*, 161–165.

Lavacic, R., & Rebmann, A. (Eds.). (1982). *Macroeconomics*. London, UK: Macmillan.

Mutula, M. S., & Brakel, P. V. (2006). E-readiness of SMEs in the ICT sector in Botswana, with respect to information access. *The Electronic Library*, *24*, 402–407. doi:10.1108/02640470610671240

Phillps, A. W. (1958). The relationship between unemployment and the rate of change of money wages in the United Kingdom 1861-1957. *Economica*, *25*(100), 283–299.

Punch. (2009). LCCI move to curb youth unemployment. *The Punch Newspaper*.

Samuelson, P. (2008). *Thought about the Phillips curve* [Electronic Version]. Retrieved from http://www.docstoc.com/docs/52714771/Thoughts-About-The-Phillips-Curve-Paul-A-Samuelson-3

Vanguard. (2009). Unemployed youth transformed to become entrepreneur. *The Vanguard Newspaper*.

Vanguard. (2010). Government to invest in Nigerias' ICT environment. *The Vanguard Newspaper*.

Chapter 6
ICT in Agricultural Development:
Its Diffusion and Adoption Pattern in Nigeria's Agricultural System

O. A. Lawal-Adebowale
University of Agriculture, Nigeria

ABSTRACT

The transformational role of ICT in the social and economic spheres of man engendered development of the communication technology in Nigeria. Based on the telecom deregulation policy of the Nigerian Government in 2001, the country witnessed an upsurge of private investment in ICT development in the country and its consequential ICT revolution in Nigeria. An empirical survey of diffusion and adoption pattern of ICT in the Nigerian agricultural system showed that mobile phone constitutes the most widely adopted ICT device by the agricultural researchers, extension agents, and rural farmers. In addition to this was computer and internet adoption by the researchers and radio transistors by the rural farmers. All eight agricultural research organisations and four extension agencies surveyed had electronic mail addresses and telephone lines for information exchanges. Six of the agricultural organisations - CRIN, FRIN, NIHORT, IAR&T, NIOMR, and NAERLS, and one of the extension agencies – LSADA, had organisational websites for hoisting generated agricultural information. Based on rural farmers' access to radio, and to some extent television, the research and extension organisations aired agricultural programmes on these electronic media for the farmers. ICT usage in the Nigerian agricultural system is however challenged by less development and exploration of communication technologies, poor electric power supply, and poor human operational skills development for effective utilisation of the information-driven technologies. This situation thus necessitates the need for pro-agriculture ICT development and human capacity for effective utilisation in the Nigerian agricultural system.

DOI: 10.4018/978-1-4666-2655-3.ch006

INTRODUCTION

Transformation of economies of the sub-Saharan countries is largely conceived to be contingent to agricultural development (IFPRI 2002). This is based on the fact agriculture constitutes the major occupation of the countries. For instance, nearly 70% of the Nigerian population depends on agriculture as means of livelihood (Popoola, 2006). In view of this, research efforts are consciously directed toward the development of agriculture in the country. Traditionally, developed agricultural innovations are disseminated to the farmers through field visits and on-farm demonstrations by the research and extension personnel. This physical method of research-extension-farmer interaction though has its attendant advantage, promptness of information flow among the research, extension and farmer subsystem remained an issue of concern. In the light of this, Kasibante et al. (2004) stress that physical method of interaction is time consuming, expensive to effect and slowed down activities between two or more units, especially those that separately exist from another. These shortcomings, as opined by Lawal-Adebowale and Adebayo (2006) could however be eliminated or reduced to the barest level through the integration of Information and Communication Technology (ICT) in the communication or interaction network of the agricultural development system.

ICT has been variously described to encompass any electronic device that facilitates the collection, processing, generation, storage, retrieval, dissemination and utilisation of data and information efficiently (Food and Agriculture Organisation - FAO, 1993; Treinen, 2004; Lawal-Adebowlae, 2009). The electronic devices ranged from simple transistor radio, television, video and audio cassette or compact disc (CD) players, camera, telephone (mobile or fixed line), telex/fax to more complex ones such as the satellite, computer, software, internet, personal Digital Assistance (PDA), Geographical Information Systems (GIS), Global Positioning Systems (GPS), remote sensing etc. Each of the ICT devices serves different purposes and as such, effective utilisation of one or more of the communication technologies is contingent to good knowledge of the technology's functional dynamics and ones' operational skills of the technologies. And given the transformation value of the ICT in information generation, management and utilisation, the technology has become widely applied in human socioeconomic endeavours.

INFORMATION AND COMMUNICATION TECHNOLOGY DEVELOPMENT IN NIGERIA

Following the deregulation of the telecommunication sector about a decade ago in Nigeria, the country has witnessed an increasing growth in ICT development and usage among its citizenry. According to the telecommunication regulating body in Nigeria, the National Communication Commission – NCC (2010), the upsurge in mobile telephone usage in the country stood at 71 million, with teledensity of 57.08 as at October 2010. This position overwhelmingly outweighs the teledensity of 0.45 at which the country was before 2001. The increased teledensity was brought about by the privatisation of the telecom sector which gave room for private investment in the sector and consequent transformation of telephony communication in the country.

The same deregulation policy prompted private investment in internet development in the country. As at 2001, the International Telecommunication Union (ITU) indicated that about 723 internet hosts were available in Nigeria, constituting about 0.3% of the total internet hosts available in Africa. With private sector participation in the Nigerian ICT development, there exist about 20 Internet Service Providers (ISP) in the country (Jidaw System Limited, online). According to the National Policy on Telecommunication – NPT

(2010), the country has internet backbone with an initial capacity of 5, 500 ports, out of which 3000 ports serves as the main point of presence (POP) in Lagos State and 1000ports in Abuja. In addition, Kaduna, Bauchi and Enugu have 500 ports. All POPs are interconnected by a 2Mb/s digital transmission link.

On a similar note, the liberalisation and privatisation policies of the Nigerian Government was not only meant to encourage private and foreign investment in computer and Internet build up, but also to have the prices of computer and ICT equipment drastically reduced, such that the devices could be affordable and accessible to intended users. In this regard was the inauguration of Computer for All Nigerians Initiative (CANI) by the Federal Government of Nigeria (FGN), in partnership with Intel Corp and Microsoft. Under this scheme, the FGN, which aims at making computers available to all Nigerians, sets the target of selling 500,000 computer units to its citizens at 50% discount cost (30% discount and 20% subsidised by employers) within the year 2006 (The PUNCH, 2006a). In order to assist the FGN to achieve this objective, Intel Corp and Microsoft, in collaboration with other local computer assemblers, worked toward selling 1million computers to homes and individuals at concessionary price of not more than N50, 000.00 (US$333 at N150 a dollar) for a full set of personal computer with an authentic operation system (The PUNCH, 2006b). To successfully achieve this, provision was made to assist individuals to obtain bank loans to pay for the computer after which deductions would be made from their salaries to repay the loans over a period of 24months (The PUNCH, 2006a). Other computer penetration initiative schemes which aimed at making computer available to people without difficulties in the country include the PC-1 and one laptop per child (The PUNCH, 2006b).

To further enhance the functionalities of the satellite-based ICT in the country, the Nigerian Government launched the country's geostationary communication satellite tagged NIGCOMSAT – 1 on the 13th day of 2006. The NIGCOMSAT – 1 is to provide bandwidth for various communication activities in the country, which may range through the telecom services, broadcasting for the radio and television, broadband multimedia and internet services, and ease of communication between various points – cutting across the rural and urban point.

ICT in the Nigerian Agricultural System

With the progressive development of ICT in Nigeria, the electronic communication devices have become integrated in the Nigerian agricultural system. A field survey by Lawal-Adebowale, (2009) showed that ICT such as computer/ laptop, internet, telex/fax, multimedia projector and mobile phones are steadily being used in related agricultural activities. Other ICT that have had applications in the Nigerian agricultural system include radio, television, radio and/or video cassette/CD players. Application of these technologies however varies across the agricultural research, extension and farmer subsystems. For instance, it was observed that all the eight agricultural research organisations surveyed in this study had computers and 87.5% of them had fixed telephone lines. Six of the agricultural research organisations namely[1] CRIN: http://www.crin-ng.com, IAR&T: http://www.iart-ng.com, FRIN: http://www.frin.gov.ng, NIHORT: http://www.nihort.org, NIOMR: http://www.niomr.com; www.niomr.edu.ng, NAERLS: http://www.naerls.org had organisational websites for hoisting their respective research-based agricultural information. All the eight research organisations had electronic mail addresses herefore indicated as – CRIN: enquiries@crin-ng.org or director@crin-ng.org, IAR&T: drart@infoweb.abs.net, FRIN: dfrin@infoweb.abs.net, NCRI: ncri@skannet.com, NIHORT: info@nihort.org, nihortinfo@yahoo.com, NIOMR: niomr@

hyperia.com, NSPRI: nspriibadan@yahoo.com, NAERLS: swoffice@naerls.org; naerls@abu.skannet.com.ng, director@naerls.org

Similarly, all the four state extension agencies[2] surveyed had computers and fixed-line telephones. Only two of them had telex/faxes. Only one of the extension organisations, namely LSADA, had organisation website with page address given as: http://www.lagosada.gov.ng. About 75.0% of extension agencies, namely LSADA, OGADEP and OYSADEP had electronic mail addresses given as LSADA: pm@lagosada.gov.ng; lagosada@yahoo.com, OGADEP: ogadep@skannet.com, OYSADEP: oysadep@skannet.com; osadepsk@yahoo.com. Common with the farmers were ICT such as radio and mobile phones and to some extent television and video players.

The agricultural organisations, though did not own radio or television stations for broadcast of agricultural information, they however take advantage of the government and privately owned broadcast media for dissemination of agricultural information. In view of this, the agricultural organisations aired a number of agricultural programmes on radio and television for education and information of the rural farmers and other would-be users of the agro-information. Among the broadcast programmes are *Lets go farming* on radio by NIHORT, *Periscope* on television by NSPRI, *Sericulture* on radio and television by FRIN and *Tree crops* on radio and television by CRIN. The entire extension organisations (100%) aired agricultural programmes on the media broadcast. Specifically OGADEP aired agricultural programmes such as *Agbe afokosoro* (crop farmers), *Agbe eleran osin* (livestock farmers) *and Agbelere* (farming is profitable) and *FADAMA S'ayedero* (FADAMA brings relief) on both radio and television; *Agbeloba* (farmer leads) by OYSADEP on radio and television, *Boluyo* (feed the community satisfactorily) on radio by LSADA and *Aye Agbe* (farmers' livelihood) on radio and *Agbe Iwoyi* (modern farmer) on TV by OSSADEP. Most of the aired agricultural programmes were through radio broadcast, and this was largely due to the fact transistor radios are readily used by the rural farmers as means of receiving broadcast information. To some extent, the farmers have access to television sets and video player which as well served as means of relaying information to the farmers.

With these ICT devices in the Nigerian agricultural system it implies that appropriate integrated use of them could bring about the establishment of a functional linkage system between the research, extension and farmers subsystems of the agricultural development process and could as well be used to fast track technical information generation, dissemination and utilisation for agricultural development.

ICT DIFFUSION AND ADOPTION PATTERN IN NIGERIA'S AGRICULTURAL SYSTEM

Based on integration of ICT in the Nigerian agricultural system, its diffusion pattern among the research, extension and farmer subsystems varied across the subsystems. While field observation showed that mobile phone was owned by all the agricultural researchers and extension agent, about 62.9% of the rural farmers[3] owned this communication device, thus making it the most widely utilised ICT device in the Nigerian agricultural system. The wide spread diffusion of mobile phone among the agricultural stakeholders cannot be unconnected with the ease and cheap rate at which a mobile line and a handset could be obtained in Nigeria. At present, field observation showed that acquisition of the Subscribers Identification Module (SIM) pack is near free. This is underpinned by the fact that a SIM pack of any of the four mobile phone service providers[4] which sold for about N100 is already loaded with N100 worth of airtime. The call tariffs have equally reduced to as low as 20kobo per second in the case of Airtel and Etisalat. In essence, subscribers on either Etisalat or Airtel lines could make calls to any other line at the rate of 20kobo per second but on the condition of daily deduction of 20kobo as administrative charges by Etisalat and 60kobo per second for the first minute call on daily basis by Airtel.

Knowledge of operation of mobile handset was also a factor. While the researchers and extension agents have the knowledge of making applications of telephone features such as text messaging, photographing and documentation of texts and voice and image recordings, the farmers, most of who are less educated, but on simple training on phone usage were able to operate the telephone set, at least for receiving calls at the click of the phone-pick button. They may however need to be assisted to locate or key-in numbers to make calls. Further interactions with the researchers, extension agents and farmers revealed that all of them conceived acquisition of mobile telephone as inevitable communication tool that facilitates their linkage with other sub-units of concern and enhanced prompt and efficient communication exchanges with the linked sub-units.

One another note, computer has a wide application among the agricultural researchers, against what it was with the extension agents. This was based on the fact that all researchers considered computer or laptop as crucial to their research activities. This electronic communication device was found essential for typesetting and preparation of research reports, documentation and retrieval of the reports, and data management and analysis, and creation of research data base for the research organisations. These applications however depend on the installed software packages on the computer system. Similarly, the internet was conceived as important by all the researchers as it afforded them, not only the opportunity of having their research outputs hoisted on their organisation's websites, but as well make it possible for them to readily source information of relevance that could be used to enhance their research activities. In the same vein, the internet creates the platform for knowledge sharing of their achievements with other stakeholders in agricultural development. The telex, as indicate by 79.6% of researchers, facilitates prompt information exchanges with other stakeholders in agriculture. Utilisation of these ICT devices by the researchers in their agricultural activities could as well have been influenced by the researchers' knowledge of functionalities and operation of the communication devices.

Unlike the researchers, only a few of the extension agents indicated that computer (9.4%), internet (16.4%) and telex (6.4%) are relevant to their extension activities. The low level of relevance indication of these electronic communication devices was due to the extension agents' poor knowledge of functionalities and lack of operational skills of the technologies. In addition, low level of computer, internet and telex/fax diffusion among the extension agents cannot be unconnected with the fact that extension services are carried out in rural areas where these ICT devices cannot be readily used due to lack of electrical power supply, and none acquisition and/or utilisation of the technologies by the rural farmers. Some of the extension agents who conceived computer as relevant to their extension service delivery, particularly in combination with the multimedia projector, had limited usage of the system in the rural communities, but good use when attending organised workshop or seminars, which normally take place in towns and hotels with electric power supply. Similarly, internet has limited applications among the extension organisations as they hardly source for agricultural information on their own but only depend on the one handed down to them by the subject matter specialists for field applications. In view of this, the extension agents mostly relate with the farmers through physical contact for field or result demonstration and through radio broadcast.

In addition to mobile phone usage among the rural farmers, about 92.4% of them had transistor radio and 23.7% of had television set. Based on this, it becomes possible to inform and educate the farmers through radio and television broadcast. Integration of the mobile phone in the electronic media broadcast helps to break the barrier of impossible feedback or interaction that was inherent in the electronic media broadcast (Lawla-Adebowale and Adebayo, 2006). By this, it becomes possible for the rural farmers to make calls on life programme for further clarification of agricultural issues being discussed. The use of video and audio players was observed among about 37.9% and 11.2% of the rural farmers, respectively. In view of this, agricultural issues of concern could be packaged in either audio or video cassettes/CD for the farmers. Depending on the content and clarity of the packaged information the farmers could playback the cassettes/CD for viewing or listening. Although, the initial cost of video/audio documentation may be expensive, as indicated by Isiaka (2006) it becomes cheaper overtime given the number of rural farmers that could be reached with the packaged messages.

CHALLENGES AND PROSPECTS OF ICT EXPLORATION FOR AGRICULTURAL DEVELOPMENT IN NIGERIA

Given the status of ICT development in Nigeria, it shows that the country has the potential of facilitating interactive linkages among stakeholders in the agricultural system for synergic agricultural development. For effective exploration of ICT for agricultural development, a number of challenges would have to be overcome. For instance, development of ICT infrastructure remains an issue of serious concern. As observed in the course of the field survey, the ICT base of the agricultural research institutes are not fully developed as the internet service functions epileptically and computers are not available to all researchers individually. These are however grossly lacking in all the out stations of the extension organisations. Apart from the mobile phone service which has had rural penetration in the country, internet facilities and rural radio and television stations or units are grossly lacking. In addition, the life style of the rural dwellers is overwhelmingly characterised by non-acquisition of formal education thereby lacking the essential operational skills for use of the modern technologies. Steady power supply needs to be ensured for effective and regular use of the available ICT. Although, in the present day Nigeria, nearly everyone rely on generator as power source but this had to be operated at extra cost and as such powered-on only when it becomes essential to do so. The implication of this is that research activities are greatly affected and information generation, flow, utilisation become obstructed.

Notwithstanding the identified challenges in ICT integration in the Nigeria's agricultural system,, efforts needs to be put in place to fully explore the potentials of the existing ICT devices, and extra efforts need to be made to incorporate the more advanced and sophisticated ICT in the Nigeria's agricultural system. For exploration of the existing ICT devices, all the agricultural research institutes not only need to hoist their organisation websites but should ensure its functionalities such that it could be readily and easily accessed at any point in time by interested users and the hoisted information should be regularly updated. The extension organisations should as well have the internet facilities installed, not just at the headquarters, but across their out stations for enhanced interactive linkage with other stake holders. The internet service provisions of all the mobile service providers in Nigeria should as well be explored for usage at the rural level since the mobile services have penetrated the rural communities. With this, ICT hub could be established for internet usage by the rural dwellers. This will be in line with some rural communities in India as reported by Meera et al. (2004), whereby a knowledgeable person is stationed in the ICT hub to source online information for the illiterate farm families and make their farm information needs known to appropriate agricultural research institutes for response.

Although, application of the existing ICT devices in the Nigeria's agricultural system might be considered appropriate (Michiels and Van Crowder, (2001), efforts need to be put in place for incorporation and exploration more advanced ICT devices. Notable ones include Global Positioning System (GPS), web-based camera servers and FieldServers (Otuka and Yamawaka, 2003; Otuka and Sugawara, 2003; Matsumoto and Machida 2002; Fukatsu and Hirafuji, 2003) for automatic monitoring of field events. By this it becomes much easier for the agricultural researchers to take agronomic and livestock productivity data accurately and thus be able to draw sound conclusion and establish efficacy of the developed technologies at the end of data analysis. To take advantage of the international market the Nigeria agricultural system needs to explore the use of ICT to create traceability record of the developed or produced agricultural crops and livestock products. In addition, the technologies create room for data banking and mining thereby creating knowledge system for transformation of the Nigeria's agriculture.

REFERENCES

FAO. (1993). *The potentials of microcomputers in support of agricultural extension, education and training* (pp. 1–173). Rome, Italy: Food and Agriculture Organisation of the United Nations.

Fukatsu, T., & Hirafuji, M. (2003). Development of field servers for a field monitoring system. *Agricultural Information Research, 12*, 1–12.

IFPRI. (2002). *Reaching sustainable food security for all by 2020: Getting priorities and responsibilities right.* Washington, DC: International Food Research Policy Institutes (IFPRI).

Isiaka, B. T. (2006). *Potentials and effectiveness of video self-learning method in agricultural information dissemination in Southwest Nigeria.* PhD Thesis, University of Agriculture, Abeokuta.

Jidaw System Limited (Online). (2011). *Internet service providers in Nigeria.* Retrieved from http://www.jidaw.com/isp.html

Kasibante, F., Mbowa, S., & Mugisha, J. (2004). Determinants of use of information and communication technologies by agribusiness firms in Uganda. *Eastern Journal of Rural Development, 20*(1), 3–11.

Lawal-Adebowale, O. A. (2009). *Information and communication technology usage in research-extension-farmer linkage system for agricultural development in Southwest Nigeria.* PhD Thesis, University of Agriculture, Abeokuta.

Lawal-Adebowale, O. A., & Adebayo, K. (2006). Information and communication technologies: Pathways to effecting feedback in mass extension method of message delivery. *Proceedings of Agricultural Extension Society of Nigeria*, (pp. 184–190).

Matsumoto, K., & Machida, T. (2002). Studies on development of work journal supporting systems with speech recognition. *Agricultural Information Research, 11*, 255–262.

Meera, S. N., Jhamtani, A., & Rao, D. U. M. (2004). *Information communication technology in agricultural development: A comparative analysis of the three projects from India*, (pp. 1-14). Agren Network Paper No. 135.

Michiels, I. S., & Van Crowder, L. (2001). *Discovering the "magic box": Local appropriation of information and communication technologies (ICTs)*. Sustainable Development Department (SD), Food and Agriculture Organisation of the United Nations SD Dimensions Special. Retrieved from www.fao.org/sd/

NCC. (2010). *Subscriber data at a glance (year 2009 – October 2010)*. Retrieved from http://www.ncc.gov.ng

NPT. (2000). *National policy on communication*. Ministry of Communication. Retrieved from http://www.ncc.gov.ng

Otuka, A., & Sugawara, K. (2003). A labor management application using hand-held computers. *Agricultural Information Research, 12*, 95–104.

Otuka, A., & Yakamawa, T. (2003). An application for insect field survey using a hand-held computer. *Agricultural Information Research, 12*, 113–124.

Popoola, L. (2006). Socio-economic potentials of the Nigerian rural landscape and ways of utilising them toward rapid economic and industrial revolution. *Proceeding of The fifteenth Annual Congress of the Nigerian Rural Sociological Association* (pp. 1–8).

The Punch. (2006a, July 7). Obasanjo inaugurates computer for all Nigerians initiative. *Punch Newspaper, Business Today*, p. 33.

The Punch. (2006b, September 25). New initiatives to drive Nigeria's PC penetration. *Punch Newspaper, Computer and IT Education*, p. 48

Treinen, S. (2004). *New ICTs for research, extension linkages: Opportunities and challenges*. Paper presented at the FAO – GTZ workshop on Effective Communication between Agricultural Research, Extension and Farmers, Laimburg, Italy.

ENDNOTES

1 Cocoa Research Institute of Nigeria (CRIN), Forestry Research Institute of Nigeria (FRIN), National Institute for Horticultural Research (NIHORT), Institute of Agricultural Research and Training (IAR&T), National Cereal Research Institute (NCRI), Nigerian Stored Product Research Institutes (NSPRI), Nigeria Institute of Oceanic and Marine Research (NIOMR) and National Agricultural Extension and Research Liaison Service (NAERLS).

2 Ogun State Agricultural Development Programme (OGADEP), Oyo State Agricultural Development Programme (OYSADEP), Osun State Agricultural Development Programme (OSSADEP) and Lagos State Agricultural Development Authority (LSADA).

3 A total of 113 researchers, 124 extension agents and 224 rural farmers were randomly selected for the study in the southwest geopolitical zone of Nigeria. The researchers were selected across the eight agricultural research institutes; the extension agents from 4 out of the six state agricultural extension agencies, and the farmers from 112 villages in the southwest geopolitical zone of Nigeria covering Lagos, Ogun, Osun and Oyo States.

4 MTN Nigeria, Airtel Nigeria, Globacom and Etisalat.

Chapter 7
Impact of Information and Communication Technology on Livestock Production:
The Experience of Rural Farmers in Nigeria

O. Anie Sylvester
Delta State Polytechnic, Nigeria

ABSTRACT

This chapter examines the impact of Information and Communication Technology (ICT) on livestock production by rural farmers in Nigeria. Questionnaire, interview, and personal observation methods were employed to elicit information on the impact of ICT on livestock production on rural areas of Nigeria. The study reveals the significance of personal characteristics of the respondents. The findings also reveal that rural farmers need to be encouraged by providing them with relevant ICT gadgets in order to enhance effective access to information on veterinary and extension services to improve productivity. It is therefore concluded that the establishment of internet facilities in rural communities should be the priority of the State and Federal Governments in order to encourage computer literacy.

INTRODUCTION

The Nigerian society is a social system compounded by contested demands on access to scarce resources especially information. Nigeria is naturally blessed because of the rich alluvial deposits and other natural composites. The fertility of the country makes it possible for the people to carry out fishing, hunting and animal rearing (Otolo, 2008).

Domestic animals are tamed and could be kept for commercial purposes and social means in Nigeria. Anyanwe and Ashaya (2007) opined that some of the domesticated animals such as

DOI: 10.4018/978-1-4666-2655-3.ch007

sheep and cow can be used for meat and milk production. Herren (2004) postulated that one of the advantages of domestic animal production is the low cost of feeding. This is because most of the domesticated animals feed on grass and less expensive grains. However, Mcnit (1995) revealed that in spite of the cheap process of rearing domestic animals, some factors still militate against intensive and extensive livestock production in Nigeria and these include inadequate veterinary services and absence of genuine veterinary drugs. But with the advent of information and communication technology (ICT) on livestock production domestic animal rearers now find solace with the application of ICT.

Most of the rural farmers are not highly educated and most of them understand and speak their dialects only. Experience shows that the ability to acquire and use information is fundamental to the development of animal production especially in the adoption of new technologies. Poor access to information is a major constraint to domestic animal production in Nigeria, and the situation is aggravated by the high level of illiteracy among the rural farmers. It is a clear evident that the ability to acquire and use information is a sine qua non for the application of ICT and agricultural development at all levels.

Fasheun (2001) declared that computer compliance by farmers have become the most important factor for productivity and prosperity. Communication is the process of imparting or exchanging information. To expose rural farmers to a better way of animal husbandry, there is need to educate them in the language they will understand better. In view of the nature of rural farmers, which are relatively isolated and the economic activities which are predominantly agriculture oriented, there is therefore the need to bring farmers to the centre of development for the upliftment of their standard of living. It therefore, calls for a communication development strategy that is democratic, participatory and productive or result-oriented. This can be achieved with the adoption of ICT in retrieving and dissemination of information on domestic animal production (Kwesiga, 2000).

Rural farmers need a wide range of agricultural information access, especially in all areas of agricultural activities such as veterinary services, prices of drugs, diseases outbreak, processing, storage facilities and marketing. The information received helps to enlighten them on the latest species of diverse animals and current prices.

The significance of this study lies in the fact that the findings will enable the rural farmers in Nigeria to be acquainted with new ideas of domestic animal production through the application of ICT.

LITERATURE REVIEW

Ogur (2003) stressed that much efforts have been made by the Federal and State Governments to enhance animal production and other agricultural activities. Few of the programmes launched by the Federal Government were "Operation Feed the Nation and Green Revolution", which focus attention measurably on students, civil servants, the police and army. These programs hardly favor the rural farmers especially the rural animal rearers.

Rural farmers are isolated from the dealers of agricultural produce, and therefore need a wide range of information to improve in their systems of farming. Information from the context of this study is defined as a processed data that is logically arranged and recorded in various forms, and is retrieved, stored and disseminated in the right format and at the right time with the application of Information and Computer Technology (ICT). Information could also be inform of recorded ideas, skills, feelings, experiences and research results that can be communicated for the improvement and development of others (Unegbu, 1999). Sanusi (2003) remarked that information, if well processed will serve as essential raw materials used in the realization of any objectives or goals set by individuals or group of persons. Rural farmers cannot and never obtain maximum production without adequate information on domestic animal production. Oyelaran (1996) opined that inability of the rural dwellers to access information via the internet is a major problem facing rural entrepreneurs such as rural farmers.

The impact of ICT in domestic animal production may not be felt as expected if the information needed by rural farmers is not well communicated. The most vital information needed by rural farmers in Nigeria is information on veterinary services, prices of drugs and diseases outbreak. Park (1997) posits it that communication of information to rural farmers must be characterized in the following ways for effective delivery. In the first place, it is good for the Federal and State Governments to supply drugs at subsidized rates to the rural farmers, but it is not good enough if the rural farmers are not guided on how to apply the drugs to the animals. If the veterinary is restricted to theoretical messages without practical demonstration for the illiterate farmers to see, the services are not complete and will not yield the needed result. Secondly; communication must be purposeful. The information being disseminated to rural farmers must relate to some pressing problems, and must be able to suggest solutions. And finally, in order to ascertain the validity of the questionnaire, extensive review of literature on information resources relevant to the study was consulted.

Objectives of the Study

This study aims at investigating the impact of IC-Ton livestock production, the experience of rural farmers in Nigeria. It also attempts to find out the effect of ICT on rural farmers and the constraints facing the rural farmers with application of ICT.

Table 1. Educational qualification of the respondents

Educational Qualification	No. of Respondents	Percent %
First School Leaving Certificate	73	24.33
SSCE/GCE	67	22.33
NCE	18	6.00
BA/B.Sc./B.ed	8	2.67
No Certificate	134	44.67
Total	**300**	**100.00**

Findings and Discussions

This section presents the analysis of the collected and discussion of the findings of the study. 600 questionnaires were distributed to the political zones but only 300 questionnaires were completed and returned. There are 300 respondents from six communities representing the six political zones under survey and their distribution is as follows; Manama in North West 40; Garkida in North East 45; Osi in North Central 50; Wasinmi in South West 50; Mgbidi in South East 65; and Isele-Uku in South-South 50.

Table 1 reveals that 18 of the respondents had formal education up to National Certificate in Education (NCE), while 8 out of the respondents made it to the degree certificate level. When they were interviewed, it was discovered that they had spent years in the city searching for job appointment but could find none, so they decided to settle down for farming. And now they are on their own with an encouraging number of livestock. It was also discovered that 134 (44.67%) of the respondents had no formal education and most of these farmers are from the Northern part of Nigeria.

The forest and mangrove zones of the country to which Akwa-Ibom state belongs have a high concentration of chicken population in Nigeria as shown in Table 2. Recent statistics on the population of livestock in Nigeria derives from the 2007 Aerial Survey and Trotting by RIM, commissioned by the Federal Livestock Department and Pest Control Services as shown in Table 3.

From the statistics shown in Table 3, it was revealed that most of the livestock are largely produced in Northern Nigeria especially in Bauchi, Gongola, Kano, Niger, Plateau and Sokoto States. Government at all levels in the country has intensified the provision of veterinary services with the application of information and communication technology.

Table 4 reveals the information needs of the rural farmers in Nigeria. Much time was spent to get the farmers well informed about their information needs by the enlightened farmers among them. To this end, their response was on a high side hence such information will help to enhance their livestock productivity. It was revealed that rural farmers in the Northern Part of Nigeria showed more interest for the information needs due to the state government's readiness to assist the farmers.

All the respondents (100%) indicated that they need information on new breeds, veterinary services, how to get credit facilities and market strategies. This indication was supported by the findings of Aboyade (2004), Broadbent (1987). 186 (62.01%) of the respondents opined that they need information on farm implements. On the application of pesticides, 199 (66.35%) of the respondents also declared that they needed information on how to acquire and use pesticides.

Veterinary agents have been the most effective source of information on the innovations and development in livestock production. This corroborates the ascertain Otolo (2008) that the major source of acquiring information on the production of livestock is through veterinary service Agents. 300 (100%) of the respondents attested to this fact. 216 (72.00%) of the respondents revealed that their major source of information on the agricultural developments and innovations is through the extension services agents. 176 (58.67%) of the respondents declared their neighbors and family members as their major source of productive information while 138 (46.00%) of the respondents opined that their source of information is through ICT materials such as CD-ROM, diskettes, internet, etc (See Table 5).

Table 7 reveals 259 (86.33%) of the respondents admitted that the introduction of ICT to the production and management of livestock farms has

Table 2. Ecological distribution of livestock population in Nigeria

Ecological Zone	State	Cattle	Sheep	Goats	Pigs	Chickens
Sahel	Parts of Yobe, Sokoto and Kaduna	1,060 (7.6)	1,105 (5.0)	1,398 (4.0)	40 (1.2)	2,580 (3.6)
Sudan	Parts of Kebbi, Sokoto, Katsina, Yobe, Borno, Bauchi, Adamawa, Taraba, and Kano	6,820 (48.8)	10,965 (47.3)	12,702 (36.8)	340 (20.4)	14,448 (20.4)
Northern Guinea	Parts of Katsina, Bauchi, Kebbi, Sokoto, Niger, Kwara, Kaduna, Plateau, Taraba and Adamawa	2,580 (18.5)	2,970 (13.4)	4,025 (12.20)	415 (12.2)	5,290 (7.5)
Southern Guinea	Parts of Cross River, oyo, Ogun, Ondo, Edo and Cross-Rivers	3,260 (23.3)	4,550 (20.6)	9,550 (27.7)	1,905 (55.9)	26,002 (36.6)
Forest	Parts of Oyo, Ogun, Osun, Ondo, Edo, Delta, Amanbra, Rivers, Lagos, Akwa- Ibom, and all of Imo and Abia	317 (1.5)	2,550 (11.5)	5,920 (17.2)	580 (17.0)	78,480 (26.11)
Mangrove	Parts of Delta, Cross- River, Akwa-Ibom, Rivers, Lagos and Ondo	10 (0.7)	464 (2.1)	900 (2.6)	130 (5.8)	4,128 (7.2)
Total		**13,947**	**22,105**	**34,495**	**3,140**	**70,928**

Table 3. Statistics on the population of livestock in Nigeria

State	Cattle	Sheep	Goats	Pigs	Chickens	Duck	Guinea Fowls	Turkey	Rabbits
Abuja	156,000	63,000	144,000	16,000	280,000	32,000	12,000	-	13,000
Akwa-Ibom	2,000	516,000	816,000	89,000	3,185,000	133,000	19,000	23,000	76,000
Anambra	65,000	425,000	1,467,000	62,000	4,306,000	251,000	6,000	8,000	3,000
Bauchi	1,731,000	2,811,000	3,466,000	66,000	2,787,000	861,000	1,003,000	1,000	533,000
Benue	48,000	737,000	1,747,000	703,000	8,368,000	1,180,000	141,000	2,000	37,000
Borno	2,727,000	424,000	3,188,000	76,000	3,292,000	466,000	198,000	-	45,000
Cross-River	10,000	117,000	351,000	68,000	1,364,000	162,000	4,000	-	4,000
Gongola	1,503,000	1,324,000	1,470,000	476,000	2,975,000	953,000	409,000	2,000	74,000
Imo	13,000	495,000	1,281,000	8,000	3,008,000	147,000	3,000	5,000	18,000
Kaduna	1,007,000	557,000	967,000	250,000	2,680,000	515,000	85,000	4,000	112,000
Kano	1,014,000	2,567,000	2,736,000	-	2,947,000	579,000	374,000	25,000	217,000
Katsina	630,000	1,696,000	2,079,000	-	2,156,000	685,000	510,000	15,000	166,000
Kwara	562,000	843,000	1,132,000	81,000	2,815,000	685,000	510,000	15,000	166,000
Lagos	3,000	57,000	158,000	25,000	2,418,000	166,000	-	-	5,000
Niger	1,165,000	752,000	968,000	81,000	2,464,000	532,000	220,000	12,000	68,000
Ogun	27,000	340,000	904,000	149,000	4,310,000	318,000	188,000	-	38,000
Ondo	9,000	589,000	1,747,000	291,000	4,492,000	582,000	5,000	1,000	32,000
Oyo	296,000	861,000	1,887,000	177,000	5,647,000	536,000	72,000	14,000	310,000
Plateau	1,054,000	964,000	1,866,000	535,000	2,941,000	917,000	166,000	53,000	189,000
Rivers	3,000	509,000	670,000	66,000	2,170,000	568,000	-	-	5,000
Sokoto	1,778,000	2,566,000	2,466,000	2,000	3,432,000	801,000	1,071,000	19,000	73,000
Total	**13,947,000**	**22,104,000**	**34,495,000**	**3,410,000**	**70,826,000**	**11,789,000**	**4,679,000**	**223,000**	**1,720,000**

Source: Ibanga, 2002

Table 4. Information needs of rural farmers

Information Needs	Masama in N.W.	Garkida in N.E.	Osi in N.C.	Wasinmi in S.W.	Mgbidi in S.E.	Isele-Uku in S.S.	Total	%
New breed Livestock	40 13.32%	45 15.00%	50 16.67%	50 16.67%	65 21.67%	50 16.67%	300	100
Veterinary services	40 13.32%	45 15.00%	50 16.67%	50 16.67%	65 21.67%	50 16.67%	300	100
Farm implements	33 11.00%	42 14.00%	47 15.67%	20 6.67%	23 7.67%	21 7.00%	186	62.01
Pesticides	32 10.67%	42 14.00%	47 15.67%	20 6.67%	26 8.67%	23 7.67%	199	66.35
How to get credit facilities	40 13.32%	45 15.00%	50 16.67%	50 16.67%	65 21.67%	50 16.67%	300	100
Market strategies	40 13.32%	45 15.00%	50 16.67%	50 16.67%	65 21.67%	50 16.67%	300	100

N.W. - North West; N.E. - North East; N.C. - North Central; S.W. - South West; S.E. - South East; S.S. - South-South.

Table 5. Frequency distribution of respondents in relation to the source of acquiring information on the production of livestock

Sources of Information	No. of Respondents	%
Veterinary Services Agents	300	100
Through neighbors and family members	176	58.67
Extension services agents	216	72.00
Through media e.g. Radio, TV, Newspapers, etc.	138	46.00
Using ICT materials e.g CD-ROM, Projector, Diskettes, etc.	110	36.67

Table 6. The effect of ICT on rural farmers

The Effect of ICT on rural farmers	No. of Respondents	Percent (%)
Positive	259	86.33
No effect	41	13.67

brought positive effects on farms (See Table 6). This opinion was corroborated by the findings of NERDC (1996) and Hawkins (2004). While 41 (13.67%) of the respondents are of the view that the application of ICT has no effect on their farm.

It was discovered that during the interview that the rural farmers are faced with a lot of constraints emanating from the application of ICT in livestock production. Table 7 also revealed that 261 (87.00%) of the respondents declared that the major constraint is poverty. This also is in consonant with the view of Otolo (2006), Ben (2006) and Kwesigu (2000). 196 (65.33%) of the respondents revealed that one of the major constraints is the high cost for downloaded information on livestock production. 183 (61.00%) of the respondents attributed their constraints to the inadequate provision of internet services while illiteracy and lack of computer knowledge also contributed to the major constraints as opined by 178 (59.00%) of the respondents.

SUMMARY AND CONCLUSION

This research work was carried out to ascertain the impact of ICT on livestock production in Nigeria with concentration on rural farmers' experience. The main purpose is to identify the impact of ICT on rural farmers; to find out if there are any remarkable changes in the rural farmers' mode of operation since the advent and application of ICT and to identify the sources of information and how efficient are the sources. The result of the findings shows that ICT has a positive impact on rural farmers in the production of their livestock. More so, ICT should be seen as an indispensable tool for effective animal production; also ICT would continue to serve as major weapons

Table 7. Rural farmers' constraints in the application of ICT

Constraints	No. of Respondents	%
Poverty	261	87.00
Illiteracy and lack of computer knowledge	178	59.00
Lack of awareness on the usefulness of ICT in livestock production	153	51.00
Inadequate contact with veterinary services agents via internet	140	46.67
Inadequate contact with extension services agents via internet	140	46.67
Inadequate provision of internet services/ centres	183	61.00
Network problems	157	52.33
High cost of downloaded information on livestock	196	65.33

in agriculture, for breaking the cultural barriers that inhibit progress in the lives of rural farmers especially now that the whole world is globally a village. To this end, ICT play a key role because rural farmers have to be aware of the requirements for credit facilities, veterinary services, conditions for the application of drugs and pesticides.

RECOMMENDATIONS

The research was carried out to ascertain the impact of ICT services to rural farmers in Nigeria. And it was discovered that the extension services agents have positively touched the lives of some rural farmers through the distribution and use of ICT materials such as CD-ROM, diskettes, films, etc. Yet urgent attention of the government is therefore needed to improve the social life and the livestock production of the rural farmers.

1. Adult education and computer literacy should be emphasized among the rural farmers.
2. Establishment of cyber café or browsing centres at rural areas should be the priority of the State and Federal Government. This will help to break the yoke of monopolistic charges by the few centres.
3. The veterinary services agents and extension services agents should be motivated and well-equipped by the governments to enhance regular visit to the rural farmers to improve their productions.
4. Credit facilities should be made available and easy for rural farmers to obtain loans to improve their productions.

REFERENCES

Aboyade, B. O. (1984). *The provision of information for rural development* (pp. 126–137). Ibadan Foundation Publication.

Anyanwe, A. C., & Maxwell, U. C. (2007). *A textbook of agricultural science for school and college* (p. 237). Benin-City, Benin: Justice Jeco Press and Publishers Ltd.

Broadbent, K. P. (1987). Mobilizing local capabilities for more effective exchange of information in developing countries: Agriculture as an example. *Quarterly Bulletin of IAALD*, *32*(2), 84–93.

Fasheun, K. P. (2001). Problems and prospects of livestock production in information age. *Delta State College of Agriculture Newsletter*, *12*(4), 23–27.

Harren, P. C. (2004). *Agricultural science and application for schools* (pp. 271–278). London, UK: Heinemann Education Book.

Hawkins, R. (2004). *Ten lessons for information and communications technology in developing world*. Washington, DC: World Bank.

Kwesiga, J. C. (2000, April). *Women in agriculture: The vital contribution of the rural women farmers to African economies and faculty sustenance farming*, (pp. 4–7).

National Education Research and Development. (1996). *Agricultural science for secondary schools* (p. 219). Ibadan, Nigeria: University Press Plc.

Ogur, S. O. (2003). *Animal production in tropical zone* (pp. 76–82). Premier Publications.

Otolo, P. U. (2006). *Agricultural information needs of rural women farmers in Nigeria: A case study of oil-producing communities in Delta State – Nigeria* (pp. 11–12).

Otolo, P. U. (2008). The impact of information and communication in extension services to rural farmers in Niger Delta. *The Information Technologist*, *5*(2), 37–38.

Oyelaran, O. B. (1996). *NISER survey of business condition, experience and expectations in the manufacturing sector*. Nigeria Institute of Social and Economic Research (NISER) publication.

Park, R. D. (1997). *Animal production and health* (pp. 38–46). Enugu, Nigeria: Moris Publishers.

Sanusi, J. O. (2003). *Special remarks on information and communication technology for the social transformation of Nigeria*. Lecture delivered at the Dinner Investor Forum organized by Leapfrog Venture Partners Ltd. The National Summit on SMIE, Federal Palace Hotel, February 15, 2003.

Unegbu, M. C. (1999). *Impact assessment of information: Agricultural technology on the performance of the Nigerian sector* (pp. 72–74). A project submitted to the African Regional Centre for Information Science, University of Ibadan, Ibadan.

Chapter 8
Marketing Strategies for E-Government and E-Agriculture Services in South Africa:
The Case of Cape Gateway Project

Blessing M. Maumbe
West Virginia University, USA

Vesper Owei
Cape Peninsula University of Technology, Republic of South Africa

ABSTRACT

The government of South Africa (GSA) has embarked on a major program for e-service delivery. The Batho Pele Gateway Project is the access portal to government information and a key driver of the e-government transformation in South Africa. Although the GSA is determined to make e-government service delivery successful, the average citizen does not fully comprehend the key transformations driving government engagement with civil society and businesses. Therefore, the Cape Gateway Project (CGP) is confronted with the immense task to develop and implement a strategic marketing program for e-government by the Provincial Government of the Western Cape (PGWC) which is not only a portal, but also a call centre and walk-in centre. Various marketing tactics have been deployed, but their efficacy in elevating awareness levels and converting that to actual sustained use of e-government service delivery remains unknown. This chapter proposes an e-government marketing framework and uses CGP experience to examine communication and branding strategies for e-government in a South African context. Key challenges facing e-government market communication strategy are highlighted.

DOI: 10.4018/978-1-4666-2655-3.ch008

1.0 INTRODUCTION

South Africa (SA) has embraced e-government and is rapidly transforming the delivery of government information and services at all three levels of government (i.e. national, provincial and municipal). In the Western Cape Province (WCP), Cape Gateway Project (CGP) is the key driver of the e-government revolution. A number of challenges arise in executing this mandate: (1) how to equitable deliver e-government services to a diverse population comprising three racial and linguistic cultures (i.e. Afrikaans, English and Xhosa) (2) how to respond to service needs without exacerbating the existing digital divide. Given that SA is a dualistic economy characterised by a modern capitalist economy co-existing alongside low income, underserved poor communities, the ability to equitably deliver state of the art e-government information and services is a major concern for the marketing unit at the CGP.

First, e-government and hence e-agriculture service delivery is a relatively new concept and for the ordinary SA citizen, conceptualizing the idea and optimizing e-government usage will take some time. Secondly, not all key public officials have the computer skills needed to effectively design and implement e-government processes. Thirdly, e-government technology is evolving and there is need for continuous learning. As civil society, businesses and other stakeholders anxiously await the widespread implementation of e-government when exactly the benefits will be realized and how these will be distributed requires a dedicated marketing effort from policy makers. The aforementioned challenges are not specific and unique to SA alone, but confront any government intending to provide e-government services.

Without doubt, CGP faces the challenge to design marketing strategies that will effectively promote citizen-centric e-government services targeting the urban and rural population in the WCP. The question of effective e-service delivery can be answered through a systematic evaluation of the performance of three channels (i.e. Walk in Centre, Call Centre and Web Portal) and how they respond to citizen service needs and aspirations. If performance is sub-optimum, the next question becomes: What alternative strategies if any, should be considered to ensure successful implementation of e-government services in both the urban and rural settings?

This paper focuses on e-government marketing initiatives being used to raise awareness levels of citizens. In this paper, e-agriculture services are considered an integral part of e-government service delivery. We present a theoretical framework for e-government service delivery and adapt that to assess the marketing of e-government services through the CGP. In addition, using CGP as an illustration, we discuss the e-government branding and highlight the uniqueness of public sector marketing. E-government practitioners at CGP have been grappling with these issues and have attempted to develop a brand that captures user needs and expectations. The CGP is keen to ascertain the effectiveness of its marketing and advertising strategies and would like to use that as a basis to justify future budgetary resource allocations and convince pessimists with negative views about advertising online public services.

Despite some of the promising starts to e-government implementation, failure to conceptualize the context and link the new initiative to the overall development agenda could result in further divides (Hanna, 2003). Crafting e-government that does not leave behind certain segments of the population is bound to yield greater success than "ad hoc and passive postures (Hanna, 2003). To avoid such failures, the CGP needs to adopt a systematic approach that is driven by a robust marketing strategy to bring all the citizens on-line. Once a systematic e-government service delivery approach has been developed, it can be used as a benchmark for deploying similar e-government initiatives in other provinces or even countries on the African continent. But, it is just on this point that there appears to be a breakdown. While the various governments may have adopted different approaches, a search of the literature does not seem to indicate the existence of a universal approach. It is this absence that motivates our study.

2.0 LITERATURE REVIEW

The delivery of quality e-government information and services on an equitable and sustainable basis remains a huge challenge for most African governments. Inherent difficulties lie in the road to universal access to digital government in SA's dualistic economy. In order to stimulate widespread adoption of e-government service, effective communication strategies are needed (Eechambadi, 1994; Galbi, 2001). Given that governments have many ways to communicate their policies, contrasting viewpoints can easily filter through to citizens if communications channels are not refined. The need for governments to develop marketing brands is now viewed as a key part of an effective communication strategy (Galbi, 2001). Also, declining budgets are forcing managers to justify their choice of advertising strategies (Woodside & Soni, 1990).

The power of advertising in raising public awareness of a given brand has been acknowledged in the literature (Miller and Berry, 1998). Early studies on advertising effectiveness in changing citizen attitudes have used measures such as return on inquiry (ROI) and cost per inquiry (COI) measures (Woodside & Reid, 1976). A key study comparing relative performance of different advertising media showed that newspapers outperformed magazines in advertising effectiveness (Woodside & Ronkainen, 1982). Much recent evidence shows that the youth niche market is relatively more responsive to digital media as an advertising strategy (Montgomery, 2000). A similar study on how adolescents process advertisements reveals that visual advertisements rank higher than verbal advertisements in elicitation of memory of more facts, inferences and emotional responses (Edens, 2000). Moreover, the study identifies adult males as being relatively poor in processing advertising information compared to females. According to Miller and Berry, (1998) "...the difference between highly recalled messages and less highly recalled ones is that highly recalled messages are much more likely to be supported by well-funded, long lasting advertising campaigns..."

The marketing theme or slogan for CGP is "easy access to government information and services." Beside the issue of whether the slogan or message is easy to recall a more important empirical question is to the extent does it create the expected awareness level. Putting the CGP message on air for years could perhaps attract greater recognition and awareness of its brand from citizens but the question of cost will also arise. While it is true that awareness comes before actual use, it is brand imagery (i.e. attribute ratings) and brand salience (i.e. mental ranking of brands) that determine citizen's customer's use of the CGP e-government brand (Miller & Berry, 1998). The CGP should strive for sustained citizen use of e-government services suggesting that the advertising strategy pursued must be cost effective and also reinforcing rather than merely persuasive.

3.0. METHODOLOGY

Governments tend to operate in traditional organizational silo[1] structures, with each silo producing and providing a large number of services. For example, the PGWC is believed to provide no less than 500 different services through its numerous departments. Typically, there is little or no coordination among the silos as far as service

Figure 1. Methodical approach to developing marketing strategies for e-services delivery

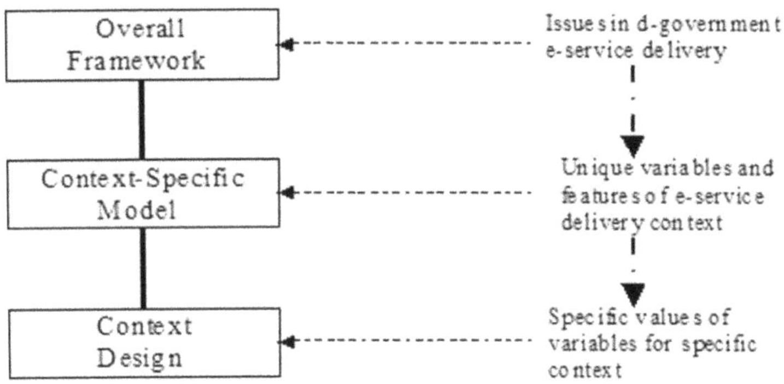

production or provision is concerned. This renders the function of government e-service delivery a huge and complex problem.

To further complicate the problem, governments are expected to provide service to different categories of service users, or consumers. Additionally, the services usually have to be made available to users through numerous access channels. Government service providing units are, therefore, confronted by the challenge of providing a *myriad of services* through *many access channels* to *various categories of service-users*, with each category possessing a set of profiles that differentiate it from the other categories. In the context of this paper, the provisioning function is that of *marketing*. To render the problem tractable, a logical approach to systems development needs to be adopted.

Our logical approach is shown in Figure 1. In marketing e-services, the use of underlying formal tools that guide the process is crucial. In this paper, therefore, we present an overall framework and a contextual model for e-government services delivery. The framework shows the key aspects in developing e-government service delivery, while the model expands on the framework to provide a description of internal organizational structure, key partners, service providers, e-government channels and users of e-government. The model considers the relationships between and among key stakeholders involved in delivery of market-led e-government service platform. In South Africa, the concept of forging partnership with the private sector, non-Governmental Organizations (NGOs), and community-based organizations (CBOs) to provide more effective service delivery has long been mooted in the literature (SA White Paper, 1997). The proposed model is intended to serve as a guide or road map for policy makers involved in designing a marketing strategy for e-government. It also enables the identification of the essential issues in marketing and distributing government e-service delivery to consumers.

Frameworks of this type are usually abstractions that apply in general. To make the framework relevant to a particular e-service delivery scenario, the variables and features specific to the marketing context must be determined and instantiated with values that are also specific to the particular context.

3.1. Overall Conceptual Framework

Our overall framework is shown on Table 1. Typically, a government unit (e.g., the Centre for e-Innovation in the case of the CGP) charged with the task of providing e-services to consumers produces the services either itself (in-house) or with the involvement and assistance of third parties ("partners"). The services are then made available to users, who normally access the services through some distribution channels.

Table 1. Overall e-service marketing framework

Government Unit variables: {unit_name}
Partner variables: {partner_name list}
E-service project variables: {project_name}
E-service variables: {information, information-based service}
User variables: {e-profile}
Marketing channel variables: {web portal, walk-in centers, call centers, digital handhelds, e-forums, traditional}
Marketing strategy variables: {distribution strategy, access strategy}
Consumer variables: {(external user-type:{name_list}), (internal user-type: {name_list})}

3.2. Contextual Model

As the name suggest, the contextual model is a model of a particular context of interest, and is derived from the framework of Table 1. It shows the e-service variable of that context and the connections between variables. For the CGP, the model is shown in Figure 2.

Note that no instance values are assigned at this stage to the variables captured in the model. Before an actual marketing of the services can be conducted, however, the variables must be instantiated with actual values. This is done in the context design step, which we only describe briefly here, because it is beyond the scope of this paper.

The last phase of our systematic approach is that of designing the actual context of e-service provision. This task builds on the context model and consists simply of instantiating the model variables. That is, the task consists of identifying the possible values of variables, choosing the values that are suitable or applicable for the context and then assigning the values to the variables.

4. MARKETING STRATEGIES

As we have found, the process of marketing the CGP to users, however, is not simply a question of developing elaborate models. The effort to make the services available to and accessible by users and getting them to actually utilize the services require extensive and effective marketing driven by strategies that target users according to specific e-profiles to create awareness of the services and, where an e-service need exist, to render the service accessible.

Marketing cannot be considered in isolation. To be effective, they must be considered in a given context. Effective strategies can be developed when distribution and access channels, i.e., marketing channels, the specific e-profiles of users, and the specific e-services that exist in the context are taken into consideration.

Marketing the CGP, therefore, involves understanding the key functions of its three main channels of information dissemination and service provision (i.e., the Walk in Centre, Call Centre and Web-based Portal[2]), the profiles of the different classes of users, and the nature of the services to be provided. The fundamental aim of our marketing effort is to build the CGP brand and raise its awareness among all the citizens of the WCP. The marketing strategies revolve around four critical goals of CGP (1) to promote the development of e-government services, (2) to enhance internal departmental efficiencies, (3) to improve the competitiveness of the province and (4) to ensure equitable and easy access to e-government information and services.

The choice of viable e-government marketing strategies depends on a number of factors including (1) cost effectiveness (2) taste and preferences of citizen customers (3) the ability to penetrate a given target market, (4) internal capacity to implement e-government services and (5) results from linkage advertising that identifies those strategies that

Figure 2. Contextual model of e-service marketing strategies

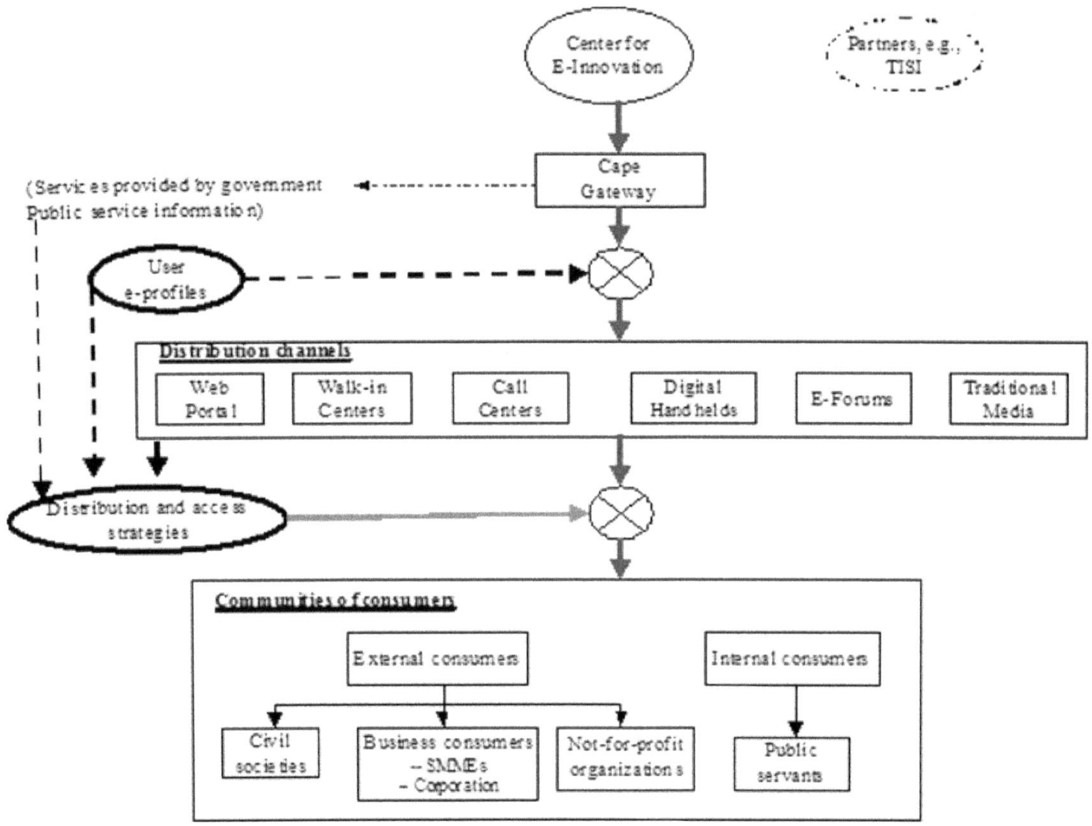

convert citizens into actual users of e-government services. In our experience we found it necessary to classify our users as either "external users" or "internal users", as shown earlier in Figure 2. Our marketing strategies (discussed next) had to be designed accordingly. Internal marketing strategy focuses on government employees while the outward looking or external marketing strategy focuses on civil society, community organizations, and businesses.

4.1 Internal Marketing Strategies

The essence of internal marketing is to obtain buy-in from management. If management adopted and internalised Cape Gateway they will automatically become advocates not only for Cape Gateway, but also for e-government. This was done by means of a "road show" to all Departmental Management meetings, where Cape Gateway were showcased as a product and more importantly an essential tool for delivering effective services to the end-user, the citizen. The potential of Cape Gateway, streamlining enquiries from citizens, cutting incoming calls to departments and at the same time marketing a specific department's services were well accepted and welcomed. Ultimately, managers had to be convinced that Cape Gateway belonged to them and that they must embrace this new form of communication, not as an ad-hoc service, but an essential and integral one.

Following the "road show", desk-side coaching was done with the Premier and all his ministers. Again, they are an important mouthpiece of Provincial Government and once they realised the value of Cape Gateway as users themselves, it put them in an informed position to market the service in their speeches, budgets, stationery, annual reports etc.

Getting the buy-in from both political and administrative heads, departmental communication officers were targeted as they are the "evangelists" of the department and often in a position of influence. Thorough training in terms of the brand of Cape Gateway as well as the service, gave them confidence to advocate the service to their citizens as much as they use it themselves. They also realised that only by taking ownership of Cape Gateway and using it as a first point of call for publishing information, it can be a success.

The rest of the staff (PGWC has a staff complement of ± 68 000) were targeted through various mediums including intranet, e-newsletters, posters in public buildings and promotional material like mouse pads, pens and coasters.

4.2 External Marketing Strategies

A very significant angle of the external marketing campaign was that it was not only designed to create awareness and a call to action for the brand, but that it has an educational focus. Market research (Webcheck, 2003) has proven that citizens know very little of government services or even who renders which government service. Brand awareness alone, was not going to encourage a citizen to pick up a phone, log in or walk in. They needed to be educated in terms of the exact services government has on offer. The educational campaign was mainly run in newspapers (especially community newspapers – free to citizens) and interviews on community radio stations. All this was done in conjunction with conventional newspaper and radio advertisements, with a stronger focus on the brand itself.

Another significant dimension of the external marketing strategy was public-private partnerships. As advertising can be very costly in mainstream media, innovative use of the limited budget was essential for effective marketing. The third of the triple bottom line for private enterprises, namely social responsibility were explored. The message that Cape Gateway was carrying was in the interest of all the Western Cape citizens and not for commercial gain. The media has a responsible role in terms of educating their readers and thus empowering them and enhancing their lives. Agreements were reached with two top media owners for a discounted advertising rate and value add in terms of exposure (e.g. Presence on their Internet Site). The partnerships have been established for the past two years already and have saved government a lot of money.

The market communication strategy for e-government information services uses diverse approaches such as digital billboards, exhibitions, printed brochures, information pamphlets, radio shows, newspaper advertisement in local press and community newspapers, listening lunches, testimonies from users and training of government information officers. The different marketing strategies are used to target specific audiences such as school children, university communities, informal settlements and other groups. Table1 summarises the various approaches that have been deployed in the past and their relative effectiveness. Radio and newspapers have high penetration and low cost suggesting that these should drive the marketing effort in the short run. Although e-community forums have a great potential to link citizens in peri-urban and rural areas, the initial infrastructural needs could increase the cost of this strategy.

CGP does not utilize television advertisement as it is relatively more costly than radio and newspapers. Moreover, the fact that not everyone has a television set in their homes makes it an illogical choice for equitable distribution of knowledge about new government information and services. Local community newspapers are a preferred advertising option because they have a wider circulation than daily papers. The fact that community newspapers are free and are published once a week provide an opportunity for individuals to be exposed to the advertisements for a prolonged period (Personal Communication, 2006). Similarly, radio is by far the most powerful source of information in developing countries (Kenny, 2002) and most people listen to radio in their homes, offices and even in cars while travelling. Another advertising media vehicle is digital billboards most of which are strategically erected by the sides of major freeways, regional link roads and in giant shopping malls. However, the effectiveness of digital signage in converting citizen to become actual users of e-government is still unknown. Nonetheless, upgrading the computer literacy skills of knowledge workers in charge of e-government should form a core part of the marketing effort designed to help with the effective delivery of e-government service. Without sustained training in e-government service delivery, SA will encounter major obstacles in modernizing its e-government services delivery in the long run.

Therefore, it is important to ask the question what are the specific challenges facing PGWC in designing and implementing their marketing strategies for e-government service provision. We dwell on this question in our discussion that follows.

5. ISSUES AND CHALLENGES FACING THE MARKETING OF THE CAPE GATEWAY PROJECT (CGP)

5.1 Branding the Cape Gateway Project

The CGP employs two critical yardsticks in its branding process--the *Batho Pele principles* of putting people first and the "*customer is king principle*" which emphasizes consumer-care ethics and responsiveness to people's needs in service delivery (SA Government White Paper, 1997). Success of the brand ought to be measured by effectiveness in the delivery of e-government services that meet the needs of all the citizens. In a desire to be citizen-centric, the portal is modelled on real life episodes and constantly undergoes content and web software updates in its bid to establish credibility and positive image for CGP.

The CGP brand has been positioned to provide a given set of attributes, benefits, values, and personality. The so called 'brand character' of the CGP conveys multifaceted dimensions including (i) maturity and credibility (ii) politeness and friendliness (iii) efficiency and reliability (iv) people centred or citizen focussed (v) easy access and convenient (vi) simple but professional. These attributes serve as a positioning platform in projecting the overall image of the PGWC. One of the challenges is to develop and project a coherent, deep, long lasting and meaningful image that imprints on the minds of citizen customers. Unlike brands that are distinguished by type of users, CGP brand equity is uniquely measured by its ability to draw upon users from diverse linguistic and cultural population groups of the WCP.

5.2 Public Sector Versus Private Sector Marketing Challenges

Marketing has long been viewed as private sector concept driven by profit motives and desire to stimulate product demand (Kotler, 1997; Markin, 1982). Public sector marketing presents unique challenges to most practitioners. For instance, flags and anthems are national government brands. Branding government information services is perceived as one way of raising awareness among citizens. But such branding effort while it consumes resources it does not translate to profit accumulation given that government is not about profit making. A number of issues and challenges arise from this:

- Central to CGP branding is the need to deliver on the "social agenda" defined by equitable and easy access to government services by all citizens.
- Although CGP has positioned itself as "different from traditional government" it is actually a government department. The question then becomes; Does the "non-governmental" image portrayed by the CGP brand conflict with real government in any way?
- As already mentioned, SA government operates at three different levels and each has its own representatives advancing the information society goals and objectives. Crafting a unique communication strategy or developing a unified voice about government e-service offering is challenging. The risk of conveying mixed messages or sending conflicting images or signals to the citizens is inevitable. Moreover, government does not monopolize expertise to deliver excellent e-services which implies the need to outsource certain functions or services offered by government.
- Currently, the CGP out-sources the Call Centre to Dialogue (Radian Consultants, 2005). With outsourcing comes the potential risk of brand depreciation which is beyond the control of the CGP. These and other challenges present obstacles in building a reputable CGP brand.
- Public sector marketing requires the need to deploy "societal marketing" principles that balances "mass marketing" and "differentiation strategies" without perpetuating the digital divide. A key strategic area that requires careful balance is the acquisition of new "citizen customers" willing to use e-government services and the retention of critical mass of loyal users of digital government. The rise in pressure on public sector to be seen as efficient providers of service has led to marketing efforts such as "this City works for you…" a slogan for the Municipality of the City of Cape Town. Although profit is not a key factor in public sector marketing, clear strategies need to be developed to the societal marketing endeavour.

5.3 Multicultural and Multilingual Racial Composition

Given the diversity in SA, the CGP brand needs to be strategically positioned to capture the benefits and values of a multi-racial society. The CGP recognizes the importance of empowering all the citizens of the WCP through ready access to government information and services easy. Multilingual content available in Afrikaans, English and Xhosa helps to promote equitable access to services. The challenge comes in balancing the width and depth of the content for each language group. The provision of e-government content based on user needs and frequently asked for information and services sounds rational. However, the approach fails to address the dilemma paused by "unconscious ignorance" among certain groups. Due to ignorance, some population groups may end up with "less" available content because they demand less information services or because they are simply not aware of what is available. Meanwhile other population groups might have more information, because they demand more services arising from higher levels of awareness and knowledge about government services. In reality, more content is available in English than Afrikaans and Xhosa on the portal (Vosloo, 2006). The promise to deliver equitable access to e-government information and services in three languages (i.e. based on user needs and frequently asked for service) requires sound judgement by experts to balance the available content. Therefore, the advertising strategy ought to raise awareness levels and rectify gaps in e-government content availability.

5.4 Marketing E-government through Multi-Channels

The three channels currently in operation; the Call Centre (i.e. 0860142142), Walk-in Centre (142 Long Street, Cape Town CBD) and Web-based Portal (i.e. www. cape gateway.gov.za) are all engaged in facilitating citizen access to information and services. Marketing e-government services through these three channels inevitably requires a unique tailor made communication strategy. Arguably, the Call Centre and Walk-in Centre share most of the burden of advancing the politeness, friendliness and or personality attributes of CGP brand. On the other hand, the web-based portal does not necessary display politeness but is expected to deliver efficiency and other non-people based attributes, benefits and values. Recognizing these differences in communication potential is critical in crafting a coherent marketing strategy. Poor coordination among the managers of the three existing channels could easily create clutter in the "marketing voice." Suffice to add that the complexity introduced by multi-channel marketing is worsened by multi-level level government services profile. CGP marketing strategy should articulate the fact that the service (or product) on offer ought to be promoted uniformly across the different channels. Failure to coordinate the marketing slogan and promote the brand equity across the various channels could confuse users resulting in the narrow perception of CGP as "complaint centre" or "tourist information centre."

5.5 Customer Relationship Management in Public Sector

In providing e-government services, customer care is a critical success factor. Like all customers, citizens develop a certain level of service expectation from their government. The challenge for government is to offer a uniform, consistent, service to all its citizens and businesses. Difficulties associated with access to e-government services, unavailability of relevant content or services (i.e. up to 500 services currently listed), poor people skills, (i.e. being unfriendly or unhelpful), slow response to citizen requests (i.e. response time currently pegged at 48 hours turnaround time) converge to raise frustration levels and thereby erode CGP brand equity. The failure to respond to citizen email requests, inadequate capacity in content development, weak coordination between departments and technological glitches are bound to reverse any potential gains made in building reputable e-government brand. According to the Advocacy Manager, one of the key challenges in achieving efficient customer relationship management is to impart knowledge and skills about CPG brand to all government workers involved in service delivery as the responsibility lies with everyone providing services to the citizens of the WCP. The tracking of perceptions of citizen customers is vital in building a citizen focussed CGP brand. Secondly, constant downtime and recurring computer glitches create negative mind set about efficiency of service delivery promised by digital government. Therefore government should allocate additional resources to capture problems that citizen customers encounter when consuming e-government services (Korsten, 2005).

5.6 Measurement Challenges and Implications for Designing E-Government Marketing Strategies

According CGP, the Walk in Centre receives an average of 700 visitors per month, the Call Centre handles about 14000 calls per month and the Portal receives 80, 000 unique visits. Measurement of usage rates, patterns of use (preferred channel by age, gender, educational levels etc.) and problems encountered is critical for designing a coherent e-government marketing strategy. In addition to quantitative usage rates, it is important that detailed qualitative information is also captured (and analyzed) about user needs, expectations, information seeking behaviour, satisfaction levels, technical preferences (i.e. navigability of websites) and views about staff willingness to help and degree of friendliness. Knowledge of these critical factors is essential in order to be able to respond better to the specific demands of users, adjust service levels, enhance quality dimensions, and make e-government relevant to diverse users. The tracking of user needs, satisfaction levels and perceptions about service offering requires the implementation of comprehensive market research study focusing on all three channels. When appropriate care is taken in conducting market research studies, analysis and interpretation, useful diagnostic information to policy makers and those approving budgetary allocations can be obtained (Siegel & Ziff-Levine, 1990). Moreover, advertising managers are under increasing pressure to become more accountable. It is expected that advertising effectiveness measurement studies will inject the necessary confidence guiding spending decisions and thus help orient the allocation of scarce resources towards better marketing campaigns (Eechambadi, 1994).

5.7 The Challenge of Reconciling Parallel e-Government Initiatives

A number of prominent e-government initiatives have sprung up in the WCP. For example the Red Door program that is managed by the Department of Economic Development and Tourism, and the Smart City Project of the City of Cape Town stand out as major e-government initiatives in SA. These programs have achieved successful name recognition in their endeavour to build an information society in the WCP. The question for PGWC is how to ensure that these programs yield positive results for the province without cannibalizing each other. Although serious policy debate on how to link these sister programs (i.e. e-government partners) has not started, it is nonetheless essential to explore mutual areas such as joint marketing, evaluations and research to inform the overall decision making process on how to effectively market e-government to citizens of the WCP. The framework and model discussed in this paper recognized the critical importance of engaging these partners in content development, research and marketing in order to advance the goals of building a strong CGP brand and excellent e-government service delivery across the WCP.

6. CONCLUSION

The paper presents a framework for e-government marketing in South Africa. E-government service delivery is a fairly new concept for both policy makers and citizens and awareness levels remain low among potential users. Based on a case study of the CGP, we examine the e-government communication strategy and its diffusion pathway in the WCP. Although various marketing tactics have been deployed in the past, radio and newspapers are the main tools currently being used to raise awareness about e-government services delivery among citizens. In the WCP, the citizens access e- government services via three main channels the walk in center, the call center and web-based portal. While additional channels are still being contemplated (e.g. Handheld PDA for SMS, MMS, etc.), the Cape Access Project has been introduced to provide internet facilities for communities in remote locations and rural town. Given presence of diverse information channels there is a need to avoid confusion, and promote a consistent communication message cross the aforementioned channels.

The CGP branding is driven by a multitude of benefit claims such as efficiency, reliability, easy access, citizen centric, simple, fast, and *Batho Pele* principles. However, its ability to deliver on the promises of the benefit elements of her brand is riddled by a number of unique challenges. Firstly, the three levels of government make it difficult for the CGP to craft a unique e-government market communication strategy across all the different public sector agencies. Secondly, there is need to develop a unified marketing voice or slogan as well as a consistent brand identity and image for all three channels. Thirdly, CGP has to devise more effective ways of differentiating its e-government marketing strategies to an economically and politically diverse multi-racial and multi-cultural population. Fourth, the CGP faces the hurdle of how to cultivate the discipline of customer relationship management within government. The fifth challenge involves lack of capacity to conduct comprehensive usability evaluations for all the three channels on a long term basis. Developing "top of the mind" awareness for CGP brand is process not an event that requires overcoming aforementioned challenges.

Once the branding effort begins to pay tangible dividends, the next challenge will be to manage the e-government transformation in ways that continually improve service quality, functionality and image. With no real competitors to talk about in e-government (besides other provincial e-government initiatives), lack of coherent long term marketing strategy that is responsive to citizen demands based on systematic usability evaluations, the same frustrations, intimidation and inefficiencies that characterized traditional government could easily return to haunt the citizens.

6.1 Future Research

As can be seen from the paper, a number of marketing initiatives to advance the goals of e-government are currently underway in WCP. In the long-term, computer literacy training for government officials will result in effective, well-coordinated and sustainable deployment of e-government services including those specific for e-agriculture. The question of effective e-service delivery can be answered through a systematic evaluation of the performance of the different marketing channels and how they respond with citizens. If performance is sub-optimum, alternative strategies should be introduced to ensure successful implementation of e-government services. The questions about effectiveness of e-service delivery require systematic evaluation a task that should be given serious consideration in future studies and policy development.

ACKNOWLEDGMENT

The authors would like to acknowledge the useful comments and suggestions that were received from Petro-Van Ryan, The Cape Gateway Project, Provincial Government of the Western Cape, Republic of South Africa.

REFERENCES

Edens, K. M. (2000). How do adolescents process advertisements? The influence of ad characteristics, processing objectives, and gender. *Contemporary Educational Psychology*, *25*, 450–463. doi:10.1006/ceps.1999.1031

Eechambadi, N. V. (1994). Does advertising work? *The McKinsey Quarterly*, 3.

Galbi, D. A. (2001). E-government: Developing state communications in a free media environment. *Journal of Futures Studies. Strategic Thinking and Policy*, *3*(1), 9–13. doi:10.1108/14636680110803012

Kenny, C. (2002). Information and communication technologies for direct poverty alleviation: Costs and benefits. *Development Policy Review*, *20*(2), 141–157. doi:10.1111/1467-7679.00162

Korsten, H. (2005). Evaluating South African government web sites: Methods, findings and recommendations (part 1). *South African Journal of Information Management*, *7*(2).

Kotler, P. (1997). *Marketing management: Analysis, planning, implementation and control* (9th ed.). Prentice Hall.

Markin, R. J. (1982). *Marketing strategy and management* (2nd ed.). New York, NY: John Wiley and Sons.

Miller, S., & Berry, L. (1998). Brand salience versus brand image: Two theories of advertising effectiveness. *Journal of Advertising Research*, 79.

Montgomery, K. (2000). Youth and digital media: A policy research agenda. *The Journal of Adolescent Health*, *27*, 61–68. doi:10.1016/S1054-139X(00)00130-0

Radian Consultants. (2006). *Cape Gateway evaluation: Report for the Centre for E-Innovation Provincial Government of the Western Cape*. Cape Town, South Africa: Author.

South Africa. (2006). *White paper on transforming public service delivery*. Retrieved from www.polity.org.za.html/govdocs/white_papers/transform.html

Vosloo, S. (2006). *Multilingualism in a South African e-government portal*. Cape Town, South Africa: Center for e-Innovation. (Unpublished document)

William, S., & Ziff-Levine, W. (1990). Evaluating tourism campaigns: Conversion vs. advertising tracking studies. *Journal of Travel Research*, *28*(3), 51–55. doi:10.1177/004728759002800312

Woodside, A. G., & Ronkainen, I. A. (1982). Travel advertising: Newspapers versus magazines. *Journal of Advertising Research*, *22*(3).

Woodside, A. G., & Soni, P. K. (1990). Performance analysis of advertising in competing media vehicles. *Journal of Advertising Research*, *30*(1).

Woodside, A. G., & Soni, P. K. (1991). Direct-response advertising information: Profiling heavy, light and non users. *Journal of Advertising Research*, *31*(6).

ADDITIONAL READING

De Ferranti, D. (2002). *From natural resources to the knowledge economy* (pp. 18–21). Washington, DC: World Bank. doi:10.1596/0-8213-5009-9

Eggleston, K., Jensen, R., & Zekhauser, R. (2002). *Information and communication technologies, markets and economic development*. The Global Information Technology Report- Readiness for the Networked World, World Economic Forum, 2001-2002.

Hanna, N. (2003). *Why national strategies are needed for ICT-enabled development*. Information Solutions Group, Staff Working Paper, No 3, June 2003.

Ikhemuemhe, G. (2004, March 3). Government and private sector partner to deliver e-government. *Vanguard*.

Onyejekwe, E. (2004). *HIV/AIDS and the digital promise – The problem and the solution*. Pearson Education.

Owei, V., Bada, A., & Aniebonam, M. (2005). Addressing the information technology skills shortage in developing countries: Tapping the pool of disabled users. *Information, Communication and Ethics in Society, 4*(2).

Sanchez, S. (2001). *Constrained global integration: A note on microenterprise in Latin America.* Washington, DC: World Bank.

Seed, A. (2005). *Report from the field: Two e-gov projects in Rwanda: PRSC II and PSCB.* Designing and Implementing e-Government: Key Issues, Best Practices & Lessons Learned, World Bank ISG Training Workshop, Washington, DC, April 26.

World Bank. (2005). *Designing and implementing e-government: Key issues, best practices & lessons learned.* World Bank ISG Training Workshop, Washington, DC, April 26.

ENDNOTES

1 The silos could be ministries, departments, agencies, or some other functional unit.

2 The portal was launched by the Center for e-Innovation in March, 2004 (Vosloo, 2006).

Chapter 9
Understanding the Information Security Landscape in South Africa:
Implications for Strategic Collaboration and Policy Development

Blessing M. Maumbe
West Virginia University, USA

Vesper Owei
Cape Peninsula University of Technology, Republic of South Africa

ABSTRACT

Information security risks are a major threat to South Africa's bid to build a broad-based information society. The integration of information security in e-government is no longer an option, but an imperative given the resulting "information overload" and the need to filter "good" from "bad" information. Unless South Africa integrates information security in its e-government development policy and practices, the acclaimed benefits of e-government will not be realized. The moral hazard problems arising from bad information behavior such as human manipulation, withholding information, unauthorized access, and violation of individual privacy and confidentiality heightens the need to combat info-security risks and vulnerabilities. South Africa's readiness to deal with the information security risks has come under scrutiny. The information security infrastructure in South Africa is also not clearly understood. This chapter examines South Africa's information security landscape and describes how institutional and agency coordination could help improve information security in e-government.

DOI: 10.4018/978-1-4666-2655-3.ch009

1. INTRODUCTION

South Africa, just like the rest of the world, is facing pressing challenges in the area of information security. As the country moves towards building an "information society" the threats from information security are also rising. The information security risks have been heightened by the growing use of e-government and e-banking as service delivery platforms (Bakari, et al, 2006). Such developments also entail an "information overload" to certain categories of users, and therefore a need to filter "good" from "bad" information. Some factors that contribute to bad information behavior are human manipulation, unauthorized access, violation of individual privacy and confidentiality, money laundering and terrorism. Recurring information systems failure also affects the integrity of the information supply chain. The information and communication technology (ICT)-induced rapid changes in information production, processing, and dissemination suggest the need to develop a sound information security "landscape" or "architecture" and simultaneously integrate information security, especially in e-government development policies and practices.

Little is being done to promote information security in Africa (Bakari, et. al., 2006). Traditional methods of protecting information such as authentication and authorization have become obsolete globally. As new security systems are being probed, an intensive debate is raging on the rationale for newly introduced information security legislation. For instance, others argue that the September 11th 2001 event has affected citizen privacy, freedom of expression and information access. As a result, concentration of focus on national and global security has led to greater control over access, secrecy over transparency, national security bias over civil liberties and surveillance over anonymity (Caidi and Ross, 2005). Further, the authors argue that the new legislation introduced has affected "what the public can know about the government not just what the government can know about the public," which is an infringement of civil society's information rights.

It is not clear to what extent the South African information security landscape is prepared to deal with the information security risks poised by the information age and the modernization of public service delivery on the e-service format. The extensive information security infrastructure in South Africa is not clearly understood by many. There is need for researchers to scrutinize the information security landscape, and examine whether or not the Government of South Africa (GSA) has aligned itself with other global agencies to combat information security threats and vulnerabilities.

The paper argues that the systematic inclusion of security in e-government projects is no longer an option, but an imperative. Information security helps to deliver information values such as accuracy, reliability, rapidity, timeliness, privacy and relevancy. Without assurances that information provision and access are secure, the aforementioned value propositions are easily eroded. E-government and its acclaimed benefits of efficiency, accountability, effectiveness, flexibility, transparency and trust among others, can also be compromised if information security is poorly handled.

South Africa is endowed with several state agencies, institutions, businesses, research institutes and various other strategic partners that could play a prominent role in arresting information security threats. But working in isolation, such organs will remain necessary but insufficient in combating information security threats. The fact that the global risk of information security seems to be growing unabated despite a proliferation of legislation globally, and renewed commitments to provide information security protocols justifies the need to probe and or transform the security landscape and enhance coordination with global partners.

Although this paper does not seek to determine the appropriate level of information security that policy makers should aim for, it seeks to examine the information security landscape and how its organization can best provide solutions that could help build secure information security landscape in South Africa. Global trends demonstrate a shift from mere information access and provision, to e-value creation which can only be guaranteed by a "state of the art information security landscape." The importance of increased coordination and collaboration in dealing with information security threats in a global information age is also highlighted.

The rest of the paper is organized as follows: The next section discusses information security in South Africa and global developments. Then we discuss the benefits of establishing secure information security infrastructure. In section 2, we present the literature review. The conceptual frame work is discussed and illustrated in section 3. Section 4, discusses e-government and information security. After that, we describe information security implementation hurdles in section 5. This is followed by a general discussion in section 6 on global information security, Internet monitoring and surveillance, and information security literacy issues. We conclude the paper in section 7.

1.1. Information Security in South Africa and Global Trends

While it is important to note that compromising information rights should not be the goal in dealing with information security threats, policy makers must guard against the under-provision of information security apparatus. Information access is a fundamental constitutional right and GSA recognizes and upholds that viewpoint through its Information Access Act of 2000 which aims to "give effect to constitutional right of access to any information" (Government Gazette, 2000). South Africa is also considering a computer misuse act that will help curtail computer related crimes.

Nonetheless, the GSA still faces the challenges of providing information security without infringing on the privacy and confidentiality concerns of its citizens. Another vital challenge is how to marshal resources that are required to build a sustainable, effective, and proactive information security landscape. Information security is a global public good and both the GSA and the private sector must coordinate their actions in reducing information security risks. In other words, low information security investment in the private sector does not constitute a risk to that sector alone, but to the public sector as well. Given the public good aspect of information security, investments in information security ought to be synchronized between the private and public sectors globally.

Following September 11, 2001 attacks on World Trade Center in New York, several security-related agencies were established and a flurry of legislation was passed around the world to respond to future threats (Caidi and Ross, 2005). Most of the legislation had direct or indirect implications for information access, storage, distribution and security. For instance, the Homeland Security and Patriot Act are examples of the legislative interventions by the Federal Government of the United States. Similar legislations were passed in Australia, Canada and the United Kingdom to name a few.

In South Africa, the Financial Intelligence Centre Act (FICA) was passed in 2001 as way to track contact details and money laundering activities using banking information. The Information Right Act advocates for "universal access to information" by South African citizens. A growing concern however is that, new security legislation appears to work contrary to some of the rights entrenched in this legislation.

Information security is now a concern not only for government, but for both businesses and individuals as well. Financial institutions bear the largest burden of computer security crimes with Internet viruses comprising major threats (Chang et al, 2006). One cannot separate information security from issues such as individual protection, privacy and confidentiality, national intelligence, and integrity of both inter - and intra- business and government relations. Since not much is known about the state of information security in Africa, research effort needs to be heightened (Bakari, et al, 2005). Part of the argument is that developed nations have the resources to put in place systems that can help guarantee high levels of information security while African nations do not have same privileges let alone the research resources to understand the scale or gravity of the information security risks.

1.2 Information Security and the e-Value Aspect of Information

Generally, people are increasingly showing concern about information security breaches (i.e. unauthorized access to computer, use of pirated software, virus or worm attacks, network failure etc.). As these breaches increase in frequency and scale, the e-value aspect of such information is eroded and the integrity of transactions that are vulnerable to security breaches is lost and maybe difficult to recover. Security breaches have long been classified into a number of areas including: authenticity, privacy, integrity, and non-repudiation (Turban et al, 1999). Authenticity involves proving beyond reasonable doubt whether the sender of the information is who they claim to be. Privacy involves exclusion of unauthorized access to information being exchanged between two parties. Integrity deals with the intentional or unintentional manipulation or modification of the contents of a given message during transmission process. Non-repudiation is concerned about disputing whether or not one has conducted a web-based transaction. While these sources of vulnerability are well known, what we still do not fully understand is how to prevent these practices from occurring.

The e-value of information has been attributed to internal, external and technical prime movers (Maumbe et al, 2006). Using this framework for e-value creation, information security falls under internal technical prime movers for e-value creation. Without enhanced information security, e-value of information is compromised. As consumers of information become more discerning in terms of demanding timely, accurate, reliable, relevant and adequate information, upholding information security contributes significantly towards information quality and high e-value credentials. In South Africa, the *Batho Pele* principles underpin the trust worthiness of e-government information and its attempt to promote a value-based system of government citizen interaction. Security remains at the core of achieving the delivery of state of the art, high quality information end users.

2.0 INFORMATION SECURITY: A LITERATURE REVIEW

The changes in the information "access landscape" attributed to September, 11, 2001 event has added a new twist to the debate on information security threats (Boxwall, 2005). The author contends that because of security risks poised by unfettered access to information, "there is a need to balance protection with openness." In this vein, proponents of right to information access would argue that people should have broad if not uncontrolled access to information, those in charge of information security risks advocate for controlled and or limited access to certain key information. The question then becomes who determines what information is right or wrong for an individual and does that not determination on behalf of someone else not violate fundamental constitutional rights.

The passing of various information-related pieces of legislation is designed to enhance the security value of information held by, and shared between government agencies. The business community, especially the banking sector seems to have battled the issue of information security earlier than government while the later seems to have under-provided for information security. The advent of e-government is set to change that status quo. The classification of information into the "sensitive" bracket by the intelligence community as required under the Homeland Security Act has created some controversy around the whole process of de-classification of information previously identified as sensitive information (Relyea, 2004).

Information security is central to the operations of many organizations including human rights groups, law enforcement agencies, intelligence agencies, and even government departments. Those concerned with information rights perceive proliferation of information laws as exceeding limits to the point of infringing on information rights of ordinary citizens (Caidi and Ross, 2005).

Relyea (2004) points out the need to prioritize a neglected area, that of "critical information protection" and argues for the need to develop federal centers for information sharing, analysis, sanitizing, and dissemination to other government agencies. In this vein, critical information protection is therefore expected to become an integral part of future strategy on security provision. Bakari (2005) cautions that e-government initiatives stand a better chance of developing secure information security systems as they are still at their early stages of development.

3.0 CONCEPTUAL FRAMEWORK

3.1 Information Security Landscape for South Africa

Information security is defined in this paper as the risks associated with the application of ICT in government, business and civil society. The information security landscape provides a description of the various role players, networks, institutions and custodians of information security. Common among these organizations, networks and role players is the critical responsibilities of 1) information security protection and 2) information sharing. Although stakeholders are easy to identify, what is unknown to us are their information security approaches and coordination strategies behind the scenes.

In terms of information security channels, GSA contracts part of its business to private firms and individuals and this makes the information security web complex. Information security needs to be established across the networks of public and private sector information users. Given the global nature of information flows and information security requirements (see Table 1), it is clear that information security landscape will ultimately be influenced by global information security network, policies and procedures. Therefore, the nature of information in particular, its diversity, coupled by the scope of the information flows and security challenges should be understood in a context that lays out the players, the multitude of networks, and the complexity of these relationships and coordination challenges within a given country, across the continent and globally.

Empirical evidence has shown that different types of institutions have different levels of information security concerns (Chang, et. al., 2006) and that affects their desire or motivation to coordinate information security needs.

3.2 The Information Security "Cobweb" Landscape Model

We envisage an information security landscape that can be described by a "cobweb" of various stakeholders that are directly or indirectly linked through information security. The Cobweb represents a landscape of various players at different levels, be it civil society, business community, national agencies, research institutes and global partners that have a keen interest on preserving the "information security culture." The cobweb illustrates that organizations in the "Information Society" face the same core challenge—information security. The information security is the oil

Table 1. Summary of global information security stakeholders, 2006

Organizational Level	Organizations (Agency, CBOs, CSO, Departments & Institutes etc)	Functional Areas
Global	INTERPOL, FBI, CIA, MI5, Scotland Yard, Computer Security Associations, etc	Information Security Law Enforcement Research & Development
Regional	SADC Organ for Security and Conflict Resolution	Information Security Conflict Resolution
National	NPA, FICA, Department of Safety and Security (DoSS), Department of Health (DOH), Department of Defense (DOD), Department of Home Affairs (DOHA), Department of Justice (DOJ), National Parliament, National Archives, National Prosecuting Authority (NPA), Emergency Services, Institute of Security Studies, National & University Library Associations, University Departments of Information Sciences, National Archives, SITA, SA National Defense Force-Electronic Warfare, National Computer Security Associations, Freedom of Expression Institute, Internet Service Providers (ISP), Private Security Firms, Software Firms e.g. Anti-virus software.	Information Security Immigration Security National Security Law Enforcement Justice Communications Research & Development Technology interoperability Health Information Systems Information Storage & Dissemination National Debates and Policy Pronouncements
Civil Society	Community Based Organizations, Civil Society Organizations, Non-Governmental Organizations, Informediaries, etc.	Information Security Business profitability Privacy and confidentiality

that lubricates the ICT wheels in the digital and networked environment. The center of the cobweb, is equivalent to the centre of gravity for the global information society.

We assume that the cobweb of information security agencies differs in shape, form and size from country to country. A typical national level cobweb is represented in Figure 1. Different size circles represent different institutions or agencies. The global information security cobweb is also expected to take a different shape and format from that of the national cobweb (Figure 2).

A critical question is how are these physically isolated cobwebs aligned in virtual space in reality? Is there a common approach in dealing with information security? How close is the core issue of information security to these global, national and even individual users of information communication technologies? Of paramount importance is the landscape ability to share information, establish credibility of information, and exclude unauthorized parties from certain key information during dissemination. Early warning systems for information threats need to be part of the configuration of this information security landscape in South Africa.

While the landscape is bound to be represented by different type of organization whether private or public, research or policy oriented, the key consideration must be the organization and or institutional orientation to information insecurity. The landscape is not exhaustive, but it is a representation of the complicated web of agencies and institutions that have strategic role in information security provision.

3.3. Hierarchy of Security Apparatus and Information Sharing

The national and global cobweb of information security suggests the existence of some hierarchy in information security provisioning. Figure 3 illustrates the horizontal and vertical information security flows and the layers of information securities filters through. However, what needs to be established is whether or not there is an under-provisioning of information security and at what level does that gap exists? Secondly, what

Figure 1. National cobweb of information security landscape in South Africa, 2010

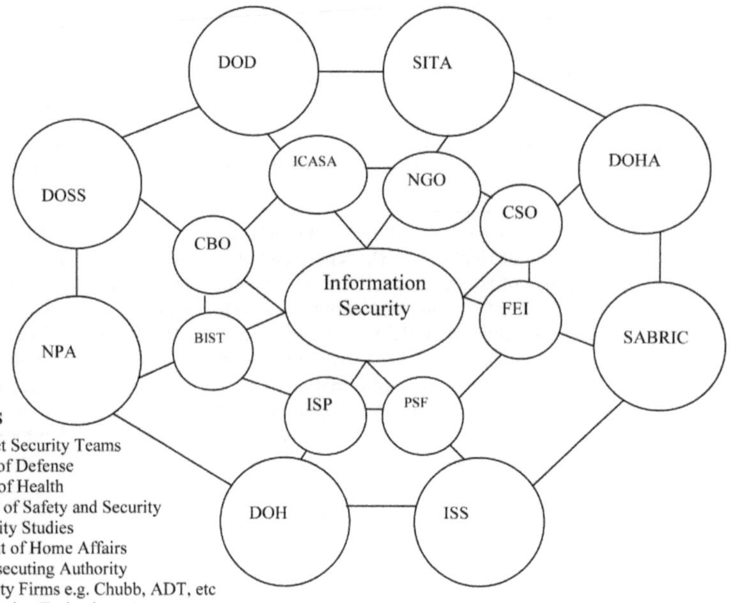

ABBREVIATIONS
BIST - Bank Internet Security Teams
DOD - Department of Defense
DOH - Department of Health
DOSS - Department of Safety and Security
ISS - Institute Security Studies
DOHA - Department of Home Affairs
NPA - National Prosecuting Authority
PFS - Private Security Firms e.g. Chubb, ADT, etc
SITA - State Information Technology Agency
SABRIC - South African Banking Risk Information Centre
SADC - Southern Africa Development Cooperation Organ for Security & Conflict Resolution

is the influence, negative or positive of global information security policies and procedures at both regional and national levels? Also not clear is the level of coordination and cooperation in information security knowledge production and exchange of that information. After identifying where the information security gaps reside it is important to establish whether global information security partners are able to support investments at regional or national levels? Unlike terrorism that is clearly a global threat, information security though viewed as global concern still awaits innovative collaborative response. Addressing security requires an understanding of the information security hierarchies, global investment and coordination opportunities and above all, realization that it's a "global public good."

4. E-GOVERNMENT AND INFORMATION SECURITY IN SOUTH AFRICA

E-government development offers tremendous opportunities for enhancing information security investments in order to protect information, secure transactions and deliver quality information services. The e-government functions are diverse and multidimensional. As shown in Figure 4, e-government functions comprise G2G (i.e. addresses information flows between government agencies and employees), G2B (i.e., covers information flows between government and business community, and G2C (i.e. deals with government interactions with civil society). All these arteries indicate the flow channels and the potential information security investments that are required to make e-government secure.

In order to understand the information security problems confronting the Government of South Africa (GSA), we need to reflect on the

Figure 2. Global cobweb of information security landscape, 2010

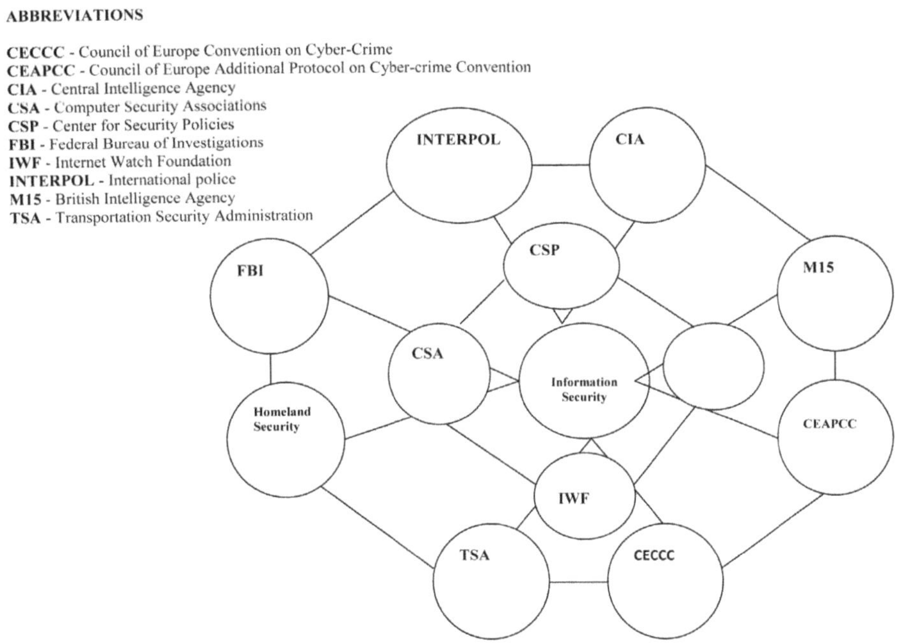

various functions and levels of e-government in South Africa. Figure 4 illustrates some of the key functional areas of government. The e-government infrastructure spills beyond individual governmental agencies (i.e. national, provincial and municipal levels) to include interactions with various other stakeholders (e.g. CSO, CBOs, and NGOs), and the business community especially those performing out-sourced government functions. E-government functions are diverse and multi-dimensional. They include e-information provision (e.g. to visitors, locals, researchers etc.), e-procurement, e-legislation development, e-health provision, e-communication within government, and with outside partners, e-learning program development, and e-filling of taxes among others. The three levels of government interactions demonstrate the degree of vertical coordination issues and the inter-agency coordination that have major implications on information security risks.

The GSA has the responsibility to protect its information by providing tight security in various information flow channels, within government agencies (i.e., national, provincial and municipal levels), and to outside partners. The security apparatus needs to be applied not only to the information flow process, but also the data storage to ensure security of stored information or citizen records.

At the second level, government deals with information security related to information exchange between GSA and the regional (e.g. SADC, NEPAD, COMESA, SACU, etc.) and international bodies (e.g. UN Agencies, G8, Non-Aligned SUMMIT, COMMONWEALTH OF NATIONS etc.). Confidential government information is exchanged, stored, shared and or distributed in bilateral and multilateral arrangements. Superimposed on this global information structure is the demand by South African citizens abroad, and visitors to South Africa for government information.

At the grassroots level, it is not clear the extent to which ordinary citizens are being made aware about information security risks. The emphasis seems to be directed at making citizens aware of available e-government information services to

Figure 3. Hierarchy of information security provisioning in South Africa, 2006

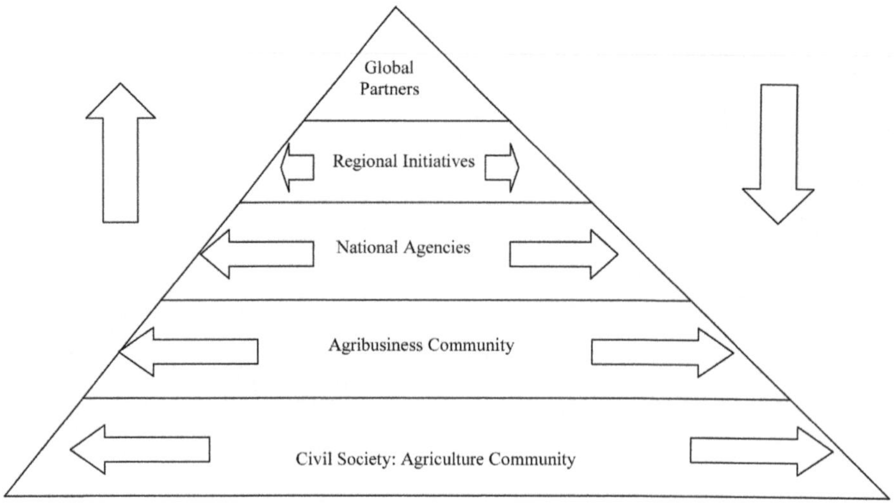

encourage widespread use. Various e-government functions can easily succumb to new abuses if there is no effort to unveil the intricacies of information security to current and future users.

Intelligence is synonymous with security. Without proper information security systems, e-government implementation cannot be expected to succeed in South Africa.

5. INFORMATION SECURITY IMPLEMENTATION HURDLES

The use of personal computer spy-ware to fraudulently obtain log-in details, card numbers and personal identification numbers are some of the information security risks and challenges facing South Africa. Specific challenges are discussed in the section below.

5.1. Coordination and Collaboration Challenges

Cyber crime is perpetrated by global syndicates. Some of the leading targets are online banking services (Chang, et. al., 2006). Recent press reports on the arrest of a South African who ran a successful international online banking-fraud syndicate which was ended by collaboration between Scorpions, Standard Bank, UK security consultancy firms demonstrates the need for strategic collaboration (www.star.co.za). With clients in diverse countries such as United Kingdom, Estonia, France, Sweden, Australia and South Africa having lost millions, information security cannot be left to individual national security advisors or institutions. To show that no organization is immune, in 1999, hackers attacked the GSA's official statistics website and spoiled a consumer price index report due for release at that time (Insecure.org). Security collaboration in the past has focused on ordinary criminals , but the emergence of cyber criminals requires a new approach to information security collaboration. These challenges lie in the spaces between nodes in the information security cobweb framework that has been presented in earlier sections.

5.2 International Harmonization of Legislations

A related issue to that of global coordination and collaboration is that of the harmonization of laws between countries. Currently, the problem

Figure 4. Overview of e-government functions in South Africa, 2006

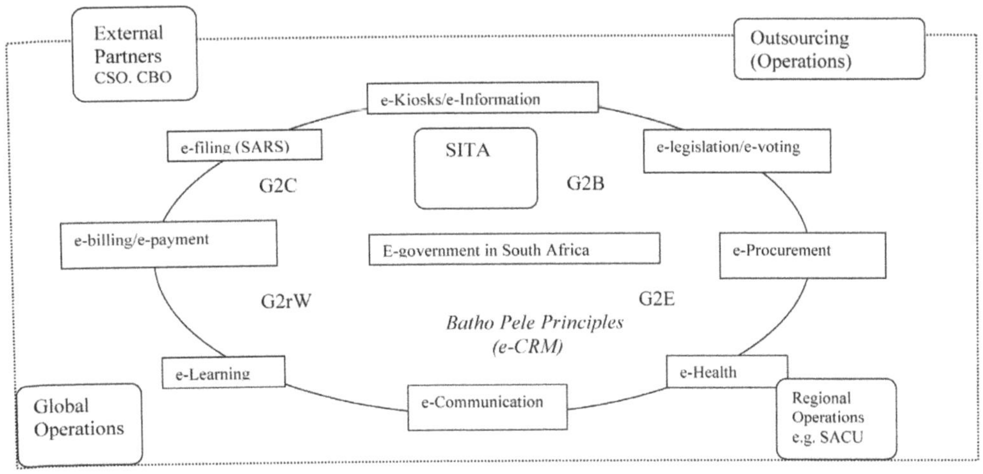

ABBREVIATIONS
G2C - Government to Community
G2B - Governmnet to Business
G2E - Governmnet to Employee
G2rW - Government to the rest of the World (i.e. regional+ global partners)
SACU - South Africa Customs Union

of arresting and prosecuting cyber-criminals is rendered difficult in cases where a criminal in one country commits cyber-crime against individuals and institutions in another country. It is usually problematic under whose laws the criminal should be prosecuted. This problem can be resolved through the multilateral and bilateral harmonization of cyber-crime legislation in different countries. Proactive initiatives need to be taken by the GSA to forestall the possibility of cyber-criminal using the legal loopholes in extradition treaties to escape arrest and prosecution.

5.3. Growing Global Trends in Cyber-Crime

Cyber-crime is intractably linked to information security. Only through greater security coordination, in both the horizontal and vertical information channels will problems begin to subside. Isolated interventions whether in public or private sector will be necessary but insufficient to curb information security related problems. As the Internet leads to more and more e-collaboration, information sharing, information systems interoperability will become inevitable. Joint policies and procedures are needed to promote system interoperability, prevent cyber-crime without compromising privacy issues. In the long run, restrictions to free flow of information will become an obstacle to development, especially if it is caused by lapses in security. The GSA cannot expect a thriving e-government program without a free flow of information. Therefore, security investments are required in both the horizontal and vertical information flow channels. The proliferation of legislation, especially in developed countries, perceived as crucial, may on one hand actually stifle information flows and even increase information security risks. On the other hand, it may violate and interfere with individual privacy. The challenge is for the GSA to craft legislation that enhances information quality and cyber-security while ensuring that the guarantee of individual privacy is not abrogated and violated. To what extent should society sacrifice aspects of its privacy to ensure security in the information age? South African policy makers should grapple with this policy question.

6. DISCUSSION

The need to understand the information security landscape and readiness of developing countries has become a major global issue of concern. Increasing attention is also turning to the Government of South Africa's readiness to combat e-government security threats and vulnerabilities associated with major world sporting events. Fueling some of this attention is the fact that the global risk of information security seems to grow unabated despite legislation and commitments to provide information security protocols suggesting the need to probe or even transform the information security landscape and enhance coordination.

6.1 What if there were no Major Global Security Threats?

As the information security discourse takes a global perspective a number of questions come to the fore. Does the rest of the world including South Africa need to wait for the USA to develop global security policies? What if South Africa faces information security risks that are unique to its soil? Would its policy response affect the rest of the region and the world? Is this the era for a national landscape for information security or is it time for a global standard? Since the post 9/11 new world order has seen the emergence of virtual communities, and information risks located in virtual spaces, maybe the world now needs one unique information security architecture. Surely, the question of resources would arise, and challenges could emerge from different resource endowment levels. However, the fact that information security risks affect virtual communities justifies the need for joint action, mutual interests, and less and less focus on security matters defined by or limited to national boundaries. Countries need to be proactive and prepare for eventualities, some of which have not been experienced before. In other words, lessons must be learnt from the global effort that was marshaled to respond to the Y2K disaster. A similar level of dedication is needed in the area of the information security risk that now confronts the global community with a potential for catastrophic consequences if not given proper attention.

6.2 Internet Security Monitoring and Surveillance?

This is a difficult proposition in a world where human freedom gains have occurred and various constituencies are fighting to preserve that. However, the vulnerabilities associated with information security threats demands some form of intervention in cyberspace. In a bid to control intelligence, some governments have considered the need to monitor Internet use, especially email and cell-phone calls. This has received public outburst. Perhaps the recent banning of Internet gaming or betting sites in the United States provide insights in what is in store for Internet users. In the same vein, South Africa has proposed the Regulation of the Interception of Communication and Provision of Communication Related Information Act (RICA) which is also referred to "sarcastically" as the "eavesdropping bill". At the time of writing, a similar bill is before the Nigerian National assembly awaiting ratification—that too has generated uproar. In Zimbabwe for instance, the Public Order and Security Act (POSA) had been described as one of the pieces of legislations that has curtailed information access and media freedom. Throughout the world, governments are confronted with the challenge of how to strike a balance between sensitive information gathering and the maintenance of citizen's privacy.

One of the key questions being asked is whether governments are not being overly incensed with information security to the point of overprotection and subsequent creation of an uninformed citizenry? The attempts so far regarding Internet monitoring and surveillance and e-mail tracking and listening in to telephone calls have met with fierce resistance from ordinary citizens and human rights groups. Although such resistance might be justified, one of its indirect benefits to policy makers is that it automatically sheds light on the key stakeholders who should contribute to the debate and force them to develop an information security landscape that is capable of securing information systems and thereby prevent security breaches that are on the rise globally.

6.3 Information Security Literacy

Information security skills are vital for the smooth operation of the information landscape. Information security is separate from ordinary computer literacy skills. Individuals can display information skills but fall short on information security knowledge. To develop information security skills and awareness, a new thrust is needed that will inculcate citizen, business and governments to become more sensitive to security issues. Without improved information security skills, knowledge and awareness, the information infrastructure can be established but will remain perpetually handicapped and will fall short on achieving intended targets. A more serious approach to information security culture should involve teaching information security as part of the curriculum on computer literacy courses in early school stages. As society becomes more aware of information security risks, that will add value to their use of computer in everyday life and in business.

7. CONCLUSION

The right of citizens to information is unquestionable. Yet that right now seems to interfere with national security concerns. South Africa has a plethora of information security laws, agencies, and policies but information security risks have not disappeared, if anything, they have heightened as a source of concern in today's information society. The development of e-government and outsourcing of government information services has also heightened the information security risks.

To achieve the desired goals in e-government, information security needs to be addressed. In building an information security landscape or architecture, attention must be given to the security value of information, information sharing, strategic coordination, and the elimination of unnecessary duplication. In South Africa's dualistic society where there is a danger from the co-existence of "information over-load" and "information paucity" and "information haves" versus "information have-nots", addressing information security requires a special balancing act.

South Africa is endowed with several agencies, institutions, and strategic partners that could play a prominent role in arresting information security threats. However, when such institutions/organizations work in isolation they fall short of meeting the necessary and sufficient conditions in combating information security threats. What this paper did not accomplish is to determine the appropriate level of information security that policy makers should aim for as that was not the initial focus. The paper starts from the premise that global trends demonstrate a shift from mere information access and provision, to e-value creation which can only be guaranteed by a state of the art "information security" landscape.

Mapping the complex information security cobweb showing information flow pathways, the configuration of a proliferation of government agencies, national archives, interest groups, libraries, academic departments, businesses, etc., and exposing the potential impact on virtual communities is a necessary first step to design innovative collaboration strategies and to improve the provision of information security. Improvements in information security require systematic institutional development and coordination throughout the "cobweb" to avoid disjointed efforts and vulnerabilities that could easily reverse the gains made at other nodes or levels in the information security cobweb. In addition, the development and alignment of information security priorities at both national and global level is a necessary requirement. In the same vein, information sharing is a critical ingredient in the establishment of a coherent information landscape, and will require rules and norms that govern that process by providing coordinates for non-disclosure to third parties.

As e-government evolves, the security architecture or landscape should be accompanied by carefully crafted e-government security policies and regulations. To prevent information security deteriorating in the future, ICT security training must become an integral part of the core curriculum in computer studies. Also, policy makers and legislators should guard against infringing on information rights by unnecessarily limiting access to key or sensitive information. Finally, information security can no longer be treated as a secondary issue; but rather it should be progressed simultaneously as part of the overall goal of building an information society with a high information security consciousness in South Africa.

REFERENCES

Bakari, J. K., Tarimo, C. N., & Mutagahywa, B. (2006). Issues and challenges to be addressed in e-government from an information security point of view. In *Proceedings of the IST Africa Conference*, 3-5th May, CSIR International Conventions Center, Johannesburg, South Africa.

Boxwall, J. (2005). The nature of geospatial information and security. *Government Information Quarterly*, *22*, 644–662. doi:10.1016/j.giq.2005.09.001

Caidi, N., & Ross, A. (2005). Information rights and national security. *Government Information Quarterly*, *22*, 663–684. doi:10.1016/j.giq.2005.11.003

Chang, I., Hwang, H., Yen, D. C., & Huang, H. (2006). An empirical study of the factors affecting internet security for the financial industry in Taiwan. *Telematics and Informatics*, *23*, 343–364. doi:10.1016/j.tele.2005.11.001

Information Security News. (2006). *Hackers attack South Africa's stats site to protest monopoly of Telcom.* Retrieved from www.Insecure.org/

Maumbe, B. M., Taylor, W. J., Wesso, H., & Erwin, G. (2006). E-value creation in a government web portal in South Africa. In Khosrow-Pour, M. (Ed.), *Encyclopaedia of e-commerce, e-government, and mobile commerce*. Hershey, PA: IGI Global. doi:10.4018/978-1-59140-989-2.ch065

Relyea, H. C. (2004). Homeland security and information sharing: Federal policy consideration. *Government Information Quarterly*, *21*, 420–438. doi:10.1016/j.giq.2004.08.007

South Africa Government Gazette. (2000). *Promotion of access to Information Act*. Pretoria, South Africa: Office of the Presidency.

The Star. (2006). *Cyber 'kingpin' nabbed.* The Star. Retrieved from www.star.co.za

Turban, E., Lee, J., King, D., & Chung, H. M. (1999). *Electronic commerce: A managerial perspective* (p. 520). Upper Saddle River, NJ: Prentice Hall.

Chapter 10
Enhanced Knowledge Management:
Knowledge Centers for Extension Communication and Agriculture Development in Ethiopia

Abebe Shiferaw
International Livestock Research Institute (ILRI), Ethiopia

Dirk Hoekstra
International Livestock Research Institute (ILRI), Ethiopia

Ermias Sehai
International Livestock Research Institute (ILRI), Ethiopia

Abrahm Getachew
International Livestock Research Institute (ILRI), Ethiopia

ABSTRACT

Knowledge management (KM) provides opportunity for organizations to achieve higher efficiency and increased output. With objective of attaining opportunities in KM at district level, Woreda Knowledge Centers (WKC) were established and used in ten districts in Ethiopia from 2005 to 2010. WKC is a telecenter used to gather, share, classify, access, and use knowledge at district (woreda) level. The study involves questionnaire, group discussion, timeline, SWOT analysis, and KM performance assessment and document review. This chapter presents WKC usefulness, establishment steps, challenges, and opportunities. In ten districts, of the 500 survey respondents, 79% and 71% agree on an increase in knowledge delivery and availability, respectively. Temporal comparison showed that WKC increased staff capacity to document, access, share, and use knowledge leading to improved extension communication. The study recommends that WKC be established by ministry of agriculture and its partners to enhance KM at district level for agricultural development.

DOI: 10.4018/978-1-4666-2655-3.ch010

1. INTRODUCTION

There are increasing evidences on the benefits of knowledge management (KM) for organizations to be more competitive, achieve higher efficiency and increased output (Hussain, et, al, 2010; 1; Sehai, 2006 ;1, Lee and Choi, 2003; 181). The benefits for public agricultural institutes from KM can enhance partner collaboration, knowledge sharing and improve extension communication leading to increased agricultural productivity. KM is a joint effort and practices used by organizations and individuals to identify, create, accumulate, re-use, apply (use) and distribute (share) knowledge (Hartwich *et.al*, 2007; 21, Hussain, *et. al*, 2010; 4). KM involves identification, creation, acquisition, transfer, sharing and utilization of knowledge.

Knowledge management is a systematic discipline of policies, processes, and activities which empower organizations to apply knowledge to improve effectiveness, innovation and quality (Sehai, 2006; 1). There is a general recognition among academics that knowledge management is a cross-functional and multifaceted discipline (Lee and Choi, 2003; 181). Knowledge management is concerned with ways of exchanging knowledge among those who can develop it and those who can use it (Hartwich *et.al*, 2007; 21). Thus, it is broader than information management (Kebede, 2010; 416). There are examples of incorporating KM in planning process to gain benefit in public agricultural organizations (Zavaglia, *et.al*, 2011;121).

Agricultural extension is a knowledge and information transfer service Gebremedhin, *et al*., (2006; 24) and KM is important to enhance development partner collaboration, improve use of knowledge in communication for increased agricultural productivity. The importance of knowledge management in agriculture sector is well recognized. Enhancing knowledge management will assist Ethiopian institutions, farmers and pastoralists to overcome challenges in agricultural extension (Sehai, 2006 ;3) and enable them take advantage of opportunities in KM.

Different empirical studies have shown that knowledge cannot easily reach farmers through extension services and development projects. In response, new way of managing knowledge have emerged across developing countries beyond the traditional farmer-extension link (Hartwich *et.al*, 2007; 21). In this case study, the use of knowledge centers or tele-centers has been examined as knowledge management enablers (Lee and Choi, 2003;198) to illustrate its' role in extension communication and agricultural development. The paper shows changes in knowledge process in accessing, sharing, transferring and using knowledge as a result of establishing and using knowledge centers at district "woreda" level.

The Ethiopian government has embarked on a major initiative for use of Information and Communication Technologies (ICTs) to improve the efficiency and impact of its sector programs(World Bank; 2003; 68). As part of this initiative, the agriculture sector needs the use of ICT application for extension programs. Recent study in Ethiopia has indicated that ICT has application in extension service as it has become more vivid in reaching the rural communities (Gebremedhin, *et al*., 2006; 7). Information and communication technologies (ICT) have great potential to contribute to community development (Gomez and Baron-Porras, 2011;1) and that is true to agriculture sector. The establishment of WKC paves the way for using ITC while addressing problems in extension communication. The argument is that knowledge cannot be simply accessed, transferred and used, when knowledge centers are lacking, which act as nucleus to enhance knowledge management. The lack of exchange of knowledge among and between farmers and those who produce farm-relevant products and services has often been regarded as the key issue in pro-poor agricultural development (Hartwich *et.al*, 2007; 22). The assumption is that knowledge center are used to eliminate some of the barriers in transfer and use of knowledge in extension communication.

In Ethiopia, lack of knowledge is considered as key impediment to development of agriculture(IPMS, 2005; 8). In the past decades in Ethiopia, effort was not paid to enhance knowledge management at district level. In the past decades in Ethiopia, much effort was not paid specifically to enhance knowledge management at district level. Need assessment shows five major problems in KM at district level. First, in most districts office of agriculture in Ethiopia, documents are not organized in a center but are kept with individual staff, departments and at office archive. Second, not all staff can access documents in archive office and staff cannot easily find relevant knowledge to address famers and pastoralists need. Third, existing schools and municipality libraries lack relevant agricultural documents. Fourth, staff at office of agriculture cannot access internet to get on line and digital documents. Fifth, staffs at office of agriculture do not share knowledge they have among themselves in a formal way because they do not have center established for that. These problems associated with knowledge access, documentation, sharing and use necessitated the need to establish a knowledge center at district (woreda) level.

Knowledge centers can help overcome challenges in agricultural extension in three ways. Firstly, the centers are used as a venue to increase access to electronic and non electronic documents at districts (woredas) office of agriculture. Secondly, establishment of the centers lead to provision of basic infrastructure for knowledge management at woreda level. Lastly, the venue (room/hall) allocated as a center, encourage culture of knowledge assembling, sharing and use among staff and with farmers leading to improved extension communication and contributing to increased productivity to both crop and livestock commodities.

In this paper, knowledge center at district (woreda) level is described as Woreda Knowledge Center (WKC). WKC is a place to gather document, share, access and classify knowledge in electronic and non electronic format to address end users (farmers) at district (woreda) office of agriculture. The term knowledge center is synonymous with the concept of tele-center (e-center or e-koiosk) as described by Masiero (2011;1). Lemma *et.al.* (2010:2) defined Woreda Knowledge Center (WKC) as an information resource center or venue that facilitates access to knowledge by providing functions which are provided by traditional and digital library, resource and meeting center. The centers serve as access point for on line and off-line documents. WKC acts as a venue to use tools of KM like seminars, meetings and trainings as described in CIDA (2003;12). The centers act as library to read and borrow educational materials. The establishment of the centers ensures establishment of infrastructure in Information Technology Communication (ITC) and non ICT tools for knowledge management at district level. At the end, the knowledge assembled, classified and shared in woreda knowledge center is to be used at grass root level with individual farmers, with group of farmers (community of practice) as described in CIDA (2003;11) and at Famers Training Centers (FTCs). The knowledge from the WKC is to be shared using training, drama, demonstration, field day, workshop or meetings. Communication method could be the form of oral communication or distributed as training materials, leaflets, manuals and other necessary materials during extension communication events. As a whole, opportunities in knowledge management can become evident in an improved extension communication and in its contribution to increased crop and livestock productivity in the long run.

This paper, therefore, is based on, Improving Productivity and Market Success (IPMS) of Ethiopian farmers' project and its partners experience in establishing ten knowledge centers at district level in Ethiopia for five years (2006 to 2010). To enhance knowledge management at district level in Ethiopia, establishment of woreda knowledge centers or tele-center lays foundation for use of both IT and non-IT technology. WKC or tele-centers enhance the processes of knowledge access, share, transfer and use. The establishment of WKC paves way to use both IT and non-IT materials to address problems in extension communication and contributing to agricultural productivity. Although there are variability in the functioning of the various knowledge centers (IPMS (b); 2010, 1) this paper draws conclusions based on general lessons learned. The paper has four sections; it begins with an introduction and continues with methodology in the second section. The third section, results and discussion provides justification, process steps, benefits, challenges and opportunities in establishing and using the centers. The fourth section draws conclusions and provides recommendations.

2. METHODOLOGY

2.1 The Approach

This case study is based on IPMS project and its' partners experience in establishing and using ten tele-centers or "knowledge centers" at district level in Ethiopia. The Improving productivity and Market Success (IPMS) of Ethiopian Farmers' project has a Knowledge Management (KM) component, which resulted in establishment and use of WKC to reach grass root level in collaboration with district office of agriculture and development partners.

In ten districts, project need assessment was conducted in 2005 and it showed poor dissemination of agricultural knowledge as key problem in knowledge management (IPMS, 2005, 8). To have functional knowledge management system in the districts (IPMS, 2005, 11) establishment of knowledge centers and use of KM tools were necessitated. Thus, the Improving Productivity and Market Success (IPMS) project and its partners established ten WKC over the past five years (2005-2010) at district level. The IPMS project and its partners have provided basic infrastructure to build ten knowledge centers (room for the center, computers, books ...) and build capacity of staff in knowledge management (i.e., training on basic and advance computer knowledge and skills) over the past five years in ten woredas (districts) in Ethiopia. Over the five years, various knowledge management tools were used to ensure transfer to knowledge from the center (woreda knowledge centers, WKCs) established at district office of agriculture to end users development agents and farmers. In this study, although there are variability in the functioning of the various knowledge centers (IPMS; 2010, 1) the importance of WKC use and its contribution in extension communication and agriculture development was assessed using KM survey, Temporal comparison, SWOT analysis and KM performance assessment in 2010.

2.2 Methods

After having established ten WKC in ten districts of Ethiopia, KM survey was conducted in 2010 in 40 Peasant Association (PAs) involving 500 survey respondents drawn from 400 farmers/pastoralists and 100 experts. The respondents represent WKC end users. PA selection has taken in to account the coverage and mix of IPMS project intervention on agriculture development. Group discussions were held with farmers, experts and development agents in the PAs. Discussion involved a group of 8 to 12 men and women to represent the diversity in terms of age, sex, socio-economic and project experience. Although measuring knowledge management performance is difficult (Lee and Choi, 2003;184),two indicators were used for measurement of performance in knowledge management (performance index) as suggested by Chang *et. al.*, (2005; 1149). The first indicator is increase in relevant knowledge availability which shows access to knowledge. The second indicator is usefulness of information received based on perception of respondents. Responsiveness of extension system was used as proxy indicator to show extension communication and reflect benefits of the WKC to reach grass root level. Temporal comparison of pre and post center establishment was undertaken to examine benefits of establishing WKC. Timeline was also used to examine progress made in establishment of typical WKC as suggested by Barry (1997; 154). As reviewed by Hermans and Thissen (2009; 809), actor analysis was used to examine actors' role in establishment and use of woreda knowledge center. The paper uses SWOT analysis as suggested by Dyson (2004; 632) to portray common pitfalls, and success factors in establishment and use of WKC.

3. RESULTS AND DISCUSSION

3.1 Overview of WKC in Ten Districts

Location of the ten districts is presented in Figure 1. Pre-WKC assessment in each district shows the existence of woreda net and school net. School nets were operational in most of the districts in 2006 while installation and use of woreda net varied form district to district (IPMS, 2005, 9 & 62). The existence of school net and woreda net reflects existence of Information Technology

Figure 1. Location of WKCs in ten districts (woredas) in Ethiopia

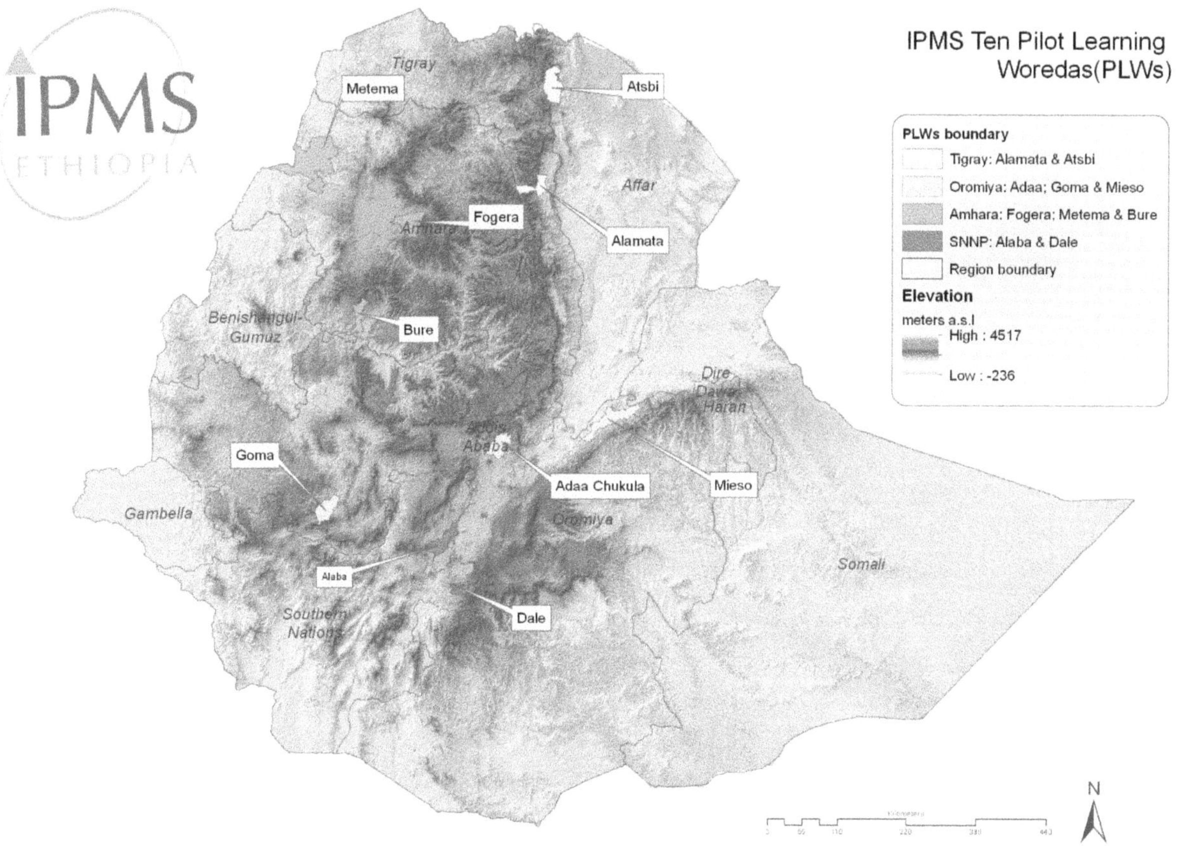

Communication (ITC) staff at district level. Ethiopian Telecommunication Authority has installed rural wireless connection in almost all districts by 2006. Depending on location of district to capital city, there is variation in availability of Internet service, involvement of private sector and use of ITC and ITC customers. In districts closer to Addis Ababa, for example Ada, the number of customers and ITC service is high. While, in district far from Addis Ababa, like Metema, there are few customers and poor ITC service. See inventory of WKC in 10 districts (Table 1).

3.2 Steps to Establish District "Woreda" Knowledge Center

The intervention to establish WKC started in 2005 with identification of gaps in KM (IPMS, 2005, 8). At district level, assessment have shown the need to establish WKC with relevant agricultural reading materials combined with modern ITC facility. In 2005 and early 2006, there were requests from office of agriculture to have access relevant agricultural knowledge to train farmers and address their emerging needs. On the bases of lessons gained, the basic steps to establish WKC are presented in Figure 2.

Initially, IPMS project and its' partners have focused on collecting (assembling), classifying and available documents for centers. Later on, over the years, depending on local demand WKC were established in larger rooms in most of the districts (See Table 2). The steps and specific procedures followed in establishment and use of WKC in ten districts varies (IPMS; 2010 (b), 1). By 2010, each WKC had a staff responsible to mange or coordinator the center. The volume and type of ITC and non-ITC materials vary from center to center.

Among the ten centers, number of computers varies from three to eleven and number of books vary from 70 to 1500 (IPMS 2010(b); 9-11). Capacity building intervention ranges from furnishing the center with computer, LCD, devices to access internet, TV sets and DVD players, books, magazine, CD and DVD materials, server…

Figure 2. Stages in establishment and use of WKC

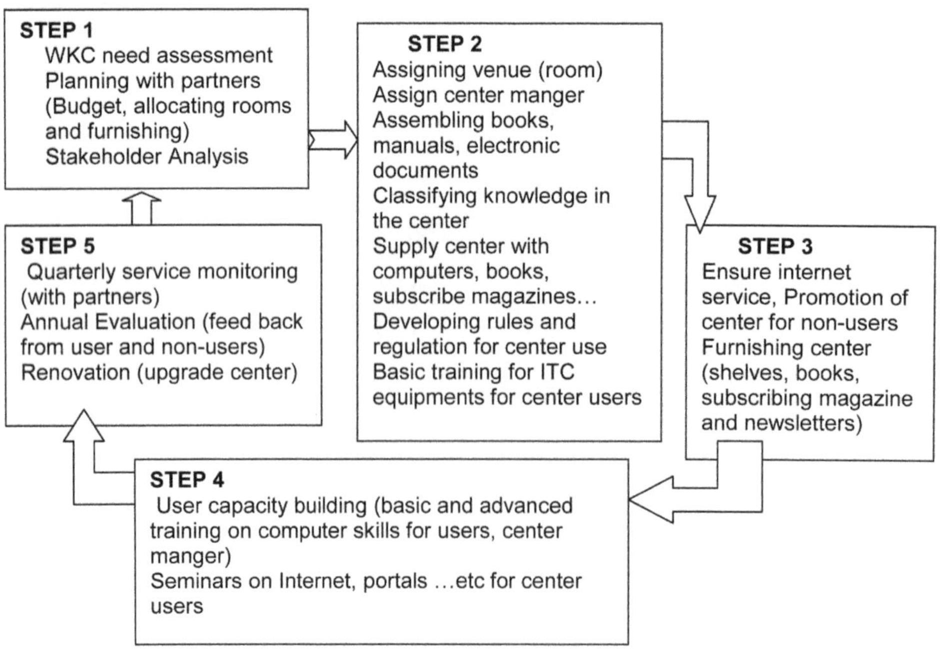

Table 1. Inventory of WKC in 10 districts in Ethiopia (2010)

WKC name	Alamata	Atsibi	Dale	Alaba	Meiso	Goma	Ada	Bure	Fogera	Metema
Number of computer	3	4	4	11	3	3	5	3	3	3
Internet connection Exist (E) Not (N)	N	N	E	E	E	N	E	E	E	N
Audiovisuals	TV, DVD	TV, DVD	TV, DVD	TV, DVD	TV, DVD	TV, DVD	TV, DVD	TV, DVD	TV, DVD	TV, DVD
Books (B)	620	881	105	677	187	375	1500	771	498	70
CD/DVD (V)	46	20	61	137	40	106	52	63	38	20
Average users/day	17	NA	18	15	NA	NA	12	NA	NA	NA
Other: Fax machine (F) Pinter (P) Satellite Receiver (D)	F,P	P	P	P,D	P	P	P	P	P	P

Source: IPMS (2010b); 9-11 NA= not available

to organizing trainings for district office of agriculture staff to make use of ITC technology in WKC (Lemma *et.al.*, 2010:4). Efforts are needed to encourage all staff to take part in knowledge gathering, sharing and reaching farmers and pastoralists as end-users.

3.3 Woreda Knowledge Center Timeline

The establishment of WKC has taken different steps in each district; however, the overall set of activity leading to the establishment of center can be summarized by looking at a typical district knowledge center in Alaba special district. Progress in the time line is governed by availability of budget, staff initiative to take the lead, role of multiple partners to support WKC and awareness created to use WKC in the office of agriculture and district.

The establishment of WKC and progressive changes to furnish the center is governed by local context. Above all the demand for knowledge (number of center users) at district office of agriculture is a driving force. In the above typical knowledge center, venue has changed initially from a small room which accommodated 3 computers (3-5 seats) in 2006 to a bigger room which accommodated 8 computers (10-20 seats) in 2009 and then finally to a hall which accommodated 11 computers (30-50 seats) in 2009.

3.4 Woreda Knowledge Center Actors

In a review of generation of knowledge management, today's second-generation knowledge management emphasizes collaboration in the management of knowledge (Hartwich *et.al*, 2007; 23) this implies the need to involve all possible actors for establishment and use of a center at all level (district, zone and region). The approach would lead to availability, creation and use all relevant knowledge through networking. In a recent study on WKC in ten districts, Lemma *et.al*, 2010:9) found that there are four categories of WKC users described as district office of agriculture staff, staff from line offices, researchers / students and others including Non Governmental Organization (NGO) staff.

KM implementation will require input in terms of people, process, tools (technology), and will operate in an existing organizational setting (Sehai ; 2006, 5). Thus, undertaking stakeholder analysis could be a starting point to see role of actors in establishment and use of WKC. Although there are

Table 2. Timeline for typical knowledge center (Alaba)

March 2006	Developed concept note on knowledge center
Aug 2006	Discussion on knowledge management
Aug -Sept 2005	Need Assessment for WKC (Identifying gap in KM)
Nov 2005	Assembling local books, documents and request partners
Jan 2006	Equipping 3 computers in a small room (3- 5 seats)
Feb 2006	Discussion with staff on WKC rules and regulation
March 2006	CD/DVDs in WKC reached 15 in number (placed in a rack)
Feb 2006	Additional books and various publications were received from partners (ILRI, EU project, ESGP project, LVIA, SARI)
April 2006 – Dec 2010	Computers maintenance and update of antivirus (as need be)
Dec 2006	CD/DVDs in WKC reached 67 in number
Jan 2007	Change in venue to bigger room (10-20 seats)
July 2007	Appointment of center coordinator(manger) by district office
Jan 2008 – Dec 2008	Promotion of WKC and increasing number of WKC users
Oct 2008 –Sept 2009	WKC furnished with chairs, conference tables, TV and its accessories, computers, a server and book shelves and generator
March 2009	11 computers, 1 server, 806 books, 173 manuals and number of 110 CD/DVDs in WKC (inventory)
Feb 2009	Change in venue to a new and bigger hall for WKC (30-50 seats)
April 2008 & Aug 2010	Dial up and wireless connection (CDMA) started
Sept 2010	CD/DVDs in WKC reached 135 in number

Source: Compiled based on review of Alaba IPMS project monitoring reports (2005-2010)

various degree of partnership between actors and variation in the role played in the process of WKC establishment, the district office of agriculture is the owner of the center. NGOs and development projects can help support establishment of WKC. Potential actors in the districts who can play role in establishment of the WKC are displayed in Figure 3.

3.5 Benefits of Woreda Knowledge Center (WKC)

The benefits of WKC are exhibited the process of accessing, documenting, sharing and using knowledge to contribute in agriculture development. Benfits of WKC are stated in IPMS (2010(a), 16).The most obvious benefits of the centers are (See Table 3):

- Source of knowledge (electronic and non-electronic documents). Electronic- documents are from web pages and off-line copies, CDs, DVDs. The non-electronic forms are the books, manuals, magazines, various reports and documents.
- Venue to share scientific knowledge in a formal way such as seminars, academic meetings, trainings, video shows, workshops and trainings.
- Relevant documents from WKC are translated in to local language to be shared in the form of drama, filed days, trainings, demonstrations and workshops.

The benefit of the centers was reflected in the increased demand for knowledge over the years as service from the center expands. This has been manifested as an increase in number of users who

Figure 3. Potential actors in establishment of WKC, 2005-2010

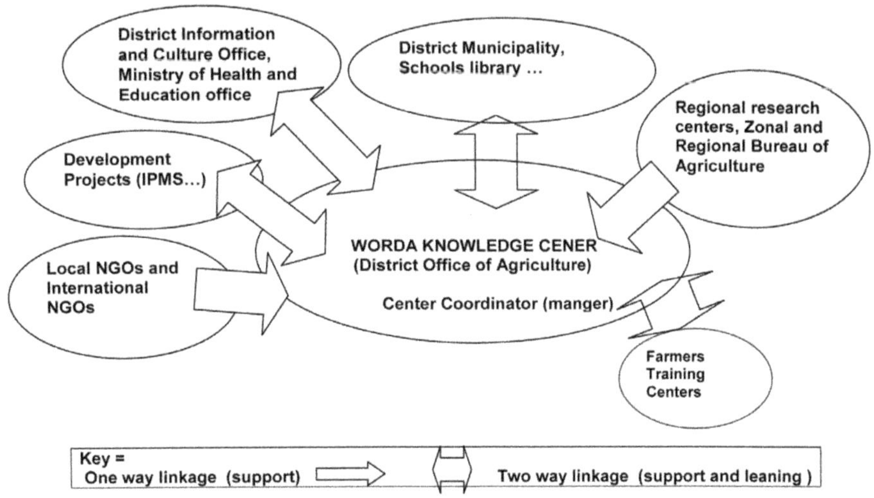

have shown interest to use off-line copies and on line documents. Users also continued to borrow books and share knowledge in the center.

Temporal comparison of WKC shows that WKC has encouraged office of agriculture staff in handling extension responsibility. The findings agree with Lemma *et.al.* (2010:9-12) in which WKC users witnessed role WKC in their daily activity. The encouragement was vivid for users involved in extension communication and documentation. The knowledge from the centers has also been shared with farmers, pastoralists and service providers by district office of agriculture staff. Since 2006, knowledge has also been shared among staff and with farmers, pastoralist and service providers using KM tools. The most common tools used are graduate seminars, staff seminar, video session, staff consultation sessions, field days, trainings, demonstration and meetings. To transfer knowledge from WKC, the above events which use KM tools were organized in all ten districts (IPMS; 2010 a, 6). Discussion with center uses shows that use of WKC made subject matter specialists (SMS) quicker in problem solving. It increased ability of SMS to guide Development Agents (DAs). SMS and DAs also became bet-

Table 3. Indicator based temporal comparison

Indicators	Pre-WKC period (Before 2005)	Post –WKC period (After 2006)
Venue for knowledge gathering	Exist as archive	Established as WKC
Venue of knowledge sharing	Did not exist	Center was established as venue
Forms of knowledge sharing	Informal	Formal (seminar, trainings, expert consultation, meetings)
Frequency of knowledge sharing	Low	High (seminar)
Knowledge sharing methods	Trainings, Meetings, Discussion, Demonstration	Diversified methods –included drama, distribution of leaflets, video show…
Internet access	No access	Access exist (dial up and wireless)
Forms of knowledge	Traditional (non electronic)	Traditional and Electronic forms
Knowledge documentation	Not organized	Organized at a center in classified manner

Source: Individual Interview and KM survey data 2010

ter decision makers, more responsive to farmers' request and it improved extension staff-farmer relations. WKC establishment reduced staff professional discouragement associated with lack of the relevant document. It means the establishment of the WKC led to an increase in the capacity of staff in knowledge sharing leading to an improved extension communication. The interventions have improved extension communication as verified in KM survey in assessing responsiveness of the extension system as proxy indicator for extension communication(IPMS, 2010 b; 25).

3.6 Assessment of Woreda Knowledge Center Use

Although assessing the impact of information and communication management on institutional performance is highly complex (CTA, 1998; 6), there are indexes for measuring organizational performance in knowledge management. Findings below attempt to assess the extent of WKC use in ten districts based on two index suggested by Chang *et. al., (*2005; 1149). The first index shows an increase in relevant knowledge availability, which reflects knowledge gathering while the second index shows usefulness of information received which reflects knowledge internalization and use. Findings of this study also agree with a recent study on WKC in ten districts (Lemma *et.al*, 2010:22) which showed that 94% of WKC users (n=336) agree that services at WKC are relevant and useful to help them undertake their duty (See Table 4).

Responsiveness of the extension system was also used as proxy indicator to show role of WKC in promoting extension communication. In group discussions held in sample of 40 PAs in ten districts, farmers acknowledged that the extension service has improved in providing relevant technical messages and in organizing periodic practical trainings from 2005 to 2010 (IPMS, 2010 (b); 25). The improvements also reflect the progressive role played by WKC in extension communication.

Project lesson shows that the establishment of WKC is not an end by itself. The impact of center establishment lies *first*, on use of center to provide knowledge for gaps identified and *second*, on reaching the farming community at grass root level using KM tools.

On the base of IPMS project experience, it is worth to focus on addressing knowledge gaps, which emerge depending on local context. The *first* gap is efficient use of the center. That can be addressed through capacity building for the center (e.g purchase of various knowledge sources materials), for users (e.g. computer skill training for center users) and for center coordinator (e.g. training on internet connection, center management). Project lesson shows that it is worth to assign staff who has background in ITC or information science as center manger or coordinator. The *second* gap could be addressing the practical need of farmers, pastoralists and service providers based on knowledge from WKC.

To describe the role of WKC in agriculture development, one can see how a particular problem can be addressed in dairy sector as an example. If

Table 4. KM performance indicators across ten districts

WKC name	Alamata	Atsibi	Dale	Alaba	Meiso	Goma	Ada	Bure	Fogera	Metema	Average
Relevant knowledge availability (% of positive respondents)	73.4	95.3	45.3	75.0	66.6	82.2	59.4	69.6	88.3	57.8	71
Usefulness of information received (% of positive respondents)	68.3	71.9	73.3	70.6	79.1	88.2	100.0	84.6	70.6	91.9	79

Source: IPMS (2010 (b); 21)

Box 1. Knowledge flow from WKC to grass root level

Farmers in Southern Ethiopia, in Dale and Alaba districts, have common knowledge on Haricot Bean (*Phaseolus vulgaris*) for decades as part of their indigenous farming system. Over the past five years, farmers in both districts have developed interest to work on improved haricot bean seeds. The farmers were determined to multiply improved haricot bean seeds (usually the red haricot beans). The farmers in both districts latter developed linkage with Melkassa Agricultural Research Center (MARC), office of Agriculture and IPMS project. The farmers can be identified as community of practice as described in CIDA (2003; 11) and Grover and Davenport (2001;8). Every year, the farmers needed external technical knowledge from subject matter specialist (SMS) and Development Agents (DAs). DAs and SMS in turn get knowledge on haricot bean from haricot bean manual (English and Amharic version) and leaflets produced by MARC which are available in WKCs. The use of knowledge from the manual and leaflets was also complemented by supply of improved haricot bean seed from MARC for farmers through IPMS project. The knowledge accessed by SMS and DAs is shared during community training and meeting at grass root level. In this typical haricot bean (commodity) development, knowledge sources were accessed from WKC and knowledge was transferred from SMS to DA and finally reached farmers at grass root level.

a problem in dairy sector is scarcity of livestock feed, knowledge should be accessed or shared from WKC (e.g., on types, amount and supply of forage materials to be developed for an area). Then, the identified forage materials (needs input supply) should reach the farmers. Coupled with this, training materials on forage management, demonstration and its planting methods can be shared and accessed from WKC with farmers for increased dairy commodity. Box 1 illustrates how knowledge accessed from the WKC can reach farmers looking at specific commodity, haricot bean seed multiplication, in two of the ten districts in the area.

3.7 Challenges and Opportunities in WKC

In spite of the perceived relevance of WKC by district office of agriculture staffs, the level of utilization is low as assessed in 2010. From the 500 respondents in 10 districts, 54% agree that WKC services has been used efficiently (IPMS (b); 2010, 9). The low level of utilizations in WKC, is attributed to challenges in KM. In a review of challenges in KM, Hussain, *et. al.*(2010; 5) describes that there are natural resistance to change in KM adoption and its successful implementation. The list of barriers in KM adoption includes: cultural barriers (distrust, lack of recognition and communication, retrenchment concerns) organizational culture, leadership, policy and enabling environment, KM tools. All these barriers have affected at different level the use of WKC in the study areas.

Analysis of internal and external conditions (Table 5) shows key factors which may influence use of WKC. There are barriers that hinder transfer knowledge from the WKC to reach grass root level. Assessment has shown that the common problems with WKC are lack of relevant reading materials in preferred language, a few numbers of printed materials on specific subject matter. Also there are problems of poor internet connection, shortage of VCD/DVD materials, poor reading culture, low level of ITC literacy, lack of budget for the center running activity (IPMS (b) ; 2010, 27). In a recent study Lemma *et.al.*, (2010:19) identified four major categories of perceived challenges associated with WKC as budget, center management, sustainability and commitment by office of agriculture.

Lessons learnt shows that using WKC at district level may create conflict between district office heads and staff because the use of WKC may overlap with time to be devoted for routine work. Thus, for proper use of the center, it is essential to develop rules and regulation. The role of WKC to address problems in extension communication and its' contribution in agriculture development depends on staff initiative to look for the right kind of knowledge when questions emerge from the community. In examining financial sustainability of centers, experience with the

*Table 5. SWOT analysis for Woreda knowledge centers**

STRENGTHS	WEAKNESS	OPPORTUNITY	THREATS
Partners exist at district level (Local and International NGOs)	Not all potential partners were involved at all level	Existence of magazine and newspaper suppliers	If there is no power supply to run WKC equipments
Existence of book stores, shops to avail materials for WKC	Annual asset inventory (not regular), Scarcity in educational digital materials (DVDs, VCDs)	Allocation of district office budget to cover running cost or employ coordinator	Lack of infrastructure (for internet connection)
Demand for internet, knowledge) exist from district staff and farmers	Poor reading culture	Existence of IPMS and other development projects	Poor internet connectivity
Data recording on WKC use exist as monitoring	Joint monitoring and evaluation of WKC is not in place	Existence of internet service provider	Books and documents can be lost (if not managed)
Venue for WKC provided by district office	Role of partners vary, lack of coordinator for partnership to work progressively on WKC	Build on cost recovery by WKC (e.g internet service charge, charge for book loans, video show)	If there are no newspaper and magazine suppliers If inappropriate venue is allocated
Existence of WKC (manger) coordinator	Initial low demand and awareness to use WKC	Existence of IT staff in the district for maintenance	If there are no IT staff to maintain computers
Effort of district staff to assemble and share knowledge	Lack of skill to browse on line document, Low initiation to translate and use materials	Existence of computer anti-virus	Imbalance in computer/staff ratio Frequent maintenance of computer

* Variation exist from one district to another

ten districts over the past five years shows that a center can run with small budget (ca 100 USD/month) to cover internet bill, magazine and newsletter subscription, assuming salary for center manger is from agriculture office budget. To ensure proper use of IT and non-IT items, to extend service duration, assigning a center coordinator or manger is essential. Usually, staff for the center coordination or management is assigned either from information documentation section or from extension communication section. As reviewed by Masiero (2011;3) it is possible to charge some of the center service (book borrowing and internet connection, video show and printing service) to recover cost and sustain the center financially

4. CONCLUSION AND RECOMMENDATION

The establishing knowledge center address key challenges in agricultural extension, mainly the knowledge and information service delivery process. At district level, knowledge management is enhanced as WKC is used to improve the process of accessing, sharing, transferring and using knowledge. The establishment of WKC laid basic infrastructure for KM technology use at district level. This study has shown that the use of WKC increased staff capacity to identify, access, assemble, document, share and use knowledge contributing to an improved extension communication and agriculture development reaching farmers and pastoralist at grass root level.

Temporal comparison and SWOT analysis showed that WKC improved culture of knowledge sharing as formal educational sessions were organized as seminars, video sessions, expert consultation, meetings and trainings. This agrees with survey result of KM in this study and recent study by Lemma *et.al*., (2010).

The establishment of WKC resulted in assembled and classified documentation in a room (center). When compared with pre WKC establishment (before 2005), in post-WKC establishment, there is an increase in responsiveness of extension communication (in 2010) as acknowledged by farmers and experts. In ten districts, 71% of 500 survey respondents agree on an increase in knowledge availability while 79% agree on an increase in delivery of relevant knowledge reflecting improvement in extension communication.

On the bases of lesson gained by IPMS project and its' partners, this study recommends that WKC be established in all districts through effort of the Ministry of Agriculture and its partners.

5. ACKNOWLEDGMENT

The authors are grateful to CIDA (Canadian International Development Agency) for funding the Improving productivity and Market Success (IPMS) of Ethiopian Farmers' project (www.ipms-ethiopia.org). The authors are also indebted to Yasin Getahun for providing location map.

6. REFERENCES

Barry, C. A. (1997). The research activity timeline: A quantitative tool for information research. *Library & Information Science Research, 19*(2), 153–179. doi:10.1016/S0740-8188(97)90041-4

Chang, K. L., Lee, S., & Kang, I. (2005). KMPI: Measuring knowledge management performance. *Information & Management, 42*(8), 1149. doi:10.1016/j.im.2005.10.003

CIDA. (2003). *Knowledge sharing: Methods, meetings and tools.* Canadian International Development Agency (CIDA) Knowledge Management Secretariat, November 2003. Retrieved from km.fao.org/uploads/media/CIDA_guide_Knowledge.pdf

CTA. (1998). Assessing the impact of information and communication management on institutional performance. In K. Batjes-Sinclair (Ed.), *Proceedings of a CTA Workshop*, January 27-29, 1998, Wageningen, The Netherlands.

Dyson, R. G. (2004). Strategic development and SWOT analysis at the University Warwick. *European Journal of Operation Research. Journal of International Management, 152*(3), 631–640.

Gebremedhin, B., Hoekstra, D., & Azage Tegegne, A. (2006). *Commercialization of Ethiopian agriculture: Extension service from input supplier to knowledge broker and facilitator*. IPMS (Improving Productivity and Market Success) of Ethiopian Farmers Project Working Paper 1. Nairobi, Kenya: International Livestock Research Institute. Retrieved from mahider.ilri.org/handle/10568/564

Gomez, R., & Baron-Porras, L. F. (2011). Does public access computing really contribute to community development? Lessons from libraries, telecenters and cybercafés in Colombia. *Electronic Journal of Information Systems in Developing Countries, 49*(2), 1–11.

Grover, V., & Davenport, T. (2001). General perspective on knowledge management: Fostering a research agenda. *Journal of Management Information Systems, 18*(1), 5–21.

Hartwich, F., Monge Perez, M., Ampuero Ramos, L., & Soto, J. L. (2007). Knowledge management for agricultural innovation: Lessons from networking efforts in the Bolivian Agricultural Technology System. *Knowledge Management for Development Journal, 3*(2), 21-37. Retrieved from journal.km4dev.org/index.php/km4dj/article/view/106/174

Hermans, L. M., & Thissen, A. H. (2009). Actor analysis methods and their use for public policy analysis. *European Journal of Operational Research, 196*(2), 808–818. doi:10.1016/j.ejor.2008.03.040

Hussain, I., Si, S., & Ahmed, A. (2010). Knowledge management for small and medium enterprise in developing countries. *Journal of Knowledge Management Practice, 11*(2).

IPMS. (2005). *Project implementation plan (PIP) submitted to CIDA.* Improving Productivity and Market Success (IPMS) of Ethiopian Farmers' Project, March 2005. Retrieved from http://www.ipms-ethiopia.org/content/files/Documents/pip/Pip%20final.doc

IPMS. (2010a). *12th progress report* (April 2010 – September 2010). Improving Productivity and Market Success (IPMS) of Ethiopian Farmers Project. Retrieved from http://www.ipms-ethiopia.org/Documents-Publications/Reporting/Progress-Reports.asp

IPMS. (2010b). *Monitoring and evaluation report of year 5* (2009/2010). Improving Productivity and Market Success (IPMS) of Ethiopian Farmers Project. Retrieved from http://www.ipms-ethiopia.org/Documents-Publications/M-&-E-Reports.asp

Kebede, G. (2010). Knowledge management: An information science perspective. *International Journal of Information Management, 30*(5), 416–424. doi:10.1016/j.ijinfomgt.2010.02.004

Lee, H., & Choi, B. (2003). Knowledge management enablers, processes, and organizational performance: An integrative view and empirical examination. *Journal of Management Information Systems, 20*(1), 179–228.

Lemma, T. T., Sehai, E., & Hoekstra, D. (2010). *Knowledge centers in the pilot learning woredas of improving productivity and market success project: Utilization, relevance and effectiveness.* Retrieved from www.ipms-ethiopia.org/IPMS

Masiero, S. (2011). Financial versus social sustainability of tele-centers: Mutual exclusion or mutual reinforcement? *Electronic Journal of Information Systems in Developing Countries, 45*(3), 1–23.

Sehai, E. (2006). *Knowledge management in Ethiopian agriculture.* Paper presented on14th Annual Conference Ethiopian Society of Animal Production, September 7th, 2006 (International Livestock Research Institute – IPMS Project). Retrieved from mahider.ilri.org/handle/10568/614

World Bank. (2003). *Higher education development for Ethiopia: Putting the vision.* World Bank Sector Study, Addis Ababa, Ethiopia. Retrieved from http://www.siteresources.worldbank.org/.../Ethiopia_Higher_Education_ESW.pdf

Zavaglia, T. T., Pierozzi, I., Rodrigues, N. P., & DeCastro, A. (2011). Knowledge management and communication in Brazilian agricultural research: An integrated procedural approach. *International Journal of Information Management, 31*(2), 121–127. doi:10.1016/j.ijinfomgt.2010.07.001

Section 3
E-Agriculture and Rural Development in Asia

Chapter 11
Situational Analysis of the Status of E-Agriculture in Tamil Nadu, India

P Senthil Priya
PSG College of Arts and Science, India

N. Mathiyalagan
PSG College of Arts and Science, India

ABSTRACT

In recent years, there is scarcely any field in the various sectors of the world economy that has not been influenced by the impact of information and communication technologies (ICT), and the field of agriculture is no exception. Within the past decade, many projects have been implemented by governments, private sector, and NGOs in developing countries to streamline the production, storage, selection of markets, and commercialization of agricultural commodities. These new ICT have opened up numerous opportunities for the farmers to increase the timeliness, quality, relevance, availability, and accessibility of critical agricultural information. The UN International Telecommunication Union (ITU) estimates that one billion people worldwide – most of whom depend in some way on agriculture for their livelihoods – still lack connection to any kind of ICT. Indian farmers face competition from global forces and the usage of ICT tools. Advances in ICT help farmers to access accurate, timely, relevant information and services, thereby facilitating an environment for more remunerative agricultural practices to meet the new challenges. This chapter provides an in-depth analysis of the existing rural agrarian digital divide in Tamilnadu, India. The chapter describes the information needs, ICT access, and new opportunities for farmers in rural Tamilnadu. The chapter also examines the status of E-Agriculture in Tamilnadu and the ICT interventions that are available to Tamilnadu farmers. The study outlines the emerging opportunities for rural farmer' participation in the knowledge society and offers some guidance on how the farmers can take advantage of ICT and utilize the information management system to derive maximum benefit out of the technology.

DOI: 10.4018/978-1-4666-2655-3.ch011

1. INTRODUCTION

Agriculture is the main source of income for around 2.5 billion people in the developing world (FAO, 2003). In most of the semi-peripheral nations, the agricultural sector is the largest and most critical economic sector. The Indian agricultural sector provides employment for about 65% of the total labor force, accounts for 27% of the GDP growth, contributes 21% of total exports and also provides raw materials to several industries (Malhan,2007). The impact of the agricultural sector is wide-ranging and extends to economic growth, food security, poverty reduction, livelihoods, rural development and the environment. Moreover, the poorest half of the population benefits significantly more from agricultural growth than growth in other sectors of the economy (UN, 2008; World Bank, 2007).

ICT can be defined as the combination of hardware, software and any other means of production that enables the exchange, processing and management of pertinent information and knowledge. ICT's include a broad array of new media technologies and methods that can effectively store, manage and process information.

Any kind of information becomes knowledge only when the relevant content is widely disseminated through a channel of communication at the right time. Thus, the information has to be streamlined and disseminated to wider audience which makes it imminent to use new media communication channels. Various new media channels of communication such as email, cell phones, pagers, instant messaging, fax and the Internet are available and these channels can be used for mass dissemination of information. ICT communication channels have become the imperative as such communications have no boundaries, no physical barriers of time and its accessibility is almost instant at a basic cost.

Information and communication technologies are rapidly transforming the face of agriculture in industrialized countries. ICT infrastructures are very strong in such countries and e-commerce, e-business, e-procurement, e-auction activities have been in practice for the past several years. High speed Internet access in rural areas enables the farmers to access credit linkages, government programs and technical assistance to farmers through direct access. Many of the activities in the agricultural marketplace are mediated by web-linked to a database that specifies prices, qualities and quantities of essential agricultural commodities. Laser technology has replaced the tractors and ancient ploughing mechanism and this new technology helps in optimizing the use of various inputs such as water, seeds, fertilizers, etc.

Agricultural business development services can be located, bought and paid for over the Internet. Often many of these services are delivered at the doorsteps of farmers. In short, the face of agriculture in developed nations has changed as ICT have become increasingly critical to farmers and agricultural planners in such countries. These nations are the predecessors of new technologies and with effective usage of ICT tools in land planning and management, their production efficiency has increased triple fold in the recent decade.

In economic terms, ICT usage in information dissemination has become so critical that it needs to be recognized explicitly as a fourth production factor in agriculture (USAID, 2003). As only 5-6% of the population is involved in agriculture and farming activities in developed nations and the supply chain is short, the economic feasibility and liberty is more for farmers. There are almost no resource poor farmers as many farmers own undivided large hectares of individual land holding. This transformation of agriculture in developed countries has taken place in a context of high literacy rates, well-functioning telecommunication systems, readily available electricity, an established and regulated credit and banking system, well-developed transportation networks, high labor costs relative to the cost of computing equipment and reasonably easy access to ICT (USAID, 2003).

Contrary to this status, in most developing countries like India, the agricultural sector is the largest employer and it effectively employs over 60% of the population. More than half of the population lives in rural areas and they are economically dependent on the performance of agricultural production. Agriculture sector remains as the largest market for labor. It provides tradable goods for foreign exchange, accelerates the national economy and contributes to government budgets through taxation.

According to Indian constitution, "Land" is listed under the state clause and hence it is the responsibility of state governments to make rules related to lands. For decades, agriculture remains as the mainstay of the state's economy and it is the responsibility of the state government to evolve a comprehensive system of land management. Hence, the state government has to motivate and instruct the farmers about new developments that have taken place in the field of agriculture through state wide extension departments.

Many farmers in the rural areas are resource poor and belong to medium, small land holders and awareness level and usage of ICT initiatives are rather less among them. Also, Land ceiling acts and acts of succession from one generation to the next has reduced the size of individual land holdings and led to land fragmentation. Fragmentation of land is widespread in India and this factor plays a major role in explaining low levels of agricultural productivity. Farmers rely heavily on subsistence farming and this factor could lead to sub-optimal usage of factor inputs and thus lower overall returns to land (Jha, 2005).

Also, In India, literacy rates among the farmers are very low and in some rural areas it is as high as 40 percent and in some tribal belts hardly about 20 percent people are literate (Chhibber, 2004). Electricity is available only for certain hours in a day, farmers do not have accessibility to easier credit facilities, do not have direct access or link to commercial markets and the farmers are highly dependent on middlemen. These middlemen purchase their products at extremely low prices and sell them at very high prices in the urban markets. The farmers have very little bargaining power for their produce. These smallholder farmers are at the bottom of the pyramid in terms of income, literacy, and access to information and services and they are the poorest and most vulnerable populations of the economy with only a little opportunity to access, use or benefit from ICT. Many of the farmers are not aware of new media technologies and their usage. Hence, there exists a tangible knowledge gap between developed and developing nations in terms of information access and dissemination.

The resource poor small-scale farmers, who contribute substantially to food production in the South Asian region, need to be empowered with such appropriate technologies to enhance sustainable agricultural productivity and production (Dar, William D et al. 2003). Agricultural ministry should create a compliant environment to support the marginal and small farmers and implement developmental activities to suit their needs. Also, to escape poverty, smallholder farmers need to enhance their skills and knowledge and the entire smallholder-dominant value chain needs to become more competitive. The combined effect of these two factors can improve agricultural productivity and raise the incomes of rural dwellers. ICT awareness and accessibility could be used to transform rural areas. Marginal and small farmers especially, can benefit a lot out of these technologies.

Efficient usage of ICT interventions in developing countries has the ability to strengthen the agricultural practices and bridge the knowledge gaps in rural economy. Internet can play a pivotal role by providing a cost-effective way to deliver information services to a large, dispersed population. It can also deliver planning tools to agribusinesses, and connects the various players in the value chain so they can conduct commerce more efficiently.

For ICT induced agriculture to benefit rural communities in developing countries, the rural digital divide must be bridged. ICT Infrastructures in rural areas have to be strengthened and access to ICT should be made affordable for them. Locally relevant digital content has to be developed in local language and circulated for use of the farmers. Otherwise, any ICT applications would remain beyond the reach of rural communities and will merely aggravate the existing rural digital divide - leading to an ever-widening knowledge gap between developed and third world nations.

The present study is divided into sections and section 1 includes the introduction, objective and scope of the study, the context and issues in agriculture, traditional farming system in Tamilnadu, India along with the Section 2 includes the review of literature, need for implementation of ICT in agriculture and the benefits of ICT in agriculture. Section 3 lists the research methodology and the needs and benefits of ICT. Section 4 provides a situational analysis of the status of E-Agriculture in Tamilnadu, the information needs of farmers and lists out the benefits and constraints of ICT interventions that are available for farmers in Tamilnadu. Section 5 gives the overview, implications of the study, conclusion and recommendations of the study.

1.1 Background to the Study

Tamil Nadu is the seventh most populous state in India with a population of 66,396,000, as of July 1, 2008 (approximately 5.79% of India's population). The state has 32 districts, 206 taluks and 17,292 villages (www.tn.gov.in). It is the fifth largest contributor to India's GDP and the most urbanized state in India. For administrative convenience, the state is categorized into districts comprising villages, taluks comprising panchayats and panchayats comprising villages. It is one of the foremost states in the country in terms of overall development. Predominantly, traditional farming in lands are the main occupation of people and majority of the farmers live in the periphery of an urban center with an average 15 miles proximity to the nearest town or city.

In recent years, the agricultural economy has slowed down in the state mainly due to migration of population and the risks associated with practicing agriculture in rural areas that lack modern amenities. Almost, all villages in Tamilnadu have basic amenities like roads and electricity connections. But modern ICT infrastructure and citizen service centers have not been established in these areas. At present, most villages in India have reliable mobile coverage and broadband Internet connectivity of 256 kbps. Traditional agricultural market information systems provided low-cost information to farmers, traders, and consumers via radios or rural extension officers, market agents have not been active participants in the information gathering process.

Majority of the population are settled in rural areas and agriculture is the main source of livelihood for greater part of the rural population. In rural Tamilnadu, the share of rural population have come down over the years from 73.31% in 1961 to 56.14% in 2000-01, due to inward migration of rural population into urban areas to look for better prospects (www.tn.gov.in).

1.2 Objective of the Study

- To carry out a situational analysis of the present status of agriculture in India on a broader framework as well as the contexts and issues in agriculture within the state of Tamilnadu.
- To provide an overview of present status of E-Agriculture in Tamilnadu and analyze the ICT interventions that was available for the farmers in Tamilnadu.
- To highlight the essential ICT factors that can be used and appropriated for rural farmers.

1.3 Scope of the Study

The insights gained from the study would serve as a situational analysis of the status of e-agriculture in Tamilnadu and it would immensely help the farmers who can access the ICT intervention. The implications of the study would give insights about the scope of the e-governance projects in Tamilnadu and how it can be effectively utilized for rural development. The overview of the study would be useful for policy formulation, identification of the right ICT initiative to be employed in a rural setting as well as to improve extension approaches and strategies in future.

The overall context of the study could be applied to any developing country as the crisis factors and literacy levels are almost the same in these regions. Hence, a broad overview of the status of agriculture in Tamilnadu will provide a general understanding of the benefits and hindrances of ICT project implementation in other third world countries.

1.4 Context and Issues in Agriculture

India has a dominant position in world agriculture, with the second largest arable land bank in the world and ranking in the top three producers of rice, wheat, coarse grains, fruits, vegetables, tea, coffee and jute. India also has some of the largest livestock populations which combined with the large farming sector ensures that agri-business remains the overwhelming contributor to the economy from both monetary and employment perspectives (Malhan, 2007).

As Agriculture is listed under the state subject, the onus is on the state government to initiate various schemes and services for upliftment of farmers. Rural India consists of 700 million people spread over 6,38,000 villages (Source: Indian Agriculture Scenario, 2008) and any developmental effort that is initiated by the government machinery has to be disseminated to all these rural production centers through extension activities.

Also, land parcels are highly fragmented in India that a 3.3 million sq km area is divided into 500 million parcels, the highest in the world. When this status is compared with countries with less land fragmentation like Australia where 7.6 million sq. km is divided into 11 million parcels of land, the Indian scenario shows too vast a difference (Second Administrative Reforms Commission, 2008). This status produces more small and marginal farmers and they majorly rely on subsistence farming for their livelihood.

Also, in India, 28 per cent of land parcels are locked in legal disputes (Paromita Shastri, 2007). Statistics reveal that 90% of land holdings in India are subjected to ownership disputes and the money spent in contesting law suits at judicial institutions constitute around 1.3% loss of Gross Domestic Product growth every year (Second administrative reforms commission, 2008). Hence, any agricultural developmental programs that are implemented in various states of India need to be tailor-made to suit micro-community requirements. Also, the average land holding size of a farmer is 1.47 hectares, and only five to ten percent of the farming population has a land holding that exceeds 4 hectares (Sharma, 2005).

The latest National Sample Survey Organization (NSSO) has observed that, an Indian farmer's household has an average debt of Rs.12, 585. Farmers from Punjab top the list with Rs.41,575 followed by farmers in Kerala with Rs. 33,907, Haryana Rs.26,007, Andhra Rs.21,965 and Tamil Nadu Rs.21,963. In fact, Andhra Pradesh witnessed highest percentage of farmers under indebtedness (82.0 per cent) followed by Tamil Nadu (74.5 per cent) and Punjab (65.4 per cent). This data further clarified that percent of indebted farmers taking loans from money lenders is the highest (29 per cent), followed by banks (27 per cent), co-operative society (26 per cent) and finally from government (3 per cent). This has created a situation of negative growth in agricultural sector and practice of agriculture is no more a viable option for farmers.

In this regard, a situational analysis of Indian farmers conducted by NSSO has brought out some highly relevant and interesting results that are:

1. An estimated 27 per cent of the farmers do not like farming because it is not considered profitable.
2. Nearly 40 per cent of the farmers, if given a choice, would prefer to take up some other career.
3. There is very low level of awareness among farmers about modern technologies as well as government programmes like crop insurance and agri-export promotion facilities etc.
4. Many farmers have reported non-availability of information regarding modern input facilities within the villages.
5. Nearly 50 per cent of farm households are indebted and the ratio as well as average outstanding loan per farm household are higher in relatively more developed states like Punjab, Tamil Nadu and Andhra Pradesh.

In recent times, the biggest problem facing the rural agrarian sector is the risks associated with practicing agriculture. Farming is not an economically a viable option and a means of sustenance for rural families anymore. The average farm-size is very small, fragmented, poverty and food-security continues to be preponderant among these small and marginal landholders. The economic condition of the farmers has deteriorated and there is a crux of migration of farmers from rural to urban area to sustain their livelihood.

Industrialized nations that lead in the deployment and use of Internet technology for agriculture has a definite economic and social edge over nations with limited use of technology. The Indian farmers have to face multiple issues, such as the severe drought, incessant floods, increase in the cost of agricultural inputs, lack of farm loans, minimal rates for the agricultural commodity, crop diseases, market manipulation by traders and decline in prices. While the farmers have no control over some of these factors, some factors can be controlled to a certain extent through planning and effective usage of ICT and agricultural information systems.

1.5 Agriculture and the Indian Constitution

The Constitution of India is regarded as a federal constitution. Unlike a unitary state, which has only the national government, India has two governments functioning at two different levels - the national government and the state governments. The states are further sub-divided into districts, which have their own local authorities of governance. The subjects of national importance requiring uniform legislation for the country as a whole are inducted in the union list. The state list includes subjects of local importance where variation in laws in response to local situations may be necessary. The concurrent list includes subjects that require legislation by parliament and by the state legislature at times. The powers of centre and state governments are distributed evenly for specified subjects under the concurrent list so as to give due importance to both local requirements and national priorities. Agriculture as a subject falls under the purview of state governments. However, keeping into the view the significance of agriculture in Indian economy, the Union Ministry of Agriculture (MoA) is responsible for policy formulation and related issues besides augmenting the developmental efforts of the state governments. The state governments are expected to put efforts to meet national aspirations while ensuring welfare of the local community.

1.6 Traditional Farming in Tamilnadu

For centuries, farmers have learned from each other. Knowledge transfer occurred rather horizontally among the farmers and they relied on each other for information sharing and cultivation. This form of communication channel sustained rural agriculture for decades among all agricultural communities in the Tamilnadu. Traditional cultivation practices prevail and to cut down on production cost, the entire family is employed for farming and cultivation in the field. Some small farmers also employ other farmers and cultivate but most often, farmers cultivated in their own field and employed members of the immediate family as many of them cannot afford agricultural laborers. The media use of farmers is mainly restricted to listening to radio programs and TV programs. Newspaper usage is limited to literates within the community. Other agricultural information sources include agricultural officer in the block, the gramsabha, newspapers, TV, radio and fellow farmers.

Agricultural extension and advisory services play an important role in agricultural development and can contribute to improving the welfare of farmers and other people living in rural areas. A range of approaches to extension delivery have been promoted over the years by the state governments in India. Agricultural department deputed block development officers and field coordinators to disseminate agricultural information among the rural community. All relevant agricultural information like cropping pattern, land use, nitrogen fixation, fertilizer use, government subsidies for farmers and rural development program were disseminated through them. However, in practice, vertical diffusion of information was restricted to communication between field officers and the village headman as it was not possible to visit every house within the village and disseminate information to each one.

However, the traditional method of dissemination of information was time consuming and slow moving. It passed over many channels to reach the farmers. Hence, on an average, there was an equitable 50% information loss during transfer. Apart from this, sometimes, the block development officer and field coordinators hosted meetings in villages and provided information about different activities taken up by the government. But the main problem in this model was that it was time consuming and less information was actually communicated as majority of farmers never attended the meetings due to heavy workload. Horizontal diffusion happened at a faster pace among farmers. Most smallholder farmers within the village got the information they need through farmer-to-farmer contacts.

After cultivation, the villagers were highly dependent on middlemen for marketing their produce. These middlemen purchased their products at extremely low prices as the farmers had very little bargaining power and sold them at very high prices in the urban markets. Also, sometimes, the middlemen lent loans to the farmers and in turn, bought the produce at a cheaper rate after harvest. As agriculture production and marketing is time-specific, most often the farmers never got the relevant information at the right time as invariably the information got delayed and it was obsolete when it reached them. This inordinate delay in information access caused various problems for the farmers and they suffered frequent crop failures and insect infestations.

2 REVIEW OF LITERATURE

In recent years, many ICT initiatives have been introduced in developing economies to reach the grass root populations and empower them with adequate agricultural knowledge and transform the traditional societies into knowledge societies. Hence we cannot overlook the role of ICT in agriculture. In this review of literature, some ICT initiatives of developing countries and some specific Indian case studies are analyzed to gain broader understanding.

Meera, Jhamtani & Rao (2004) studied the impact of ICT projects in rural community and pointed out that that when ICT projects for agricultural development were set up, along with technical infrastructure, equal attention should also be paid to efficient usage of ICT tools in information collection as well as information dissemination to rule out any imbalances. Rahul De (2003) concluded that though there was direct evidence of women participation in rural ICT projects, the participation of dalits and other marginal groups of the society was rather limited. The author pointed out that capacity building of government departments should include decentralization of power centers and provision of citizen services in rural villages rather than district head quarters or taluks. Only when this process of reengineering was established at easily accessible nodal points of proximity, citizen use would increase in developing countries. Ballantyne et al., (2010) have assessed that farmers and scientists were utilizing a wide range of devices and platforms to share agricultural knowledge and in future, networked communities would accelerate with much more advanced technologies and establish global communities of knowledge. Chauhan (2002) analyzed the role of information technology in the lives of rural farmers and remarked that the Indian farmers' needed much more diversified information from the extension officials and usage of satellite based information and communication technologies would only be able to provide the specific information that the farmers need. Adhau (2009) reviewed the current scenario of agricultural development in India and pointed out that the country was moving towards an agricultural emergency mainly due to lack of adequate attention to agriculture, insufficient land reform projects, defective land management policies, insufficient price for agricultural producers, inadequate investment in irrigational and agriculture infrastructure within the country. The author proposed that the solution to the problem should be initiated at the grass root population through an effective use of ICT for agricultural purposes. Walsham (2010) concluded that ICT should not be seen as 'silver bullets' for instant development of the country but rather, as a potentially important contributor for human development through wider socio-technical interventions.

2.1 Need for ICT in Agriculture

Knowledge is power and it is based on information. In this era, management of information is power. Knowledge that can be codified is available in two forms: *explicit knowledge* that can be codified and articulated in formal language and *tacit knowledge*, which is personal knowledge that is obtained or embedded in experience (Polanyi 1966). Traditional corporate and development theories generally focus on developing and diffusing explicit knowledge. In traditional farming, tacit knowledge was disseminated through vertical communication and explicit knowledge was rather limited within the community.

In the 21st century, agriculture is moving rapidly away from an artisanal, labor extensive, traditional activity towards a more sophisticated, mechanical, information intensive sector of the global economy, access to information and modern communication technologies has become a necessity for the world's farmers, especially in developing countries (FAO, 2003). Traditional societies are being transformed into knowledge societies all over the world and ICT play a key role in improving the availability of agricultural production and market information in developing countries. ICT-based market information systems have a proven track record for improving rural livelihoods in middle income developing countries where they have been introduced.

There are twenty million agriculture laborers and 8.7 million agriculture families in Tamil Nadu (Financial Express, 2010). At present, the ratio of the farmers to the extension worker is 2500:1, which is rather inadequate for effective agricultural information diffusion. It is almost impossible for the extension workers to meet every farmer within the belt and disseminate agricultural information. Although the appointed village extension workers disseminate the relevant information in regular meetings, most often, the farmers are unresponsive and do not attend meetings. This factor has created a need to evolve a new mechanism wherein the farmers themselves have access to relevant agricultural information services and are motivated to access such services.

Hence, the need for ICT intervention tools is necessary at this juncture as the rural farmers have to be educated on the uses and potential of ICT to compete in a global environment. In recent years, sustainable food production relies more and more on agricultural information management and communication technologies.

Innovation in ICT offers immense opportunity for the farmers to transform agricultural marketing and associated services especially among rural communities. Through this access, links to commercial markets and supply chains can be shortened and farmers can directly access the traders. It also offers a communication platform transcending all traditional physical barriers and backwardness within the community and creates wider reach for the community surpassing all social and gender bias. Access to these relevant, ICT based market information system and community based transaction chains offer a more reliable communication platform to the farmers. It has the potential to empower the farmers and make them self-reliant in information.

2.2 Benefits of ICT in Agriculture

ICT have the power and potential to help farmers by improving their bargaining power and by enabling direct sales. ICT can be used to integrate small-scale farmers of a particular geographic location and thereby increase their bargaining and negotiation power. ICT can also be used for the improvement and strengthening of agriculture sector in India by disseminating timely information on weather forecasts, natural calamities, better and spontaneous agriculture practices, better marketing exposure, better pricing strategy and enhanced income, reduction of agricultural risks, better awareness, exposure and accessibility of information, improved networking and communication, facility of online trading and e-commerce and collective representation at various forums. It can play multiple roles and create a viable environment for rural farmers to practice sustainable agriculture and enable rural communities to interact with other stakeholders, thus reducing social isolation and bringing about a participatory mode of communication.

3 RESEARCH METHODOLOGY

The present study gives a situational analysis of ICT interventions in Tamilnadu. Qualitative analysis of ICT interventions were done and secondary data was extensively used for the study and the source of secondary data were websites, journals, magazines, web publications, newspaper articles, government reports, white papers, seminars and the Internet. Previous research studies done by researchers and papers presented at various conferences have also been analyzed for the study.

3.1 Status of Internet Penetration in Rural India

In India, Internet awareness and usage has been on the increase with more concentration on urban cities. Urban areas experienced a higher rate of Internet growth till the year 2000. On an average, there has been more than 30% increase in the Internet adoption among urban population every year. But, lately, high Internet growth has been witnessed in non-metros and rural parts of the country are also slowly rousing to this interactive medium. It is reason enough to convince us that ICT penetration is slowly taking roots in the rural parts of our country (See Figure 1).

Also, availability of Internet access points serve as crucial for Internet usage in rural India. There is a less likelihood of rural citizens owning personal computers and they rely on facilities provided by either public or private entities. As

Figure 1. Internet penetration estimates of India from 2008 – 2010

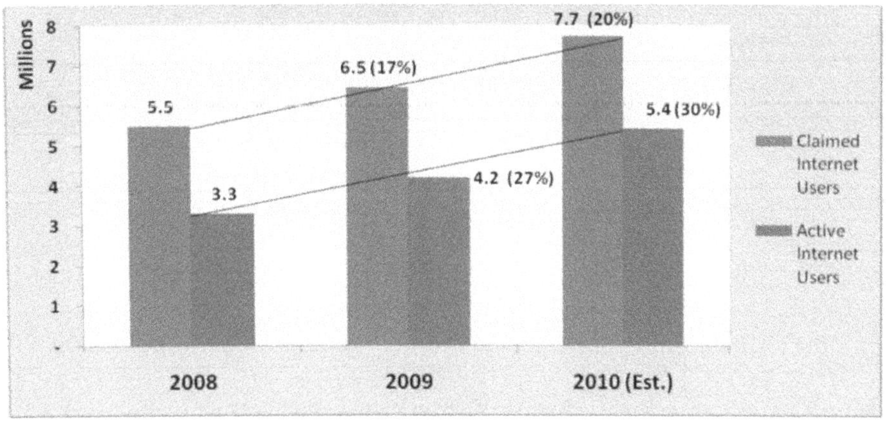

per the IMRG survey, Claimed Internet users, who have used Internet at least once, in rural villages, have grown from 5.5 Million in 2008 to 6.46 Million in 2009 - a 17% increase in a year.

The number of active Internet users has also risen from 3.3 million in 2008 to 4.18 million in 2009, exhibiting nearly a 27% increase. Although an increase in usage has occurred, the numbers are not high considering the large population based in rural villages of India.

3.2 Purpose of Internet Access in Rural India

Rural users access Internet for communication. Email is the dominant purpose of Internet access in rural India. It is interesting to note that after communication, 67% of the rural users access music and video over the Internet. About 48% of the people claim to have used the Internet for educational research and 42% claim they used it for general information search (See Figure 2).

With the proliferation of e-agricultural initiatives, rural people have started using the Internet appreciably for agriculture-related aspects. About 13% of the people use the Internet for knowing more about latest farming techniques and 8% of the people use the Internet to find more about fertilizers and pesticides. This is an interesting fact considering the importance of farming in rural India. It proves that if there are Internet related initiatives with farming as the major focus, it could certainly trigger Internet literacy to rise faster in rural areas. Awareness has to be generated among the rural population for more efficient and prudent usage of such technologies.

4 E-AGRICULTURE IN TAMILNADU

"E-Agriculture" is an emerging field in the intersection of agricultural informatics, agricultural development and entrepreneurship, referring to agricultural services, technology dissemination, and information delivered or enhanced through the Internet and related Information and Communication Technologies. More specifically, it involves the conceptualization, design, development, evaluation and application of new (innovative) ways to use existing or emerging information and communication technologies (ICT). IT seems to be a promising medium through which relevant information can be communicated to help in farming and post harvest processes.

4.1 Agricultural Information Need of Farmers

Rural farmers in Tamilnadu are faced with multiples challenges as most of them are poor, illiterates and lack the basic skills to operate computers. Although there are a few who have these

Figure 2. Purpose of rural internet access

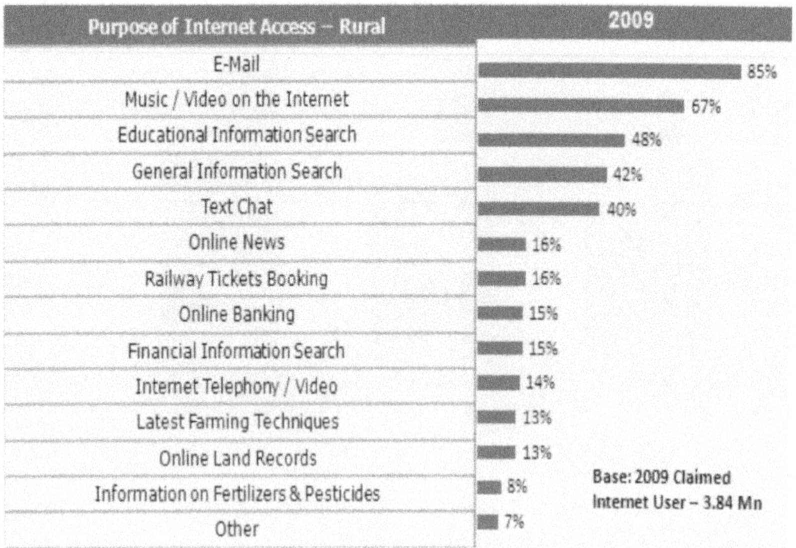

skills, they are in the most cases the exception rather than the rule. Information needs of farmers can be categorized based on the three main phases of Agriculture. Pre-Harvest, Harvest and Post-harvest are the three important phases of agriculture and farmers need to receive relevant and timely information to pass through these three phases (See Figure 3).

The above mentioned information needs of the farmers must be addressed through an easily accessible, up to date, and accurate information management system. The content has to be indigenous and relevant, pertaining to the local needs of farmers. It should also be available in the local language and within reach as only then the farmers would be influenced to use it.

4.2 ICT Interventions Available for Farmers in Tamilnadu

Today, technology plays a key role in changing people's lives by improving access to information and sharing of knowledge. Rural livelihoods would be greatly enhanced by improvements in areas such as access to agricultural markets, improved agricultural practices, information on weather, and web based marketing communication tools. ICT would not only to provide tools and valuable cropping and management information, but also very crucial linkages with expert's support as and when the farmers require it.

However, availability of any ICT technology does not guarantee the utilization and usage of the same by the farmers. Only when the infrastructures are laid in place and support structures are established to facilitate citizen usage, level of usage would increase in rural areas. To provide effective services for rural farmers, an efficient back-end system of operation as well as rural service delivery network has to be established. Awareness is most likely to stimulate the demand for information services that is essential for ICT to be used to their full potential. Locally relevant information, in a local language portal would facilitate more usage.

In a vast country like India, with 127 different agro climatic zones, the need for personalized services are stronger and only an effective utilization of technology could bring a change in the lives of farmers. In Tamilnadu, many such ICT interventions have been initiated to harness the power of ICT in rural areas. These ICT initiates

Figure 3. Agricultural information need of farmers

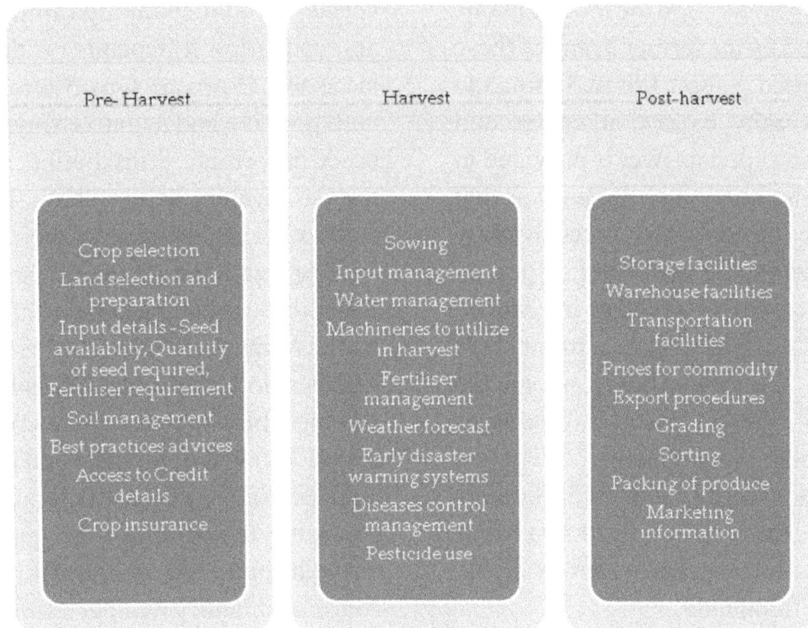

have been implemented by the extension center at Tamilnadu agricultural University, Coimbatore and an overview of the ICT interventions are provided below.

4.3 Mobile Based ICT Interventions

India has 525 million mobile phone users and fewer than 150 million people with Internet access, so mobile-phone technology can be used as a tool of e-governance. The growth in mobile-telephone technology demonstrates that the digital divide is shifting, and the focus on development effort in the field of agriculture must change with it.

With the exponential growth of mobile telephony in rural India, the mobile phone has become one of the best channels to reach the rural farmers. Penetration of mobile phones are very high in these areas and farmers often use mobile phones to keep in touch with block and district level officials, basically to check on market information and related activities.

Hence, mobile phones can be used as an effective medium of ICT based agricultural interventions. Cell phones have the flexibility to allow consumers, traders and farmers to search for market information when, where and how they want. Researches prove that cell phones are a particularly effective means of providing such information to market participants, and are being quickly adopted in low-income countries (Aker, 2008).

4.3.1 Nokia Life Tools Project

Nokia Life tools project was mainly introduced by the E-Extension department of Tamilnadu Agricultural University in the year 2009. The aim of the project was to primarily reach the farmers who had difficulty in using the computer based ICT technologies within Coimbatore region. The project has been test- implemented at two belts and it is still in Research and development stage. Initially 30 farmers were selected and provided with a Nokia handset mobile with life tools services. Through this feature, the farmers can launch query based agricultural search and service facility and consult the experts with decision based queries. When confronted with an agricultural query, the member farmers can either text or voice-out a query through their mobile phone. Audio, Text and video messages could be sent through the mobile phone and any query raised by the farmer in any of these formats is transmitted as text file to Tamilnadu Agricultural University extension center and within 48 hours, the expert answer is provided to the farmer. Also, the query and the answer that was sent from one life tools user is received by all other service users at the same time.. If proven successful, this intervention would be introduced for all farmers in Tamilnadu. At its current stage of implementation, this project has overcome the computer illiteracy of rural farmers in these areas and established instant, cost effective, cell phone based communication facility between the farmers and experts. This is the only mobile or text based intervention of Decision Support System available for the farmers of Tamilnadu.

With the widespread use of mobile phones in rural areas, the mobile phone manufacturer Nokia has developed an information service to provide farmers with agricultural advice, weather forecasts and market data in Indonesia and Africa. The company developed the service, known as Nokia Life Tools, after a consultation with users who seemed interested in an add on service that delivered up to date, relevant information directly to their mobile phones. At present, similar services are implemented in some developing countries like Indonesia and Africa and Nokia Life Tools facility delivers customized information to farmers based on their location, language and crops. Analysis of the feedback in these countries proved that the farmers were willing to pay a small subscription fee for a service that provided information to meet their specific needs and this tool can be effectively used to provide wide array of benefits to the farmers.

4.4 TNAU Agricultural Weather Network

The linkages between agriculture and climate are pronounced and often complex. Agriculture is most sensitive to extreme climatic events such as droughts, floods and hailstorms and to seasonal variability and changing rainfall patterns and as an occupation it depends on the vagaries of the monsoon. Crops are sensitive to climate change in both positive and negative ways. Hence, weather based information prediction is very necessary for the farmers as they have to plan ahead for cultivation and storage of their produce.

The weather network of Tamilnadu circle is a 150 million project that has established 214 Automatic Weather stations (AWS) within Tamilnadu weather circle that updates weather information of 365 local blocks every hour. AWS are established in various parts of the state and information about humidity, moisture content, rainfall, maximum and minimum temperatures are gathered, processed through the server and updated on the website. Automatic, hourly weather information is provided on the agricultural website for farmers to access. Using this information, the agricultural officers could also develop weather based agro advisories for farmers at block level.

This project is one of its kind in Tamilnadu as the Indian meteorological department has only 167 blocks for weather display at national level. Weather prediction network helps in monitoring the global warming and climate change impacts on Tamil Nadu Agriculture as well.

5. BENEFITS AND CONSTRAINTS

5.1 Farmer Information Portal

This information portal acts as a data warehouse that stores information of plot wise agricultural information of member farmers. Details of the farmer, his area of cultivation, soil information and other farm related details are stored in the computer system under particular identity numbers. Any farmer can send a query through accessing the "ask the expert facility" within the portal. Decision support services are provided to the farmers through the e-extension department of TNAU and the farmers can call in with any kind of queries. On receiving a query from the farmer through his mobile phone, the research extension officer directly accesses the details of the farmer based on the identity number of his individual land plot through the web server, identifies the problem and instant solution is provided to the farmer. At present, the project is in data entry stage.

5.2 Online and Offline Expert System

This type of ICT intervention uses artificial intelligence to provide solution to farmers and the process do not require human intervention. Once, the farmer types out a problem based query, the system automatically tracks the relevant details and information stored in the master database, analyses the systematic and possible solutions for the problems and fetches it as a search result. The farmer can view the search details and track the solution that best suits his query. Disease patterns and photographs of over 2000 crop melodies are scanned and stored in digital format to produce such an instant solution to agricultural queries. Off-line ICT intervention is accessible to the farmer community that is remote and access of Internet is rather limited. The farmers could search the static content and photographs available in the off-line website and references can be viewed to find solution to any of their queries. This project is still in Research and Development phase and not launched for public use.

5.3 Dynamic Market Information

This project is implemented by Ministry of Information Technology, Government of India and funded by CDAC, Hyderabad at a budget of 8000 million. This website is a community based website that offers a facility to receive agricultural price information through the mobile phones. The service offers market information to all farmers by sending out an sms about agricultural commodity prices of different markets within Tamilnadu. Market information of almost 13-20 markets is available at present.

Field extension support officers, collect the relevant details of agricultural and horticultural products based on the quantity of arrival from each individual markets in the state and the rates are fixed for each market. The market price information is updated on the web and the farmers can view the price information via e-mail or as sms sent through mobile phones. Every day before 3 pm in the afternoon, the required price information is sent to the farmers via the communication channels.

This facility could be accessed by the farmers by uploading their personal information and needed details in the related website. Once the personal profile is created the farmer can choose the markets and the commodity for what he seeks the information and he can register to receive the details either through an email or through his mobile phone. Any user interested in agricultural market prices can receive real-time information via SMS messages on their mobile phones or as an e-mail sent to them about the prices and availability of fruit and vegetables, in any of the regulated markets in one or two states.

This ICT facility provides farmers with cost of agricultural commodities at all regulated markets in the state to help him choose and sell his commodity for a better rate at any of these markets. This also leads to a price transparency of agricultural commodities across the country.

The farmers get the price information of the particular commodity through their mobile phones, contact the traders directly and fix up the commodity rate based on the DMI data of any regulated market in their vicinity. The farmers also establish trade contacts at all regulated markets through their associations and collectively send their produce through shared transportation facility from their fields. In the weekend, the farmers travel to the markets and collect the cumulative sum of their produce. Market transaction costs have been brought down by high level of coordination between the farmers and market vendors. The speedier collection and dissemination of price information on commodities enables farmers to realize better prices through direct access of agricultural markets without the intervention of middlemen.

5.4 DEMIC (Agmarknet)

Market information is a key ingredient for small farmers striving to increase their incomes. For small producers trading in rural areas in Tamilnadu, the challenges that they face include unavailability of commodity storage information and queries of where and when to sell the produce. In the past, the farmers relied on middlemen to provide market information, but it had its own disadvantages. The middlemen did not give them cultivation and storage details and just paid for their produce and controlled the marketing aspects with other traders. Also, public services and government extension departments were not able to serve the rural farmers due to several constraints. Transaction chains were long, while the volumes of goods were often small and of varied quality. Prices too were highly unstable.

AGMARKNET service portal facilitates generation and transmission of prices, commodity arrival information from agricultural produce markets, and creates a web-based dissemination to producers, consumers, traders and policy makers transparently and quickly. It aims at improving the decision-making capability of the farmers and strengthening their bargaining power.

Launched in March 2000, this project has linked all important agricultural produce markets spread across the subcontinent with the State Agricultural Marketing Boards and Directorates for effective information exchange. 810 nodes, 735 agricultural markets and 75 state marketing boards/directorate of marketing and inspection [DMI] offices were linked in the first phase and 2,537 additional markets were networked in the next phase. A commodity base, comprising of more than 300 commodities and about 2,000 varieties has been evolved. The commodities are being categorized into various groups: cereals, pulses, oil seeds, fruits, vegetables, spices, fiber crops, beverages, forest products, drugs and narcotics, dry fruits, flowers, forest products, livestock/poultry, and so forth to facilitate easy retrieval of market information.

5.5 Agrotech Portal of Tamilnadu Agricultural University

Despite the increase in the use of mobile phones for disseminating information, they do have some limitations. Their information storage and retrieval capacity is small, particularly in relation to the large size of files in video format. Hence, Internet should be developed as an alternative medium for communication among the farmers through extension activities.

The emergence of the Internet and especially the World Wide Web is regarded as an important development among all other ICT technologies because it offers connectivity at a lower cost and from many varied sources. Also, the information available over the Internet is instant and it transcends all physical barriers of communication. The Web also possess both sound and picture abilities, which can be used to further address the problem of illiteracy faced by rural farmers. The Internet remains as a cheaper and more reliable option for the foreseeable future, as it allows the transmission, storage and retrieval of large volumes of information. Governments should therefore do more to directly encourage private operators to provide reliable Internet facilities in rural areas of India.

Only 5% of the rural farmers are literates in Tamilnadu and their media uses are restricted to newspaper and agricultural magazines. Even the literates, though they are aware of Internet, are not familiar with computer usage and do not access the agricultural portals for information. Small and marginal farmers in rural areas are not aware of Internet technologies and their usage. Group extension activities and association meetings are used to disseminate agricultural information and technology transfer demonstration among the farmers.

However, almost all farmers belong to different farmers associations and the usage of Internet and ICT tools among the associations are very strong. Some tech-savvy farmers have Internet availability and their communication access is very strong in Tamilnadu. Hence, once the associations are informed of the web services, secondary dissemination of information happens at a faster pace. Group communication access is very strong and the agricultural information dissemination happens at a faster pace and in reasonable time the information reaches all farmers in the community.

5.6 Video-Conferencing Facility and Touch Screen Kiosks

There are 60 extension centers of TNAU within Tamilnadu and all the centers have e-connectivity between them. TNAU is used as a source hub for technologies. At any point of time, the main extension cell at Coimbatore, Tamilnadu could convene a video-conferencing facility with all the other extension centers within the state and farmer interactions, technology transfer, question and answer sessions happen through the e-conferencing facility almost every week. Farmer-Scientist interactions take place and this facility has evolved out to be very beneficial for the farmers. All queries of farmers are addressed by the experts and moderators of the project.

Touch screen kiosks have been provided at three districts like Chennai, Villupuram and Dindivanam to provide agricultural information access to farmers.

6. IMPLICATION OF THE STUDY

Availability of ICT technology perse, does not guarantee the usage of the same by the farmers. Any technology is only as good as its usability, adoptability and penetration among the rural farmers. Technologies remain useless if the farmers don't find it adoptable. Hence, e-agriculture ventures have to be easily adoptable and available to them. Technological innovations that can best raise agricultural productivity are those that are easily adoptable by end users and are cost-effective.

For these innovations to be adoptable, the rural farmers must find it suitable, acceptable and usable. Only then, the process of adoption cycle would begin among the users. One of the best ways to ensure that this happens is by actively engaging the views of the stakeholders in the process of the content creation and an assessment of their needs should be taken into account to effectively make the technologies adoptable.

There are a wide variety of new ICT tools for agriculture in Tamilnadu at present, but the tools need to be placed in an overall context of agricultural information and communication needs of the farmers. Only when the technology tools are placed within the suitable context and made available in local language there could be a prevalent usage of it among the farmers. This way, the ICT technology becomes adoptable without any push once it is fully developed.

An analysis of all the ICT interventions established at present shows certain benefits and constraints of usage.

One challenge facing mobile phone projects in any developing country is that a Short Message Service can carry only a limited amount of information and it definitely requires basic literacy. Though, voice-based solutions are complicated to develop it is more effective in rural areas. Studies in Africa have shown that voice based delivery systems remains by far the most promising platform, as it can be customized for any language and is readily accessible. It also offers direct responses to specific questions and it has got its own limitations as it cannot offer detailed information or pictures, as websites can. Hence, farmers can be prompted to use mobile sms facility as a first stage of evolution but it has to progress to usage of Internet and online forum at a later stage.

Weather information portal is very useful for literate farmers and rainfall prediction and other related predictions are by far the most accurate in Tamilnadu. Weather alerts are available and accessed by the farmers only through the web portal. The weather prediction facility helps only the literate farmers who are Internet aware. But for poor, illiterate farmers it has not brought in any change of status.

Farmer information portal uses novel ICT facilities to provide instant solution to farmers. As land fragmentation is huge in rural villages and every farmer changes his cultivation within the span of three to six months, the data of farmers become obsolete very soon. But, eventually new details have to be collected from them and updated after the end of every cultivation cycle. This would be a very tedious process. Though this facility leads to quick intervention and instant solution, the back end mechanism and data entry have to be done at a very faster pace. Only then this ICT intervention would be successful. But a master database of such kind would be very beneficial for all kinds of agricultural and rural development works.

On-line and Offline portal is very beneficial for educated farmers who can type out a query and find a solution. But, for uneducated farmers this facility might not bring in any change of situation.

By far, Dynamic Market Information service has evolved out to be the most successful of all ICT interventions introduced for farmers in Tamilnadu. This ICT innovation offers farmers a fair price for their farm produce, and it gives them a flexibility to select different markets based on their convenience, cuts out the information search and transaction costs and thus raises the net income of farmers. Farmers are able to access price information details of agricultural and horticultural commodity and conduct transactions. However, the benefits of prices have not been channelized to consumers. Many a times, the mobile SMS sent quotes a lesser price than the market price at retail shops. The retailers or shop keepers receive the commodity for specific prices from the farmers, but the prices that they sell the product to the consumers vary significantly. Hence, the benefit has not been accessed by the common man who buys the agricultural commodities from the retailer or nearby vegetable vendor.

DEMIC portal also gives market forecast information about what commodities to keep in stock and what to sell for the next three months. It effectively disseminates the market and price information to the farmers well in advance for planning, production and holding stocks of agricultural commodities. The constraints are that the information is available only in English language and it is not the language of the mass in India. The portal is delineated due to illiteracy and lack of awareness.

Internet and related information technology access and usage is imminent and relevant in rural areas as it reduces costs of acquiring information, is relatively easy to use, and can provide faster access. It cuts out all barriers of communication and establishes instant access to information. Information on the Internet is often not available in local languages, which seriously constrains rural farmer's access to relevant information.

Video-Conferencing facility acts in contrast to the old extension system of field visits and training. All queries and demonstrations are effectively done through video-conferencing and this process has evolved instant grass root and participatory communication of farmers with scientists.

6.1 Overview of the Study

In the past decade, there has been a remarkable progress in the use of ICT in agriculture, especially in the area of farmers' access to market information. Various projects have been developed that integrate ICT into the dissemination of agricultural information to farmers. Farmers Information Services at both the national and regional level are a promising new field of research and application in the emerging field of e-agriculture. The agricultural sector requires a well-organized learning community in the form of group extensions, farmers' associations, cooperatives etc.

A whole array of e-agricultural services are available for the farmers at present, but the services are rather more related to pricing of agricultural commodities and market prediction inputs. Some decision support services are also offered to farmers through video-conferencing facility project at present, but, it is not been used widely by the farmers due to lack of awareness among them. Only when the farmers know of the technology and have credibility in using it, the process of adoption would begin. Hence, adequate publicity measures should be used to bring about awareness within the farming community.

Rural farmers are getting adoptable to usage of simpler technologies like mobile phones and are learning to explore the facilities offered by them. Many of the mobile users, access mobiles for various activities like listening to FM radio, watch movie clips, capture images and videos and simultaneously transfer them via infrared or Bluetooth to other mobile users. They explore the functions available in their handsets and evolve their own mechanisms to establish communication within the community. Availability of mobile technology has created a need to use it and this need has facilitated a kind of learning environment for the farmers. Equipped with only the basic literacy, these farmers have harnessed their skill by adequately mastering the facilities available for communication through the mobile phone channel and are able to use blue tooth technology and infra-red portability without any inconvenience.

The adoptability of mobile phone usage among the farmers has in fact, created way for other newer ICT technologies. The farmers have adapted the usage of mobile phone technology by exploring through a self-learning process and likewise, when an opportunity is created for them to use computer technologies, effectively combined with their need to learn the usage of computers would propel them in the path of technology adoption process.

Need based learning is the best teacher for mankind. Some farmers, who are only basic literates, have begun to use Internet and learnt to operate the system, send and receive e-mails and have also begun to adapt to voice and video based online messenger services to communicate with their children and grand children settled in other countries. As this was the only the cost effective and suitable medium for communication, need based self-learning has occurred among them and the farmers have learnt to use the ICT tools by themselves.

Many rural communities do not participate in any ICT based initiatives at present because of the lack of awareness, lack of demanded content in local language and detailed information on governmental programmes in rural areas as well as region-specific information on agricultural insurance, agricultural credits facilities, e-commerce facilities, agricultural inputs and storage details. When this type of information is provided, farmers will feel more encouraged to participate actively in information sharing activities. A major focus has to be not on just collecting the right information, but on delivering it in a format which is accessible and will be accepted by farmers.

7. CONCLUSION

Information and Communications Technologies can be used to shift from the old tedious, less profiting approaches to a more modernized, convenient and potentially profitable approach. But a comprehensive plan for ICT adoption in agriculture should be adopted and a wide array of services has to be made available for the farmers through a common portal. *The level of awareness among professionals and decision makers in the region about the role of ICT in development is generally low.* E-illiteracy, lack if ICT infrastructure, lack of budget, lack of citizen service centers, high cost of ICT equipment, shortage of trained manpower who can assist the farmers in operating computers, as well as lack of awareness on ICT role are the serious impediments to any ICT expansion strategy within the rural community. *Awareness has to be created within the community and among the opinion leaders and decision makers.*

Now, the challenge before the state government is how to broaden awareness, acceptance and usage of e-agricultural services as the whole. The state government has to execute a roadmap that covers the broadband infrastructure in rural areas, a rural service delivery infrastructure that delivers a wide array of services that enables rural, agriculture-based transformation through one single portal. At present, the major concern is that many of farmers do not avail the services as they do not feel the need and dependency to use it.

Farmers have to see the tangible benefits that technologies could offer through awareness drive. ICT have to be integrated in the developmental agenda of rural India, only then the adaptability would eventually happen. Hence, when the need and access to the Internet is inevitable and at their doorsteps, access and usage can also be forced on the farmers in due course by presenting the agricultural information content in local language. An active collaboration among a complex system of stakeholder functions can turn this vision into reality.

Lack of appropriate local content and diversity in the Internet like local language, local problems and local needs has posed the greatest challenge at present and micro information portals in local language have to be created for usage among the local community.

Computer literacy among the farmers is low and the common model based on individual computer access in most cases is unfeasible due to high computer costs and lack of energy resources. One way to achieve this is through creation of Dynamic Market Information) in all villages of rural India. The Indian Government had approved a CSC scheme to create 1, 00,000 CSCs in 6, 00,000 villages in the country. These centers would serve as front-end delivery points for the rural citizens in the country. After these centers are established, need based learning will occur among all rural communities.

Awareness level among the farmers about the ICT interventions remains relatively low. Publicity measures have to be mooted to develop awareness about the ventures in all rural areas. *Connectivity and access at an affordable cost in the rural areas in particular is still a problem.*

Online access to financial services, online access to land record copies and online access to commodity procurement sites need to be developed to create the need for information access. When tangible benefits that are attached to the projects are more, the utilization of ICT services would improve. Internet centers and information kiosks should be made available to the rural community.

In 10 years time, computers and the Internet will not be a new technology, they will simply be part of the environment in which people live and work. The technological environment will help shape the skills and values of the next generation. To support this behavior, public and private ICT initiatives could be an effective solution. However, in addition to the accessibility, targeted and niche services and relevant content need to be provisioned for the farmers.

The motivation and the incentive to use such traditional means of disseminating information is low at present, emphasis have to be laid to create a comprehensive value added agricultural service portal along with citizen service centers in all rural hamlets would definitely go a long way in increasing the adaptability. Only when this is done, the country and the economy will reap the full benefit of ICT.

7.1 Recommendations

Efforts are being accelerated to eliminate areas without broadband access and regions without mobile phone reception by the end of 2010 to promptly create an environment in which all members of the public can access broadband services and enjoy the conveniences they provide.

1. Awareness level of many of the ICT interventions remains as the main constraint of adoption and efforts have to be made by the government to publicize the ICT interventions.
2. Farmers can be prompted to use mobile sms facility as a first stage of evolution but it has to progress to usage of Internet and online forum at a later stage.
3. Weather alerts could be sent to farmers as an sms through their mobile phones. It would help them to better anticipate natural calamities.
4. Farmer information portals should also have back-end supports and constantly updated information only would be useful for the farmers and scientists. Hence, back end operations have to be developed to enter new and store new crop information details every three months.
5. A comprehensive farmer's portal should be created with an integrated facility to access e-procurement, e-financial services and provision to receive online land record copies.
6. A comprehensive master database of farmer information in all villages and districts of the state should be created as it would be very beneficial for all kinds of agricultural and rural development works.
7. Region-specific details needs to be provided for the farmers. As what works in the neighboring state might not work well for Tamilnadu.
8. Rural telecenters have to be created to facilitate e-agricultural transactions and provide complete agricultural services to all farmers.
9. Internet kiosks and information kiosks needs to be established at taluk offices to provide agricultural information access for the farmers.
10. Area-wise farmer information portal can be created with farm histories of every plot to provide instant interventions.
11. The need for farm and crop insurance schemes has to be stressed among the farmers to help them overcome the risks associated with agricultural practices.
12. Approaches need to be developed to overcome these constraints and the value of local knowledge and micro issues need to be emphasized with focus on small scale farmers and rural communities. It has many advantages as a medium of information and knowledge exchange. But limited access and poor connectivity constrain many individuals, particularly in rural areas of Tamilnadu.

REFERENCES

Adhau, B. P. (2010). Contribution of agriculture in the development of India. *International Research Journal, 1*(17).

Aker, J. (2008). *Can you hear me now? How cell phones are transforming markets in Sub-Saharan Africa*. USA: Center for Global Development.

Aker, J. C. (2008). *Does digital divide or provide? The impact of cell phones on grain markets in Niger*. BREAD working paper number, 177.

Assadi, M. (1998). Farmer's suicide –Signs of distress in rural economy. *Economic and Political Weekly, 23*(14), 747–748.

Ballantyne, P., Maru, A., & Porcari, E. (2010). *Crop Science Society of America, 50*(10).

Chauhan, N. (2005). Information technology for agricultural development in India. In Dipak, D., & Jirli, B. (Eds.), *Communication support for sustainable development*. Jangamawadi Math, India: GangaKaveri Publishing House.

Chhiber, N. (2004). Enhancing developmental opportunities by promoting ICT use: Vision for rural India. *The Journal of E-Government Policy and Regulation, 27*(3).

Dale, P. (1999). *Is technology a blessing or a curse in land administration?* Presented at the UN-FIG Conference on Land Tenure and Cadastral Infrastructure for Sustainable Development, Melbourne, Australia, October 1999.

Dale, P. (2000). *The importance of land administration in the development of land markets - A global perspective. Land markets and land consolidation in Central Europe.* TU Delft – Report of UDMS 2000.

Dar, W. (2003). *Technology for poverty reduction: Working towards a South Asia strategy*. Intermediate Technology Development Group-South Asia (ITDG-South Asia), British Council, New Delhi, 10-11 October, 2003, South Asia Conference on Technologies for Poverty Reduction, New Delhi.

De, R. (2005). E-government systems in developing countries: Stakeholders and conflict. In Wimmer, M. A. (Eds.), *E-Gov - 2005, LNCS 3591* (pp. 26–37). doi:10.1007/11545156_3

De, R. (2006). The impact of e-government initiatives: Issues of poverty and vulnerability reduction. *Regional Development Dialogue, 27*(2), 88–100.

De Silva, H. (2010). *The Govi Gnana Servic: An ICT info structure to fight agricultural poverty in Sri Lanka. Report of E Development Labs*. Public Interest Program Unit.

De Silva, H., & Ratnadiwakara, D. (2010). *Using ICT to reduce transaction costs in agriculture through better communication: A case-study from Sri Lanka*. LIRNEasia, Colombo.

Division of Agricultural Economics, IARI. (2006). *Agricultural Economics Research Review, 19*.

Dwivedi, S. (2008). *Challenges of e-government implementation in India: Emerging technologies in e-government* (pp. 210–215). E-Governance Publication.

Ewing, J. (2007, September 13). Upwardly mobile in Africa. *Business Week*. GSMA Development Fund. 2008.

FAO. (2003). *Statement at the ministerial conference of the WTO.* Fifth Session, Cancun, 10-14 September, Doc No: WT/MIN(03)/ST/61.

Gates, B. (2008, May 8). Promoting technology adoption—and prosperity—Through public-private partnerships. *Wall Street Journal Asia*.

Jensen, R. (2007). The digital provide: Information (technology), market performance and welfare in the south Indian fisheries sector. *The Quarterly Journal of Economics, 122*(3), 879–924. doi:10.1162/qjec.122.3.879

Malhan, I. V. (2007). Agricultural knowledge transfer in India: A study of prevailing communication channels. *Library Philosophy and Practice,* 2007, February.

Meera, S. N., Jhamtani, A., & Rao, D. U. M. (2004). *Information and communication technology in communication development: A comparative analysis of three projects in India.* Agricultural Research & Extension Network, Paper no 135.

Ministry of Agriculture, Indian government report, New Delhi. (2008). *Indian agricultural scenario*.

Mohanty, K. (2007). Power of information in the Indian rural set up – An analysis. *International Journal of Rural Studies, 14*(1).

Nath, V. (2003). *Digital governance models: Moving towards good governance in developing countries.* United Nations Development Project Report.

Okello, J. J., Okello, R. M., & Ofwona-Adera, E. (2010). Awareness and the use of mobile phones for market linkage by smallholder farmers in Kenya. In Maumbe, B. M. (Ed.), *E-agriculture and e-government for global policy development: Implications and future directions*. Hershey, PA: IGI Global. doi:10.4018/978-1-60566-820-8.ch001

Raghbendra, J. (2005). *Land fragmentation and its implications for productivity: Evidence from Southern India.* ASARC Working Paper 2005/01.

Rahul, D. (2005). *The impact of Indian e-government initiatives: Issues of poverty and vulnerability reduction and conflict*. Bangalore, India: Indian Institute of Management.

Sharma, D. (2005 June 11). The business of hunger: ICT and millennium development goals. *Motion Magazine.*

Shastri, P. (2001, May 1). Digitizing land records key to reform. *Hindustan Times*. Retrieved from http://www.hindustantimes.com/News/business/Digitising-land-records-key-to-reform/Article1-219609.aspx

UNDP. (2004). *E-governance- UNDP practice area: Democratic governance synthesis of lessons learned.* Human development report 2001, No.15, April 2004.

UNESCO. (2003). *Newsletter, 1*(106).

UNESCO. (2005). *E-governance toolkit for developing countries*. New Delhi, India: UNESCO Publishers.

United Nations. (2008). *Trends in sustainable development: Agriculture, rural development, land, desertification and drought*. New York, NY: Department of Economic and Social Affairs of the United Nations. Retrieved from http://www.un.org/esa/sustdev/publications/trends2008/fullreport.pdf

USAID. (2003). *Future directions in agriculture and information and communication technologies (ICT)*. USAID/Economic Growth, Agriculture, and Trade/Agriculture and Food Security.

Walsham, G. (2010). ICT for broader development of India: An analysis of literature. *The Electronic Journal on Information Systems in Developing Countries, 41*(4), 1–20.

World Bank. (2007). *World development report 2008: Agriculture for development.* Washington, DC: World Bank. Retrieved from http://siteresources.worldbank.org/INTWDR2008/Resources/WDR_00_book.pdf

Chapter 12
Information and Communication Technology and its Potential to Transform Indian Agriculture

Saurabh Singh
College of Agribusiness Management, India

ABSTRACT

The role of Information Communication Technology in enhancing food security is well established and has been recognized at worldwide forums like World Summit of Information Society (WSIS) 2004-2005. Information Technology includes use of computer, internet, geographical information system, and mobile phone, besides traditional media. Every sphere having to do with agriculture has scope for information technology to make significant contribution, but in a few areas, its impact can be instantaneous. The present era, also known as the information era, requires creation of information expressway interlinking farmers, markets, corporate, agricultural scientists, and buyers of agricultural products. This may help producers to realize maximum share from the price paid by end consumer. In the present agricultural marketing system, middlemen share a significant portion from price paid by end consumer, and what the farmer gets is residue. Thus, the two major areas where information technology intervention can create progress are creation of information expressway for farmers; and strengthening farmers' linkages to supply chain. These two aspects have been discussed with the help of initiatives like AGMARKNET, NDDB Procurement Process, e-Choupal, and EID's Parry's Corner. The effort is to facilitate learning with real life examples as contrasted against pure and abstract discussion based learning.

HISTORICAL OVERVIEW OF INDIAN AGRICULTURE

India, which always enjoyed the envious position of being 'Global Seat of Power and Trade', was forced to suffer tyranny and oppression, at the hands of its foreign conquerors, for nearly four centuries. Preceding line is important in the context that the structure of India's present day economy is not just of current making; it has its roots steeped in history. Particular reference is being made of period when India was under British rule, which lasted for almost two centuries, before India finally won independence on August 15, 1947.

DOI: 10.4018/978-1-4666-2655-3.ch012

This was a period, when the renaissance of Britain as well as Europe was shaping up in all the spheres of human activity after its long unwelcome tryst with dark ages, spanning centuries, approximately from 600 AD to 1600 AD (Hodge 1982). Besides Britain, few other nations that harbored imperialistic agenda were Spain, France and Portugal, and all being nearly contemporary. Fascinated by wealth and developed trade in orient, British with intension to trade reached the Indian shores. After a brief period of trading relationship, these traders started to ruin the fortune of India's Trade and Development, as they turned greedy, invaders, conquerors and exploiters. It seems pertinent here, before deliberating ahead about British India and Indian Agriculture; to provide a glimpse of Agriculture, as it was, in India during Pre- British era, and the same could be found in the words of a seventeenth century French Traveler to Bengal (Uppal, 2006).

"(Bernier, 17th Century) the knowledge I have acquired of Bengal in two visits inclines me to believe that it is richer than Egypt. It exports in abundance, cottons and silks, rice, sugar and butter. It produces amply – for its own consumption – wheat, vegetables, grains, fowls, ducks and geese. It has immense herds of pigs and flocks of sheep and goats. Fish of every kind it has in profusion. From 'Rajmahal (Palace)' to the sea is an endless number of canals, cut in bygone ages from the Ganges by immense labour for navigation and irrigation". In India, Bernier came under the protection of Daneshmand Khan (Mullah Shafi'a'i, a native of Yazd), an important official at the court of Aurangzeb. (Wikipedia).

Economy of India prior to being forced turned as a colony of British imperialists, was an independent one. Though agriculture was the mainstay of livelihood for majority of Indians, the nation's economy had made enough strides in manufacturing industry, to enable it to be called as developed. Prominent among the industries were handicrafts, cotton & textile, metal and precious stone works etc. The economic policies pursued by the colonial government in India were concerned with protection and promotion of economic interests of their home country than with the development of Indian Economy (Uppal, 2006). Such policies transformed Indian economy and country into just being a supplier of raw material and consumer of finished industrial products for Britain.

INDIAN AGRICULTURE SECTOR POST INDEPENDENCE

The situation of agriculture sector as well as economy of India witnessed a phenomenal growth when compared with what India was at the time of independence. Things improved by a speed much higher than what was expected. The production of agriculture sector, when compared with 1950s, shows a significant increase in agricultural productivity. This phenomenon did not alone make India, self – sufficient but also, a net exporter of variety of food grains. All this can be attributed to the remarkable and surprising success exhibited by agricultural sector in 1968, which marked the beginning of Green Revolution'. Somehow these days the same sector gives an impression of stagnation, which perhaps may be due to the reasons of nothing great having been achieved by the stakeholders comprising scientists to farmers, over last few decades. Arabi (2009) opines that technological breakthrough brought about by the Green Revolution in terms of yield were unsustainable after the sixties and seventies in twentieth century; as they were limited to specific crops – wheat and rice and in a limited geography. Geographical areas covered by Green Revolution were restricted to Punjab, Haryana, Western Uttar Pradesh, and parts of Andhra Pradesh and Karnataka. Last decade has witnessed a fatigue in the green revolution with growth rate of food grain falling below population growth. Swaminathan (2009) states that today the human numbers are increasing at a rate higher than the nation's present capacity to make the goal of food for all a reality. Due to inadequate purchasing power, which is a manifestation of famine of jobs or livelihood generating activities, as a consequence of poor growth of employment opportunities for employment in rural non – farm and off – farm sector, is not allowing consumption to go up and is leading to a famine of food at household level.

Even the growth in agricultural productivity when compared with 1950s, could not succeed in accommodating majority of farmers, when it comes sharing the benefits of the growth. The deep cause, underlying the phenomenon, could be traced as inefficient land holdings (Annamalai & Rao, 2003). Burden of housing and other socio – economic needs of the growing population has been and is still being catered by sacrificing agricultural land in favor of Commercial, Industrial and Residential projects. Sharma, (2010) is of the view that other issues which have further added to the misery happen to be the problem of shrinkage in agricultural land as a consequence of Soil Health deterioration leading to land degradation, land acquisition for urbanization, infrastructural activities and special economic zones. These have made their impact significantly felt, by the way of leading to shrinkage in Net Sown Area (NSA).

Even today, when India has nearly succeeded in implementing a fast but controlled process of development and growth, importance of agriculture sector could not be ignored. The importance of and contribution from 'agriculture sector' to the Indian Economy is of the nature, that it has earned adjectives like, being backbone (Chand 2004); main stay (Moni 2004); and vital etc. to convey this relationship.

The inter sectoral linkage of agriculture could be better understood by learning that Agriculture on supply side does not alone provide cereals, fruits, meat, egg, vegetables etc., which are seen on food plates of both urban and rural consumer; but also raw material to several industries like sugar, jute, cotton textile, oil, milk products, fruit and vegetable based industries in processing sector etc. It is certain that abundance in agricultural production would bring prices down, due to inherent character of agricultural produce and also the need of producer like perish-ability, need for money, lack of storage facility etc. This phenomenon will result in less than proportionate (compared to production) increase in the income of farmers, but it has fair chances to result in increased real income of net buyer of agricultural produce. Consequently, both farming community as well as net buyers of food shall have increased disposable income to spend on non agricultural products. At the same time availability of raw materials at lower costs will lead to higher production in secondary and tertiary sector (Mohan 2007). Thus development of agricultural sector influences the performance of other sectors directly.

The role of agriculture sector as mainstay in the rural Indian economy gets re-established further by above arguments. This happens to be a sector around which socio-economic privileges and deprivations revolve and, any change in its structure is likely to, have a corresponding impact on the existing pattern of social equity. Sustainable Agricultural production depends on the judicious use of natural resources (soil, water, livestock, plant genetic, fisheries, forest, climate, rainfall, and topography) in an acceptable technology management under the prevailing socio-economic infrastructure. Food and Agriculture Organization (FAO) define sustainable development in case of Agriculture, Forestry and Fisheries as in double quotes, "Sustainable development is management and conservation of natural resource base and orientation of technological and institutional change in such a manner as to ensure the attainment and continued satisfaction of human needs for the present and future generations. Such sustainable development, (in the agriculture, forestry and fisheries sector) conserves land, water, plants and animal genetic resources, is environmentally non-degradable, technically appropriate, economically viable and socially acceptable (FAO 1989)." Various research studies and policy papers highlight that the Indian Agricultural sector faces resource constraints in many spheres. It is, therefore, essential to combine natural resources, capital resources, institutional resources, technological resources and human resources (i.e. to optimally utilize the agricultural resources).

Indian government has embarked on the developmental process in big way which has resulted in numerous developments that have already taken place providing more avenues of fruitful employment. Somehow this has resulted only in marginal net fall in the parentage of people dependent on agriculture for their livelihood and employment. This can be attributed to the reasons that the developments and progress made by nation have been offset by the growing population. Thus agriculture sector still remains the source of direct or indirect livelihood earning for approx 67 per cent (Mohan 2007) of population.

INDIAN AGRICULTURE SECTOR POST REFORMS

In India a new phase of economic reforms got initiated with New Government taking Charge having Mr. Narsimharao as Prime Minister and Dr. Manmohan Singh as Finance Minister of the country in 1991. The new government launched a new economic policy known as New Economic Policy 1991 (NEP) also known as Narsimharao Manmohan Singh Economic Reforms. This was a time when a significant worldwide trend in public policy had been to disengage decades of direct government involvement in the agricultural sector. Price supports, input subsidies, and publicly owned agriculture-related institutions were increasingly paving ways in favor of private market determination of prices and other incentives (Saurabh, 2008). On the similar lines, deregulation, liberalization, and reduction in government involvement in market activities were basic pillars of the NEP 1991.

Much of the reforms initiated, though, had no direct impact on agriculture sector, but the sector due to being vulnerable to a number of factors could not be saved from the indirect heat that it had to face. Indirect heat was generated due to events like devaluation of exchange rates, liberalization in favor of external trade and dis-protection of the industry. These economic reforms were soon followed by a new international trade accord and WTO, making opening up of domestic markets for external trade mandatory. The two decades preceding the new economic reforms failed to witness any major government intervention in agriculture sector. The interventions done in this era by government in agriculture sector were limited to 'price interventions' and that too were limited in coverage (Chand, 2005).

All this created a scenario of felt void in context of any formal document on agriculture in form of policy, which could address the new challenges thrown up by and in the years since initiations of economic reforms in 1991. Consequently, pressure started building on the government to guide the sector in face of new set of development, as regards to new challenges and opportunities, which had started making their presence felt, both on agriculture sector as well as farmers. Finally government of India on July 2000 announced a New Agricultural Policy.

The new agricultural policy aims to attain output growth rate in excess of 4 per cent per annum in agriculture sector based on efficient use of resources. It seeks to achieve this goal in a sustainable manner and with equity.

Thus, in nutshell, besides many other things, it could be agreed upon, that focus of New Agriculture Policy 2000 was on:

1. Efficient Use of Resources;
2. Sustainable Agriculture; &
3. Equity

SUSTAINABLE AGRICULTURE AND ROLE OF INFORMATION TECHNOLOGY

The objective behind sustainable agriculture rests in providing enhanced food security. The Recommendations of United Nations Conference of Sustainable Agriculture and Rural Development (1992) in its Agenda 21 recommend major adjustments in agricultural, environmental, and macro – economic policy to create conditions for sustainable agriculture and rural development. The Agenda 21 prescribes, issues mentioned below, as 'information for decision making':

1. Development of indicators for sustainable development
2. Promotion of global use of indicators for sustainable development
3. Improvement of data collection and use
4. Improvements of methods of data collection and analysis
5. Establishment of comprehensive information framework
6. Strengthening of capacity for traditional information
7. Production of information usable for decision making
8. Development of documentation about information
9. Establishment of standards and methods for handling information
10. Establishment and strengthening of electronics networking capabilities
11. Making use of commercial information Sources

Utilization of information resources over internet and/ or intranet and development of information system is a matter of gaining strategic lead over other countries in the world. Information network plays an important role in the flow of information from implementation level to planner at macro (National) level to macro-meso (Regional Level covering more than one state) level to meso (State) level and finally to micro (District, Block and Village) level(Moni 2004).

Role of Information Communication Technology and Tools in enhancing food security has been well established and has got recognized at worldwide forums. The same was officially endorsed at the World Summit of Information Society (WSIS) 2004-2005. Information Communication Technology includes use of computer, internet, geographical information system, mobile phone besides traditional media such as television and radio.

Informatics for agricultural development requires coordinated inter sectoral approach and application of appropriate information technology tools, in nearly all the areas related to agriculture, like Agricultural Research, Agro-meteorology, Agricultural Marketing, Food Processing, Agricultural Extension, Crop Production and Protection. Fertilizer and Manure, Irrigation and Drainage System, Rural Development, Live Stock, Animal Husbandry, Dairy Development, Fisheries, Environment and Forest, Watershed Development, and Wasteland Development etc..

Contributing to this trend toward free market economies are the growing global efforts to liberalize trade by lowering tariffs, harmonizing other standards, and facilitating foreign direct investment. While the short-term adjustments to increased competition are recognized as potentially painful, it is generally agreed that medium to long-term gains more than compensate. Nevertheless, in a very real sense, the adjustment to removal of both domestic support structures and barriers to external competition may be especially painful for smaller producers, a group which is often the target of poverty alleviation measures. Indian food-processing sector is undergoing rapid transformation. As the economy of the country moves on the path of development, agricultural sector evolves from traditional subsistence level farming to commercial agriculture producing high value and processed products. There is high potential demand in the domestic as well as export market. On the supply side, with vast arable land and agro-climatic diversity, India has the potential to be a food basket of the world. With this interaction of demand and supply forces the food processing sector has the potential to be the driver of economic growth and enhance rural incomes. Food Processing Industry is widely recognized as a 'sunrise industry' with a vision to make India, the Food Basket of the world, achieves what has been achieved in the field of computer technology (Saurabh, 2008).

Avenues Where it Can Add Tremendous Value

Though in every sphere having to do with agriculture, agricultural process, agriculturists, and agricultural markets etc., information technology can make significant contribution, but in few areas identified here, its impact will be felt instantaneously. Present era, which is also known as information era, requires information technology to create an expressway of information interlinking farmers, markets, corporate, agricultural scientists and experts, and buyers of their products. Information technology has a responsibility to ensure that producers gets maximum share from the price paid by ends consumer to purchase the produce. This can be done easily and in much smoother manner, as it requires eliminating structural inefficiencies of agriculture produce marketing system. In present system a significant portion of the price paid by the end consumer of the agricultural produce is taken away by middlemen and what farmer gets is residue.

Thus, the two major areas that have been identified here, for further deliberations, where information technology can create a miracle, happen to be as below:

1. Creating information expressway for farmers;
2. Strengthening farmers' linkages to supply chain. The deliberation further will focus on these two points and will be explained with help of various initiatives taken by government, public and private sector in the area.

CREATING INFORMATION EXPRESSWAY FOR FARMERS

The Agmarknet Story (http://www.agmarknet.nic.in)

It is a very well established fact today that any growth, development. Achievement, plan and policy etc., can never prove a success without involvement of the stakeholders of agriculture sector. The old axiom that "India Lives in Villages" even today is as true as it was when won independence. Agriculture still remains vital to all, i.e., India, Indians and Indian economy. This can be attributed to the reason that agriculture sector has the responsibility to feed more than one billion people besides providing the source of livelihood to approximately 66 per cent of them. Its contribution to Indian GDP still remains significant at 23 per cent. Though the contribution of agriculture when seen in percentage terms may give an impression of its reducing share but it's simply an illusion, as agriculture still remains the largest sole contributor to Indian economy. In such scenario where such a large population is dependent on agriculture for livelihood, the activities of agricultural production and marketing of agricultural produce assume prime importance.

The existing regulations that prohibits the farmers from directly engaging in trade with end consumers and advising the farmers to route their produce by the way of selling the same at government regulated agricultural produce marketing mandi by the way accepting the highest bid made at mandi by any trader having license to do so from the concerned state governments. Though the provision has been made to safeguard the interests of framers, but it also adds volatility to the activities/ process of agricultural production and marketing of agricultural produce. Since the farmers' do not have much exposure to developments around the globe turn to be easy prey for rumor mongers, who may play with emotions of the farmer to help themselves achieve some malicious hidden agenda of theirs. The regulated mandies specifically created for this process are managed by an elected committee called as "Agricultural Produce Marketing Committee (APMC)" and the regulation of mandi and trading activities as regards to marketing of agriculture produce is regulated by Agricultural Produce Marketing Committee Act (APMC Act).This act was designed to ensure that farmers get, if not the best, then too, a proper price for their produce. These mandies are in normal course supposed to cater to the farmers falling in a radius of 25 to 50 Kilometers.

However despite of best of the structures created with all good intensions cannot be claimed to be perfect one. These imperfections get noticed gradually and based on same corrections are incorporated in the earlier structure. Thus despite all the efforts of government some structural and/ or efficiency related problems find some via root to enter the scene of action. One of such deficiencies prevalent in mandies happens to be the lack of transparency. Though some scholars also feel weighing too as a problem area, the author does not share the same opinion. This is just for the regions that the weighing instruments are open for public scrutiny due to being stationed at the counters allotted to licensed traders, anyone and everyone is free to raise the issue. Further if someone is willingly getting cheated then none can rescue the individual. The priority area needing attention of policy makers happens to search the ways and means of adding transparency in auctioning process.

The limitation of farmer in overcoming the problem of transparency in auctioning process, that fixes the price of his produce, is mainly due to information asymmetry. The farmers often, due to working in more or less isolated manner coupled with absence of any tool of getting the real time information that could inform them of prices prevailing in nearby mandies and other issues of the same nature are not able to be in a position to have price discovery prior to selling the produce. This information asymmetry works against the farmer and in favor of the trader.

Any democratically elected government cannot close its eyes in such a scenario. It is supposed to work in a manner so as to minimize the chances of suffering a loss by the farmer. Perhaps the government of Union of India, is the first government in world to have embarked on the process of creating an information highway, through the length and breadth of the country to minimize this information asymmetry. This certainly is an herculean task, given the length, breadth, topology, population, numerous agro climatic zone and a great variety of agricultural products being produced and marketed through mandi channel. Due to being aware of the fact that socio-economic transformation of farmers would not take place till they do not get equipped with latest agricultural and allied knowledge besides the real time information of what is happening in the market.

Thus, with commitment of huge funds and riding on the developments made by information communication technology, the Government of India launched an ambitious scheme named "AGMARKNET" (See Figure 1). This scheme got initiated by Department of Agriculture and Cooperation under aegis of Ministry Agriculture of Union of India with active participation of National Informatics Center (NIC). The full name this scheme is "Central Sector Scheme of Agricultural Marketing and Research Network" popularly called AGMARKNET Scheme (See Figure 2).

The Objectives of the Scheme as detailed under Ninth Five Year Plan and then Tenth Five Year Plan were as mentioned hereunder respectively.

The Objectives of Scheme under Ninth Five Year Plan

1. To establish a nation-wide information network for speedy collection and dissemination of market information and data for its efficient and timely utilization;
2. To computerize data on market fees, market charges, total arrivals, arrival by agencies, prices (variety-wise and quality-wise), storage, dispatches with destination, method of sale, payment, weight, grading facility, quantities graded, market personnel (trained/ untrained), market functionaries, market finance, development programs, infrastructural facilities, constitution/ composition of market committees, state marketing Boards and Departments;
3. To ensure flow of regular and reliable data to the producers, traders, and consumers to derive maximum advantage out of their sale and purchase; and
4. To increase the efficiency in marketing by effective improvement in the existing market information system.

Objectives under Tenth Five Year Plan

1. To establish a nation-wide information network for speedy collection and dissemination of marketing information and data for its efficient and timely utilization (See Figure 3).
2. To facilitate collection and dissemination of information related to better price realization by the farmers. This would cover:
   a. Market related information such as market fee, market charges, costs, method of sale, payment, weight, handling, market functionaries, development program, market laws, dispute settlement mechanism, composition of market committees, income and expenditure, etc. (See Figure 4).

Figure 1. Home page of AGMARKNET portal

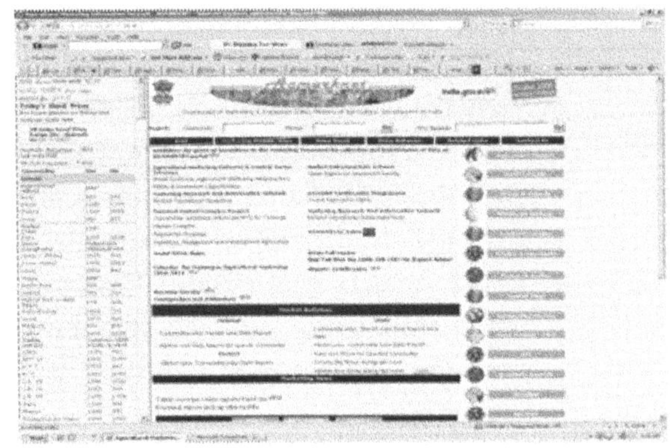

Figure 2. Site map of AGMARKNET portal

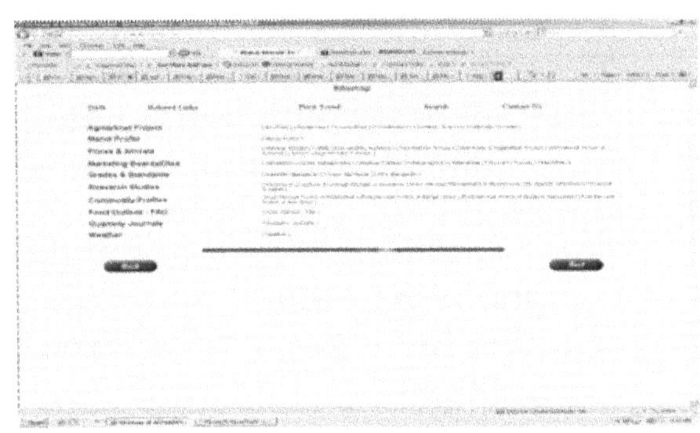

Figure 3. Total number of regulated markets networked

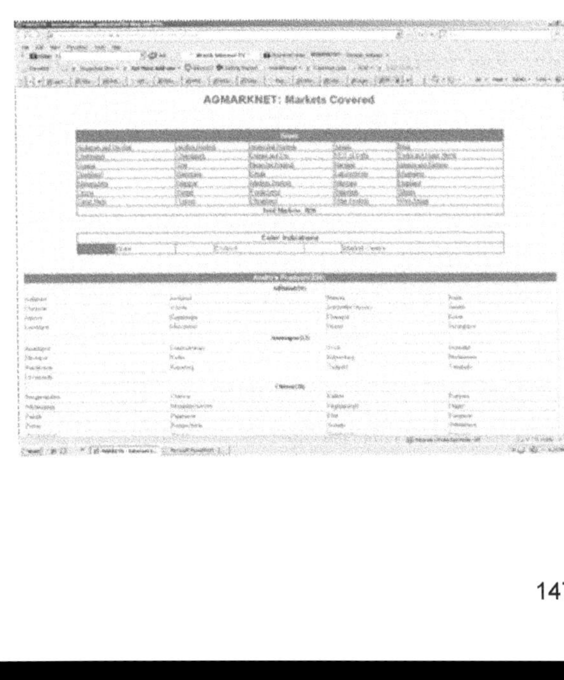

b. Price related information such as minimum, maximum and modal prices of varieties and qualities transacted, total arrivals and dispatches with destination, marketing cost and margins, etc.

c. Infrastructure related information comprising facilities and services available to the farmers with regards to storage and warehousing, cold storage, direct markets, grading, re-handling and repacking, etc.

d. Promotion related information covering accepted standards and grades, labeling, sanitary and phyto-sanitary requirements, pledge finance, marketing and new opportunities available in respect of better marketing (See Figure 5).

3. To sensitize and orient farmers to respond to new challenges in agricultural marketing by using ICT as vehicle of extension.
4. To improve efficiency in agricultural marketing through regular training and extension or reaching region-specific farmers in their own language.
5. To provide assistance for marketing research to generate marketing information for its dissemination to farmers and other marketing functionaries at grass-root level to create an ambience of good marketing practices in the country.

Figure 4. Listing of Mandies as shown on portal

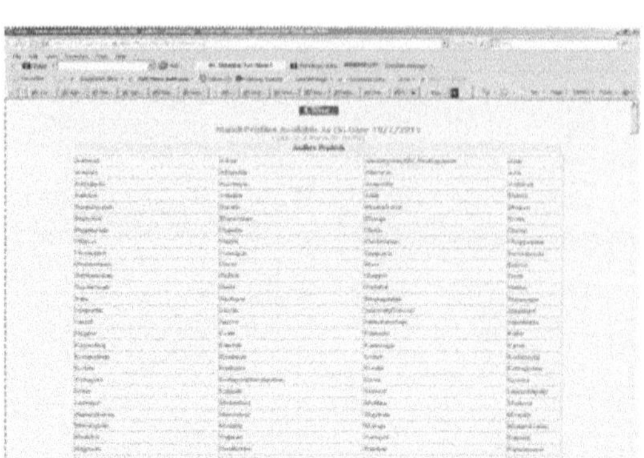

AGMARKNET Model

The Physical Setup of Model

Each AGMARKNET node has been equipped with the following hardware and system software tools:

1. Hardware
   a. Client computer system
   b. Printer supporting Hindi, English and a local language
   c. Line interactive UPS system with battery back up
   d. Modem to enable dial up based communication.
2. Software
   a. Windows 98 Operating System
   b. Lotus SmartSuite Millennium
   c. AGMARK application package

The Implementation Plan

The plan of implementing the MRIN project is as follows:

1. The agencies involved in execution of the scheme are Directorate of Marketing & Inspection (DMI), National Informatics Centre (NIC), and State Governments through State Agriculture Marketing Boards

Figure 5. Page providing information on standards

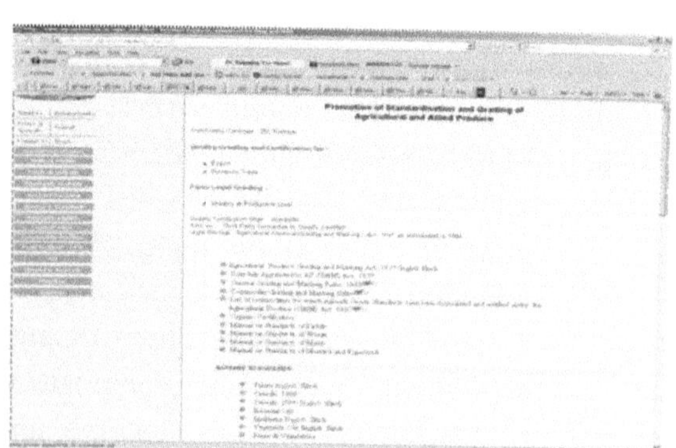

(SAMB) / Departments, National and State level institutions and individual market committees/authorities wherever applicable in the country.

2. NIC has been assigned the following responsibilities:
   a. Make available computer hardware
   b. Develop Software
   c. Train market personnel in handling computer hardware and software
   d. Update the software package from time to time
   e. Provide internet connectivity
   f. Develop and commission State level portals
   g. Harmonize/integrate software package developed by the State with AGMARKNET to bring uniformity in database
   h. Manage AGMARKNET portals
3. It was planned that State Governments will provide to the DMI the list of markets to be covered.
4. The selected markets are envisaged to provide site for installation comprising
   a. Dust free computer room
   b. Power requirement
      i. Electrical points-3, input to be protected by 6 amp MCB
      ii. 220/230 volt power supply
      iii. Line, neutral and each connection on the specified socket terminals
      iv. Proper Earth pit copper wire earthing (with earth to neutral voltage less than 3V) with an exclusive phase
   c. Computer operator
   d. One telephone with STD facility, wherever required.
5. Marketing Committees/Controlling authorities of AGMARKNET node at market level are required to collect relevant data and information, feed it and transmit it to the State level and AGMARKNET portal.
6. The SAMB/Departments are to nominate a nodal officer to coordinate functioning of the scheme.
7. The State level nodal officer will have to guide and monitor the market level officers.
8. The State level servers are to be preferably located at the Headquarter of the State Marketing Board/Directorate implementing the scheme.

To fulfill the objectives of the Central Sector Scheme of Agricultural Marketing & Research Network, a user-friendly software package *AGMARKNET*, has been developed to facilitate organization and transmission of market data and it has been implemented in the markets. In order to strengthen interface with farmers and other beneficiaries, AGMARKNET portal *(*http://agmarknet.nic.in*)* has been evolved. Hundreds of markets are regularly reporting price related data, which is being disseminated through the portal. The AGMARKNET portal also serves as a single window for accessing websites of various organizations concerned with agricultural marketing. It provides weekly trend analysis for important markets in respect of major commodities. It also provides links to various Online Commodity Exchanges of India providing future prices in respect of oilseeds, fiber crops etc. International price trends of various agricultural commodities available on FAO website are also accessible through the portal. The portal is constantly being enriched.

The Present State of AGMARKNET

The AGMARKNET Scheme has turned to be a very successful scheme. The efficiency of the Government organizations involved in the process could be gaused by the fact that against the given target of networking 1676 regulated mandies through the length and breadth of country, from total number of such mandies available, during ninth and tenth plan put together, the actual number of mandies that got networked came at 2379, way ahead of the target. The Scheme is still operational in the eleventh five year plan and aims to network all the government regulated mandies. The farmers can get a information on nearly everything they could wish starting from good agricultural practices to prices of various commodities prevailing in the madies throughout India, besides prices prevailing in futures markets and much more (See Figure 6).

Further it also provides information or linkage a wide variety of information like information on International Prices, Agriculture Marketing Journals etc. Many National Governments may learn lesson from this initiative (See Figure 7).

The Story of National Dairy Development Board (NDDB) of India: Use of IT in Procurement and Payment of Milk

Founded in 1965, the National Dairy Development Board has a mission of organizing poor milk producers and to establish dairying as a potent tool bring about economic development specifically in Rural India. It was vision shared by then Prime Minister of India Shri Lal Bahadur Shashtri, who wanted to extend the success of Kaira Cooperative Milk Producers Union to other parts of India and benefit the people involved there in dairying.

Figure 6. AGMARKNET portal linking to FAO

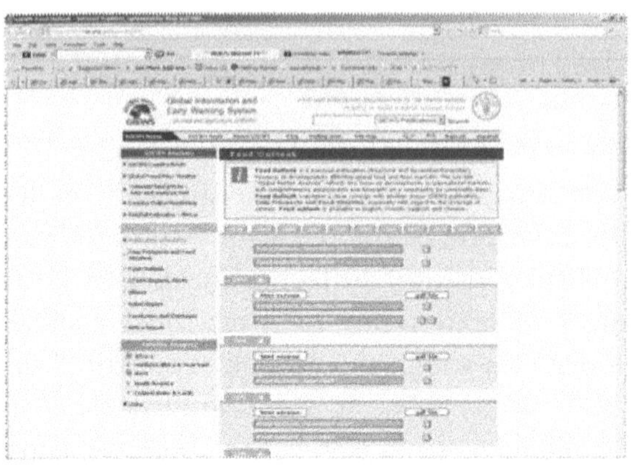

Kaira District Co-operative Milk Producers' Union Limited was the First Co-operative Milk Producers' Union started in Gujarat, formed in Kheda district, whose founding chairman was social worker, Tribhovandas Patel. Its success, also known as Anand Pattern led to creation of District Co-operative Milk Producers' Union at each district, created Amul (Anand Milk Union Limited) in 1946

The Amul Co-operative in Anand, Gujarat, has today become a role model and benchmark for any dairy development project that is launched in India. This model showed that an integrated approach along cooperative lines could enhance production, procurement, processing and marketing of milk.

NEED OF ICT INTERVENTION

The conventional system of milk collection though was functional and working well, but possibilities existed in few areas where efficiencies could be added with development in technology. These can be mentioned as below:

1. The technology was now available in market to reduce quality variation among sellers, which was really a long and testing problem in efforts of NDDB.

Figure 7. Page linking quarterly journals

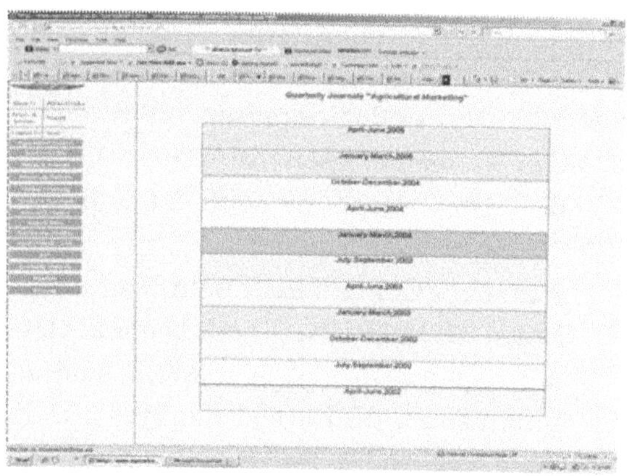

2. The technology had given birth to new weighing machines that could measure the quantity with great accuracy as compared to manual system.
3. Numerous accounts management and data processing software were there in the market to make the recording of transaction a easy task while simultaneously adding efficiency to it. Now it was in possible limits to pay the farmer the value of his contribution on same day as contrasted against existing 10 days interval.
4. Since the payments were to be made based on the content of fat in milk, the process had to be completed before making payments.
5. The compromise done with hygiene by putting the milk in plastic bucket for testing was possible to be done away with and technologies were making entry to make new tools available for replacing the complex Gerber Method of fat testing.

Solution on all above fronts was being provided by a new Entrepreneurial Venture M/s Shree Kamdhenu Electronics Private Ltd. (SKEPL). SKEPL was founded in 1996 with less than US$11,000 to develop IT-based tools that could increase the efficiency and productivity of the Indian dairy industry at a grassroots level. Under the brand name of AKASHGANGA the SKEPL provides integrated solutions that automate the milk collection process at local dairy cooperatives. The AKASHGANGA system not only minimizes handling and increases efficiency, but also increases transparency, and creates a basis for improving the quality of the milk produced.

Business Model

SKEPL's business model is centered on providing technology-based products and services to help milk cooperatives become more efficient and productive. The company provides complete IT-enabled solutions that automate the milk collection process at local milk cooperatives. Its high-end system, selling for about US$3,300 (Rs 151,800), incorporates an electronic weighing system, a milk analyzer to test milk quality, a personal computer, and accounting and management software. Compared to earlier manual procedures, the AKASHGANGA system is faster, more accurate, and more transparent. That means milk can be sent on to the cooperative union for processing more quickly, reducing spoilage; farmers can see for themselves the weight and quality of their milk via a display and printed receipt, increasing their trust in the cooperative process. In addition, farmers are paid immediately, rather than sometimes days later as under manual procedures; and local cooperatives need fewer employees and have better records and reports for planning purposes. SKEPL places an emphasis on delivering quality products and services as well as responsive and efficient after-sales service.

In just a few short years of operation, the AKASHGANGA brand has become quite popular in the Indian dairy industry, especially in the states of Gujarat and Maharashtra, where the bulk of the company's 600 installed systems are located. The company and its founders have received wide recognition for their efforts. Moreover, the company has been consistently profitable, and has recently raised additional investment to enable it to expand more rapidly. With only a small proportion of India's 96,000 local milk cooperatives using automated collection systems today, SKEPL's vast potential market is scarcely tapped. The company has also recently formed a strategic alliance with ICICI Infotech Ltd, a large software consulting firm linked to ICICI Bank Limited, the second largest bank in India in terms of asset size. ICICI Infotech is in the process of developing an integrated supply chain management software system that will seamlessly connect milk societies, milk unions, and milk federations on a single technological platform, and has chosen SKEPL as its development partner. ICICI Infotech will integrate its system with AKASHGANGA's solution at the local society level.

With the success of its basic model, SKEPL plans to incorporate Internet-based information and communication technology (ICT) into its products to facilitate online information exchange between local cooperatives and milk unions. In addition to featuring access to AKASHGANGA's dairy portal, the upgraded system will also enable farmers to exchange e-mail and obtain information in local languages about market prices of milk and dairy inputs, as well as general access to information about hospitals, government offices, educational institutions, and market prices.

Automating milk collection has resulted in accrual of a number of benefits to farmers and local dairy cooperatives by the way of adding towards increasing efficiency, transparency and fairness, and speed of payment. Moreover, it enables faster processing of perishable milk, preventing spoilage, and provides the mechanism for capturing the historical information base that farmers and local cooperatives need in order to plan more effectively and make improvements in quality and yield (See Figure 8).

Figure 8. PC based automatic milk collection system

Such technological Interventions will play critical role in sustaining the competitiveness of the Indian dairy industry and possibly the survival of its cooperative structure as it faces a growing threat from more efficient foreign producers. Transformation of IT-enabled automatic milk collection systems into networks that provide Internet-based information and communication services may, in the future, also help farmers improve their productivity and gain better access to government and commercial services.

Strengthening Farmers' Linkages to Supply Chain

The comprehension may turn easy and involving both, if in place of putting too many abstract numbers and serious text, some other attempt is made. The author feels that if any new initiative is presented in form a story, the same may prove interesting to both, reader as well as learner.

The Story of ITC Ltd'S E-Choupal Initiative (http://www.echoupal.com & http://www.soyachoupal.com)

Sometimes there are issues that seem very dissimilar when seen in isolation. It does not require a magician but a simple individual who could think of new possibilities to answer old problems that created miracle.

The issue being discussed here is stunted growth that was being witnessed by both, the farmers on one hand and a corporate house on the other. In reality these happen to possess very dissimilar attributes and thus normally not expected to share the same platform even.

Who could have thought of in even wildest imagination that same tool and model will reignite the growth process and the affected will chart a new growth story (See Figure 9).

Figure 9. Home page of E-Choupal

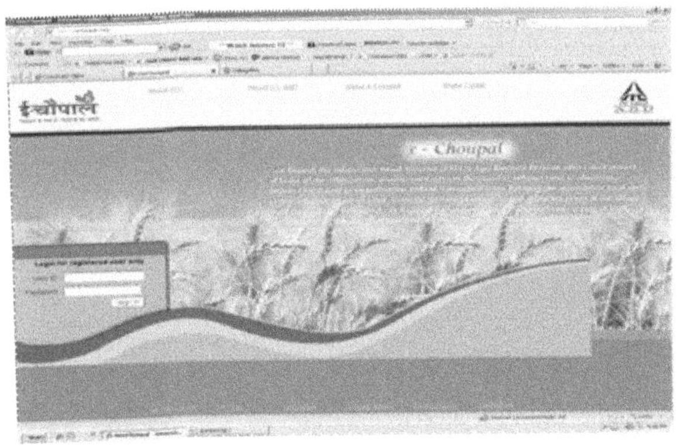

The Farmers' or Soybean Cultivators' Story

On one side of the story were the Soybean cultivators from Madhya Pradesh; a state geographically located centrally in Union of India. That particular area, i.e., rural Madhya Pradesh is popularly also called as 'Soybowl' of India as it contributes 4 million tons in total 5 million tons of Soybean produced in India.

The Farmers here too, very much like their counterparts in other provinces in India, were practicing the same methods of farming and marketing as were used by their elders. This probably was the basic reason behind carrying the legacy of poverty and exploitation faced by their elders in their generation too.

The Issue of Information Asymmetry

To have a real feel of information asymmetry, a thorough understanding of agriculture produce marketing system in prevalent in India is prerequisite. This phenomenon leads to market inefficiencies and market failures, which manifest as sub-optimal outcome for farmers in their both roles, i.e., as customer of agricultural inputs as well as sellers of agricultural produce or output.

Prevalent Agricultural Produce Marketing Setup

The marketing process starts with farmers having harvested their produce. The produce is then uploaded in bullock – carts or tractors depending upon the volume of produce and farmers capability to hire one. The produce is then taken to nearest mandi (market yard) which is normally located within 20 to 30 kilometers. If it happens to be season, farmers on an average may have to wait for approximately two to three days to even get entry into mandi. The auction process starts after the produce has been valued by a government appointed bidder and the initial bid has been placed. Now it is turn of buyers licensed by government popularly known as commission agents to place bid upwards to the level until the produce is sold. Probably only character found inactive during whole process of sales that takes place at mandi happens to be producer of the produce being auctioned or sold. The real suffering in greater part was a resultant of time taken in selling the produce in mandi as they depend a lot, on the timely receipt of the value of produce sold, for their subsistence. This process is culminated by the farmer taking his produce to the shop of commission agent and gets his produce weighed. Normally it is expected that commission agent will pay the price there and then itself but as per settled practices commission agents ask farmer to visit in a few days again and get the money due to them (See Figure 10).

The farmers, more often than not, are not in a situation to refuse the prices offered by commission agent. This could be attributed to four important factors. The first being the time consumed in making it to auction, which will have to be reinvested if the price is refused; second being the transportation cost incurred in bringing the produce to mandi, that will have to be taken care

Figure 10. Pictorial depiction of traditional agricultural marketing system

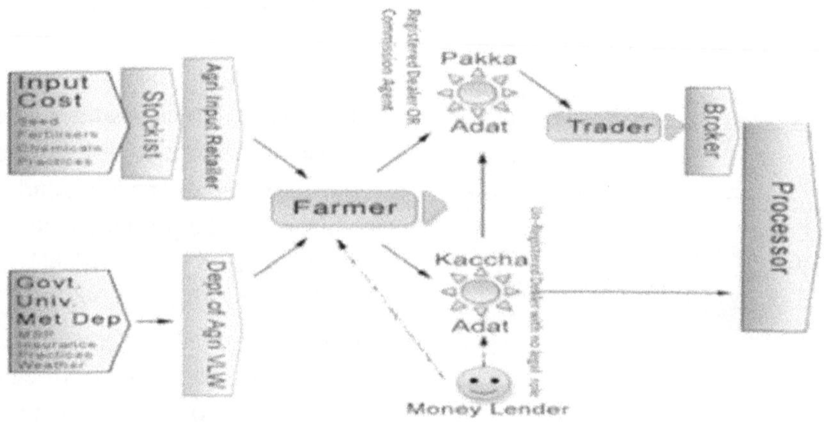

of again; third is lack of storage capacity with the farmer and this storage capacity itself has two inbuilt issues, one being non-availability of physical place for safe storage like godown or warehouse etc., and other being the capacity in financial sense, i.e., taking care of living as well as input cost for next crop without realizing the value of produce produced in previous season; and finally fourth happens to be lack of any information, as to when the prevailing prices in mandi will benefit the farmer.

Farmers normally cultivated and sold their produce almost in isolated manner. The situation was further compounded by information asymmetry created due to absence of telecommunication tools. This phenomenon resulted in price discovery at the end of whole process, i.e, at end of growing and selling cycle.

The International Business Division of ITC's Story

The ITC Ltd. is a well settled and known corporate house in India having very diversified interests ranging from lifestyle products to hospitality business to export of agricultural produce and their derivatives to tobacco business and many more areas. ITC Ltd.'s International Business Division (IBD) happens to be its exporting arm for agricultural commodities. The two third of total agricultural commodities exported by IBD comprised Soybean and its derivatives (See Figure 11).

The financial results for the financial year 1997-98 stated that IBD had grossed Rs 450 crore. This value indicated that IBD was failing in making its impact felt in total grossed by other divisions of ITC put together. ITC Ltd in same financial as whole grossed Rs 7701 Crore, for which Rs 450 Crore were nothing but a marginal addition. This also conveyed that IBD was not witnessing growth and its contribution to total turnover was even less than 6 per cent.

S. Sivkumar who was heading IBD arm of ITC took it as a major challenge. He himself set a target for himself and that was beyond a normal individual's imagination. He had set a target to gross Rs 2000 crore by 2005 from Rs 450 Crore in 1998. This was followed by a brainstorming session of ITC management team in Palancheru near Hyderabad. The team was well aware that to achieve this target they need to kill inefficiencies and reduce on costs incurred along the current supply chain. This clearly meant that current supply

Figure 11. Home page of ITC Ltd's international business division

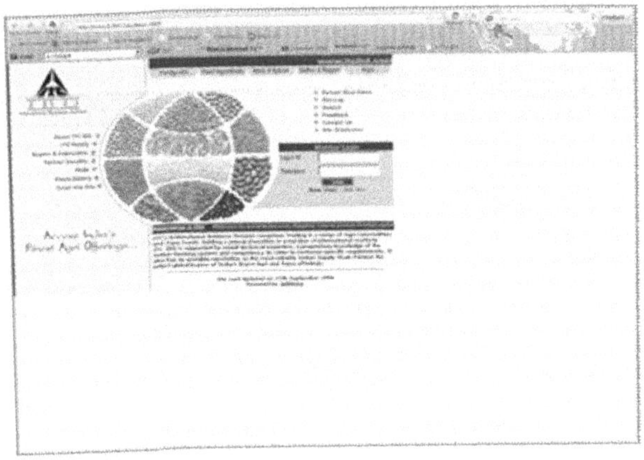

chain needs to broken and its time for inventing a new supply chain.

The recent developments made by information technology came handy in reworking the supply chain and this process succeeded in adding "e" to the supply chain. S. Sivakumar and team worked on issue of eliminating the information asymmetry to the extent possible. Now they were geared to reach farmers with this "e" in supply chain. They planned to empower the Soybean farmer with this "e" in supply chain so as to enable them in making informed decision through whole Soybean cycle, i.e., from sowing to harvesting to selling the produce. They wished to help farmers in price discovery not at the end of process but before they planned to visit the mandi.

Choupal

There could never have been a better way than to visit the village choupal to share this new development that would work to the benefit of the farmer. They were creating a win-win situation. Comprehending the discussion further requires an understanding of what is meant or conveyed by the word "Choupal".

Choupal is a place where normally villagers gather while coming back from their farms or work in the evening, while on the way to home. It often happens to be outer part of the house which is located at a crossroad often used for going and coming back from work. Often house located at such place belongs to Sarpanch/ Pradhan/ Mukhiya of village (various terms used to denote the elected village head) or opinion leader of the village, but is not a necessary pre-condition. The space at Choupal is good enough to handle gathering of score of villagers.

Choupal is a great place of learning for villagers as at Choupal they share a knowledge and information on a wide variety of subjects and issues. These range from weather to launch of new hybrid seed, to shortage of fertilizers, to misdeeds some rogue element in village to where to get the wards admitted for good schooling to who should be elected as new mukhiya/ pradhaan/ sarpanch of the village. Thus there could be no better place than Choupal for sharing any new intervention that could benefit farmers.

E-Choupal Model

The Physical Setup of Model

Two basic components out of which one being the installation of required hardware, software and up linking facility at the place of farmer who hosts the setup, also known as 'Sanchalak' and other being the creation of a physical network of Processing Centers, Collection Centers etc., in a manner that as per location they should be, if not nearer to the mandi, then also at least equidistant, at the other end layout wise these should offer better facilities when contrasted with mandi while giving the farmer a feeling of being at home (See Figure 12).

The Handing Over and Installation of E – Choupal Kit

This model involved an initial investment on the part of ITC Ltd., as computer kits with the acces-

Figure 12. E-Choupal setup VSAT mode

sories mentioned hereunder were to be supplied to villages.

1. One desktop computer with Operating System (MS-Windows OS was used in this Case)
2. Power supply comprising UPS and solar powered battery backup
3. A Dot-Matrix printer
4. Internet connectivity through either telephone lines (if present) or by VSAT.

Total setup cost incurred by ITC per e-Choupal has somehow not been consistent in the cases authored by various authors. Where as in one case study it is quoted to be around Rs. 170000.00 (US$3762) and another Rs. 100000.00 (US$2213) people, travel, communication, software and training (Upton and Fuller, 2004), another quotes it as Rs. 40000 to Rs. 60000 in terms of setup cost (Sukhpal, 2004), the other one puts this figure at US$3000 to US$ 6000 as one time set up and US$100 as maintenance cost per year (Anupindi, R, 2004) (See Table 1).

Table 1. Hardware installation cost of Soya-choupal

Cost of Choupal without VSAT	***Rupees***
Computer with Multimedia	42,000
Dot Matrix Printer	07,000
UPS with Battery	08,000
Solar Charger Panels	09,600
Earthing	04,500
Painting the Choupal Wall	01,000
Insurrance & Warranty	03,500
Plaque, Mousepad, Wiring & Misc	02,000
Keyboard	01,500
Total	***79,100***
Choupal Cost with VSAT Installation	***Rupees***
Basic Cost of Choupal	79,100
VSAT	90,000
UPS with Battery*	08,000
Solar Charger Panels*	09,600
Total	***186,700***

*When installing VSAT at any location, as additional UPS with Battery and Solar Charger Panel are necessary.

Source: (Kumar, R., 2004)

The Location and Layout of Collection and Processing Centers

The ITC Ltd. proceeded to create this setup which for convenience of understanding may termed as Hub and Spoke Model. While this e-Choupal initiative was conceived ITC Ltd, already had forty four contact points in Madhya Pradesh, where a farmer could bring his produce or to be specific Soybean. These contact points were in form of its five processing units and thirty nine warehouses. This setup was pitted against five large soybean mandies present in Madhya Pradesh regulated by government in place. These collection points of ITC required farmers to travel a distance from 5 Kilometers to 20 Kilometers, depending upon their location.

ITC already had 1695 e-Choupals in MP which provided access to 8400 farmers comprising 80 per cent of the total Soybean farmers in the state. At the time of conception it was planned at corporate office to have a physical architecture in a manner, so that any farmer, comprising target farmer, should have an e-Choupal web kiosk within walking distance of five kilometers and a hub within driving distance of not more than 30 kilometers. To meet this criterion ITC added three more processing hubs followed by setting up of thirty six warehouses, once the project got launched in full swing. This ensured that distance within Farm to Hubs or Farm to Mandi almost turned equidistant.

The Business Model Based on Physical Setup

Thus the term "e-Choupal" got coined as consequence of connecting the village choupal with help of advances made by electronic and telecommunication technologies. The efforts further are to detail the e-Choupal model. In this model technology was at the crux and it transformed the fate of soybean cultivators of soybowl of Madhya Pradesh in India. The ITC Ltd has also launched new projects, which in pilot phase, in three more states of the country. It is working for wheat farmers in UP, coffee cultivators in Karnataka and for sea food in Andhra Pradesh. These new projects are in pilot phase.

The intentions here are not to go in details of these three but explain how the model works. The model drives its strength by impacting on two aspects of the agriculture produce marketing and information supply chain. They can be listed as under:

1. *Eliminating the information asymmetry* to the extent possible by helping in real time price discovery; and
2. *Minimizing the transaction cost* thereby increasing the farmers' share in the price paid by end consumer.

This set up is established for use by the farmers, free of charge. The farmer who hosts the setup is called a sanchalak and he has to incurs some operating costs. Sanchalaks have expressed normally three motivations for assuming their role, first being will to help their community, second a profitable business proposition for themselves and third and final a means to get access to a functional computer. The Sanchalak is obligated by a public oath to serve the entire community and thus benefits from increased prestige. The farmers can use the computer to access daily closing prices on local mandis, and also the global price trends or find information about new farming techniques—either directly or, since a significant number of farmers are illiterate, they seek the help of sanchalak (See Figure 13).

E-Choupal also serves as a marketing channel, i.e., for selling of inputs or other items being produced by ITC Ltd. Thus it turns in a tool for penetrating to serve the demand existing in rural markets. The villagers can put their orders for seeds fertilizers and other products such as consumer goods from ITC or its partners via Sanchalak (e-Choupal), at prices lower than those available from village traders. The sanchalak typically aggregates the village demand for these products and transmits the order to an ITC representative.

At harvest time, ITC offers to buy the crop directly from any farmer at the previous day's closing price; the farmer then transports his crop to an ITC processing center, where the crop is weighed electronically and assessed for quality. The farmer is then paid for the crop and the transport fee too is reimbursed, which is not the case in traditional agriculture produce marketing system. "Bonus points," which are exchangeable for products that ITC sells, are given for crops with quality above the norm. In this way, the e-Choupal system bypasses the government-mandated trading mandis. Farmers benefit from more accurate weighing, faster processing time, and prompt payment, and from access to a wide range of information, including accurate market price knowledge, and market trends, which help them decide when, where, and at what price to sell. Farmers selling directly to ITC through an e-Choupal typically receive a higher price for their crops than they would receive through the mandi system, on average about 2.5% higher (Anupindi, R, 2004).

The total benefit to farmers includes lower prices for inputs and other goods, higher yields,

Figure 13. Pictorial depiction of e-Choupal model

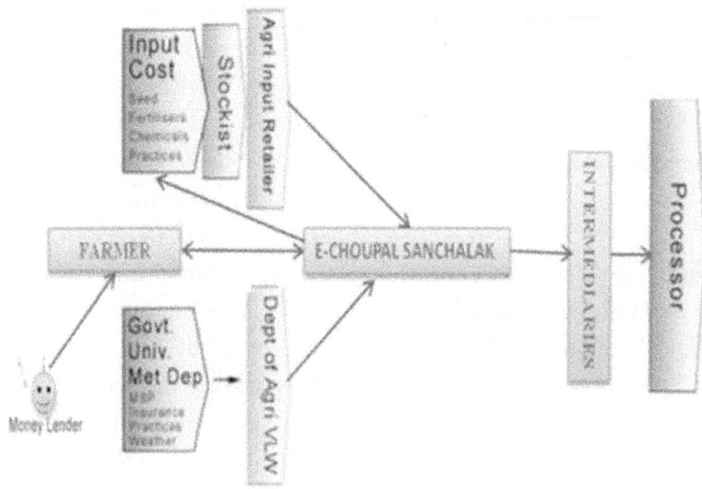

and a sense of empowerment. The e-Choupal system has had a measurable impact on what farmers chose to do: in areas covered by e-Choupals, the percentage of farmers planting soy has increased dramatically, from 50 to 90% in some regions, while the volume of soy marketed through mandis has dropped as much as half. At the same time, ITC benefits from net procurement costs that are about 2.5% lower (it saves the commission fee and part of the transport costs it would otherwise pay to traders who serve as its buying agents at the mandi) and it has more direct control over the quality of what it buys. The system also provides direct access to the farmer and to information about conditions on the ground, improving planning and building relationships that increase its security of supply. The company reports that it recovers its equipment costs from an e-Choupal in the first year of operation and that the venture as a whole is profitable.

E-Choupal: As Marketing Channel for Company's Products & Support Services

The investment initially made by ITC in providing and installing the e-Choupal kiosks, from the very start has lead to anticipation of future stream of profits, i.e., significant return over the investment made. A number of synergies got realized once the channel to procure Soybean, directly from the farmers got in place. Where on one side e-Choupal was seen as connecting farmers to ITC, on the other side the farmers were getting automatically connected and exposed to another world that was networked with internet connecting them. As stated earlier that 66 per cent of Indian workforce was rural and depended on agriculture for livelihood, this in terms of size of market turns out to be a bigger market than few smaller nations put together.

In mid-2003, e-Choupal services reached more than 1 million farmers in nearly 11,000 villages, and the system is expanding rapidly. ITC gains additional benefits from using this network as a distribution channel for its products (and those of its partners) and a source of innovation for new products, while simultaneously permitting channel partners also sell their products through this network. Considering at moment only about amount of money invested by farmers in purchase of farm inputs, figures reveal that fertilizer market, when market for Urea, Diammonium Phosphate and Muriate of Potash are put together, alone has a turnover of Rs 26,000 Crores (US$ 5.7 Billion). The next is market for agrichemicals at Rs 3,500 Crores (US$ 774.5 Million), which is closely followed by market for Seeds at Rs 3,000 Crores (US$ 663.9 Million). Cumulatively inputs alone result in market having turnover of Rs. 32,500 Crores (US$ 7.2 Billion).

Initially when ITC started with e-Choupal network, it had only four organizations partnering with it. Today the story is very much different. Now there are 160 companies riding this network to make their presence felt in the rural area and these include likes of Maruti Suzuki, Colgate Palmolive, Eveready, Syngenta etc. Around 50 per cent of the e-Choupal business comes from ITC and its related divisions while the rest is contributed by other channel partners. ITC is also exploring partnering with banks to offer farmers access to credit, insurance, and other services that are not currently offered or are prohibitively expensive.

Present Status and Expected Future

The launch of e-Choupal initiative began in year 2000 with six installations, and the number of e-Choupals touched 6,500 mark in 2007. This was a period when ITC was making six new e-Choupals functional per day. The target that ITC had setup for itself was to have 20,000 e-Choupals, covering 1,00,000 Villages by the Year 2010.

This was the Year from when the E-Choupal initiative has been put on hold. Virginia House, Kolkata, ITC's Corporate Headquarter had decided to freeze the expansion plan. This happened as a consequence of government action, which was made by government for welfare of citizens of the nation. The government had to restrict the export of agricultural produce to curb the inflationary pressure. The company at such a juncture found it futile to expand to kiosks when it was barred from buying the commodities in large quantities.The company after its decision to stop expansion of e-Choupal Kiosks has decided to reach the farmers of 34000 odd villages within the ten states, which have not been directly linked to e-Choupals.

The ITC has now launched M-Choupal, where mobile phones are to play a crucial role. The focus of how to reach is shifting from engaging them through kiosks to reach them where they are with help of mobile technology. This has also saved ITC from the money that it had to invest on setting up of e-Choupal Kiosks.

ITC has of now tied up with the job portal Monster India and has launched its job portal through e-Choupal Model involving sanchalaks. This portal has the domain name as rojgarduniya.com. This initiative got launched initially as a piolet project in few villages of Uttar- Pradesh. The same has now been extended to 4000 villages in Uttar – Pradesh, Madhya Pradesh, Maharashtra and Rajasthan.

INDIAGRILINE: STORY OF EID PARRY (http://www.indiagriline.com)

Indiagriline is another effort by a corporate to reach the farmers for disseminating the relevant and timely information to them. These portals in physical form are known as "Parry's Corners". These kiosks were called Parry's Corners, named after a famous landmark in Chennai. Thomas Parry, a young Welsh trader who came to Madras (now renamed Chennai) in 1788, founded Parry & Company. Now called EID Parry India Ltd., the company is well diversified with interests in fertilizers, pesticides, sugar and ceramics. The fertilizer and sugar businesses comprise 70 per cent of turnover (See Figure 14).

Thomas Parry was the force behind the company until his death in 1824. The company started its sugar operations in 1803 with a cane sugar plantation near Chidambaram. The first sugar distillery at Nellikuppam in Tamilnadu came up in 1845. In 1906, it set up the Presidency Manure Works to manufacture super phosphates in Ranipet. It got into ceramics when it began making acid

Figure 14. Home page – Indiagriline

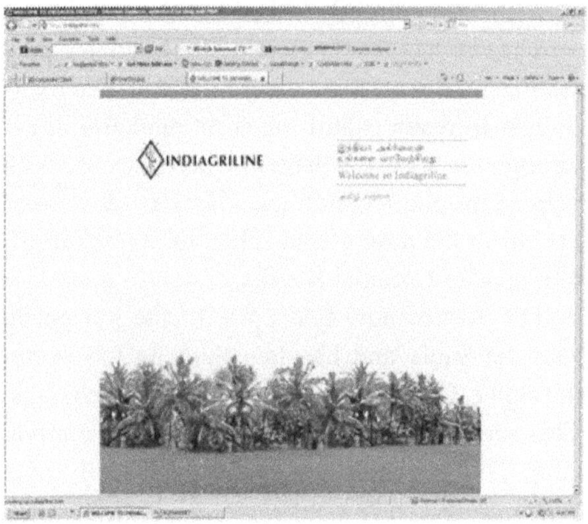

Figure 15. First page when one enters from home page

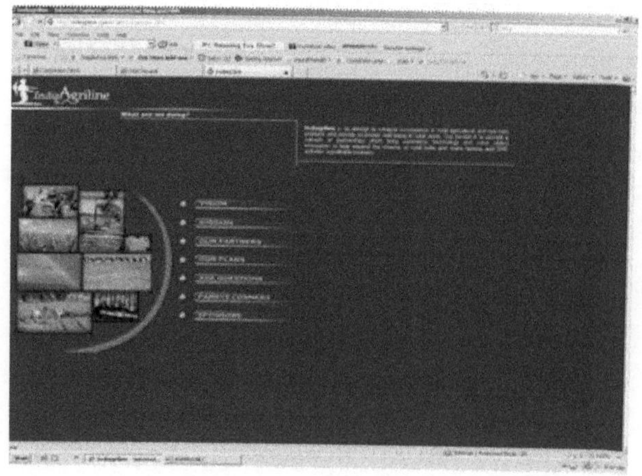

jars in 1908 for the sulphuric acid produced in the super phosphate factory. In 1975, the company was registered under the Companies Act as EID Parry (India) Ltd. From the late sixties onwards, the company started showing signs of decline (See Figure 15).

In the Year 1981 – 1982 A private corporation owned by the Murugappa Group, having interests in farm inputs, sugar, confectionary, building materials, abrasives, bicycles, plantations and finance, took over the EID Parry. At this point of time the situation of EID parry, and specifically at its Nellikuppam factory, was in dire state. The takeover by Murugappa group improved the situation by lifting up both morale and productivity of the employees. The group also installed modern weighbridges and digital displays and made the weighing process unambiguous.

IT in Corporate Business: A Challenge in Competition

Recognizing the significance of information and communication technology (ICT) as a powerful tool for bridging the infrastructure gaps in rural India, EID Parry regarded the Internet as the next logical medium for delivering its farm extension services. Indiagriline is an effort to provide an end-to-end solution addressing the needs of the farming community in South India. EID Parry launched this project in early 2001 by setting up Internet kiosks in 16 villages around its sugar factory in Nellikuppam. These kiosks were called Parry's Corners, named after a famous landmark in Chennai and were intended to be business hubs of their respective villages-a one-stop shop that acted as a storefront for buying farm inputs, a market for selling goods, and an Internet café for communication and information services.

Indiagriline: The Vision

Strategic goal of the Murugappa Group was to ultimately develop the following capabilities:

- **Distribution Infrastructure:** This infrastructure would be capable of supporting bidirectional distribution of products and services into and out of rural India. Therefore, developing a low-cost channel for rural distribution was a key goal.
- **Trading Infrastructure:** This would serve as the foundation to a platform for trading agricultural commodities and rural industry manufactured goods.

EID Parry's Kiosks' Business Model

Towards achieving these ends, EID Parry forged and facilitated partnerships among a wide range of organizations, including Tamil Nadu Agriculture University (TNAU) and its research stations, Tamil Nadu University for Veterinary and Animal Sciences (TANUVAS), National Horticulture Board, AMM Foundation, and Murugappa Chettiar Research Center, to create the agro-portal Indiagriline. www.indiagriline.com was developed by using in-house expertise.

Stakeholders in EID Parry

The stakeholders in EID parry system can be classified in three major categories, first being rural communities and households, second rural service providers and third policy makers and their advisors.

Rural communities and households comprised farming and non farming community. The farming community had its stake due to having access to Cane farms using the services of the Parry factory, where as farmers not in sugarcane used the Parry's Corner for learning about crops other than cane. Non farming community, specifically students found it useful for accessing academic information and also training etc., whereas others found it as a convenient and cost effective communication tool.

Rural service providers having stake included the field staff employed by EID Parry to help and guide the farmers. Content of the website was also managed by company officials; franchisees running the Parry's Corners; banking sector participants disbursing the loans from the Parry's Corners and other stakeholders like subject matter experts from TNAU and many more (See Figure 16).

The project by itself does not take policy makers in its ambit, but there has been found to be considerable interest and encouragement from local administration on the work being executed by Parry.

Figure 16. Stakeholders in Parry's Corner

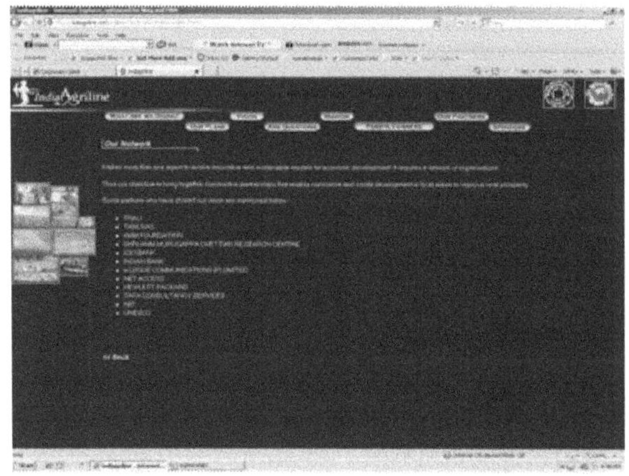

Franchisee Based Business Model

Muruguppa was also having the intentions to help farming community, especially sugarcane farmers, by developing the spirit of entrepreneurship in them.

So very much unlike the e-Choupal concept, here the Inadigriline portal was being operated in Kiosks, owned and operated by franchisee trained to operate the system. In this particular case EID Parry allows the franchisees to use its brand, procure commodities on its behalf, and sell its products or services and by adapting this model the company has been able to keep its fixed costs low. Although EID Parry covers the cost of establishing the infrastructure for voice and data connectivity, each franchisee invests approximately Rs. 50,000 to cover the cost of the computer and all related equipment (Prahlad. C. K., 2004). The operating costs of running the kiosks, such as electricity and connectivity charges, are covered by the franchisees. For EID Parry, franchising is a cost-effective way to build a chain of many kiosks all operating under their banner. This cooperative relationship builds a win–win situation for both parties involved. It provides the incentives each party needs to contribute to the other's potential success.

The Scale and Physical Spread of EID Parry Kiosks

EID Parry has restricted the indiagriline.com project to the areas in which it conducts its sugarcane operations. The project Covers 275 villages around the Nellikuppam factory with 40 EID parry's Corner Kiosks franchised in the area. One kiosk on an average serves 2000 to 3000 farmers who come from 5 to 6 villages. Besides Nellikuppam plant the project is being extended to Parry's other sugar factories. Few have already been setup around the Pudukottal factory and sites where further installations have been planned are undergoing technical evaluation.

Benefits to Farmers

The Farmers use Parry's corners facility for different purposes like farming methods, technical information, computer education of children, entertainment, and international news. However, the primary focus remains to cut down on the farmers transaction time and cost of doing business with parry. A sugarcane management system happens to be the one of the most used features in these centers. Farmers using the system are being considerably benefited by getting information on Sugarcane, Paddy and Banana cropping pattern, besides having access to updated farming information, speedy help from farm experts to tackle diseases and pests in farming, and easy transaction on sugar related issues.

Possible Sources of Payback to the Franchisee

At an individual kiosk level, the capital invested and the risk borne by the franchisee in setting up the Parry's Corner kiosk is greater. It was therefore essential for EID Parry to identify ways to make this a profitable venture for the franchisee. Sources of revenue fall into the following categories. Although some franchisees do not charge the farmers anything for some of these services to promote use, they are listed here because these services could be charged for in the future.

The sources of revenues projected by EID Parry happen to be many and these look very promising too. On the other hand it seems that these promises will take many years down the line in really providing significant earning. At present the total earnings of kiosks operators range between Rs. 800/- to Rs.2000/- per month. (Singh, A. 2004).

The revenue generating activities as identified for EID Parry kiosks are mentioned hereunder (Prahlad, C.K., 2004; Bowender, B. & Yadav, Y. 2003; Singh A, 2004).

Procurement

Relaxation of restrictions on buying, selling, and holding of certain commodities including varieties of rice by the Indian government has opened up a whole new opportunity for EID Parry. EID Parry expects to leverage the following strengths:

- Farmer relationships. EID Parry has a strong relationship with the farming community in the Nellikuppam area. Paddy is one of the other crops grown in that region. EID Parry saw an opportunity to leverage this strong relationship with the community it had built over the years.
- Agricultural know-how. EID Parry, before entertaining any plans of entering the rice market, had accumulated significant know-how in paddy farming by virtue of its farm extension services in the region. It saw an opportunity to increase paddy productivity by improving farming techniques and by providing the farmer with certified seeds and quality inputs. Closer inspection of the rice value chain revealed that EID Parry could eliminate inefficiencies and reduce costs. By bringing in transparency to the

process, through unambiguous pricing, prompt payment, and accurate weighing, it knew it would be able to provide the farmer with an alternative.

- Synergies. EID Parry also expected to derive synergies from its sugar business by leveraging the brand, packaging facilities, stocking points, and distribution channels. In the pilot phase, certified ponni variety paddy seeds were sold to 500 farmers who reported a 20 percent increase in their yield. In their very first year, it procured 27 metric tons of ponni from these farmers. The franchisee gets a commission for his services in the procurement process.

Marketing Products and Services

The same network that brings produce from the village can be used to transfer goods to the villages. EID intends to leverage this network to sell products to rural India. The franchisees help identify products and aggregate demand. By disseminating knowledge on relevant agriculture technologies, EID Parry creates a demand for agricultural inputs and services. Access to state-of the- art agronomy coupled with quality farm inputs, such as high-yielding seed varieties and fertilizers, can improve the farm productivity. The franchisee gets a commission for the products sold through this channel. The franchisee in Maligaimedu were also selling sugar in retail to the villagers, in addition to fertilizers and pesticides.

Desktop and Publishing Services

Desktop and publishing services are a key source of revenue for the franchisee. The franchisee is given a free hand in deciding what services can be provided and at what cost. At present, EID Parry does not partake in the revenues generated for such services. As part of project e-inclusion, EID Parry, in partnership with Hewlett-Packard, has deployed all-in-one print/scan/fax/copy devices and digital cameras. These devices enable the franchisee to provide a range of services for which villagers would otherwise have to travel great distances.

Communication Services

A Parry's Corner kiosk doubles as the local Internet café. Franchisees usually charge customers an access fee for surfing the Web. Information accessed at these kiosks relates to current affairs, education, health, entertainment, and weather. The telecommunications infrastructure deployed in these villages uses a wireless local loop (WLL) technology called corDECT, which provides sufficient bandwidth (70 kbps) to allow for simultaneous voice and Internet connectivity. In addition to a data line for Internet access, the kiosks get a telephone line as well.

Education Services

EID Parry has used its Parry's Corners to launch adult literacy and computer education programs. In partnership with NIIT Ltd, a pioneer in the field of IT education, EID Parry organized computer education programs. These programs were tailored to teach the basics of computer operation and applications to people who had never before seen a computer. In partnership with Tata Consultancy Services, it also conducted an adult literacy program to educate illiterate women. Besides the initiatives organized by EID Parry, franchisees were free to conduct programs tailored to local needs. One franchisee was also conducting computer classes for school children between Grades 1 and 5.

Farm Extension Services

Agricultural know-how and expertise exists within India, but the lack of infrastructure has impeded their dissemination. The farming community is not only fragmented, but also scattered throughout the country. Providing farm extension services is central to EID Parry's business model. The extension services provided in the Cuddalore district have focused on paddy, banana, groundnut, tapioca, and cashew. For the cane farmers in the region, EID Parry provides the following farm advisory services:

- Expert visits and crop seminars
- Soil sampling and analysis Land preparation
- Arrangements for labor and machinery
- Nutrient management
- Irrigation mechanisms
- Crop diagnostics
- Advice on farm inputs
- Harvesting techniques

Through effective use of technology, farm extension services now are available from village kiosks. Farmers can gather information directly from the kiosk or communicate with an agronomist to get specific, customized advice via e-mail. The typical turnaround time is a day. Services such as crop diagnostics actually can be performed remotely. The franchisee can use the digital camera to take a picture of the crop to be inspected and e-mail the image to the agronomist. The agronomist then will be able to follow up with his diagnosis. All this can be done without the farmers leaving the village.

Business Processes at Parry's Corners

Cane Management System (CMS)

CMS is EID Parry's enterprise resource planning (ERP) system that helps manage sugarcane procurement. About 5,000 tons of sugarcane are crushed daily at the Nellikuppam factory. Forty divisional officers manage the procurement process. They work with the sugarcane farmers on an ongoing basis and provide comprehensive assistance with all aspects of farming throughout the lifecycle of the crop.

CMS enables registered sugarcane farmers to access and maintain their transaction records with the company. This application also is designed to track the progress of crops from sowing to harvesting; every aspect of cane farming is managed here. Conventionally the farmers would have to go up to the divisional office to access CMS. This system now has been Web-enabled. Instead of going to one of the factory's eight divisional offices, farmers now can go to a kiosk in their own village to leverage CMS services.

EID Parry's future plans include eliminating their physical divisional offices and carrying out the cane management processes from its Parry's Corners.

CONCLUSION

The four stories about intervention done by taking the advantage of progress made by technology have proved their worth. Thus information technology has the potential to transform the agricultural landscape anywhere in the world, if it's proving as a life saving tool for small, marginal and fragmented farmers in India. What is required on part of Government and Corporate as well is that they should promote testing of new technology in the field while promoting all possible innovations to derive synergy and milestone solutions.

REFERENCES

Agrawal, G. K. (2008). Village knowledge centers/common service centers, information technology in developing countries. *Newsletter, 18*(1). Retrieved on February 9, 2011, from http://www.iimahd.ernet.in/egov/ifip/feb2008/gk-agrawal.htm

Ahmed, R. (Ed.). (2009). *Globalisation and economic development –Part – 1*. New Delhi, India: Mittal Publications.

Annamalai, K., & Rao, S. (2003). *What works: ITC's E – CHOUPAL and profitable rural transformation web – based information and procurement tool for Indian farmers*. Retrieved February 6, 2011, from http://www.bdsknowledge.org/dyn/bds/docs/292/ITC e-choupalIndia.pdf

Anupindi, R. (2004). *Going direct to farmer: ITC's e – Choupal initiative*. Retrieved February 6, 2011, from http://www.mba.tuck.dartmouth.edu/digital/Programs/CorporateEvents/SupplyChainThoughtLeaders/session6.pdf

Arabi, U. (2009). State of Indian agriculture sustainability under changing technology: Emerging policy agenda. In Ahmed, R. (Ed.), *Globalisation and economic development –Part – 1* (pp. 49–71). New Delhi, India: Mittal Publications.

Assisi, C., & Gupta, I. (2003). ITC's rural symphony. *Business World*, *22*(34), 30–37.

Badgaiyan, B. D. (2002). Avoiding crisis in rural areas through information technology. *Information Technology in Developing Countries, 12*(2). Retrieved February 9, 2011, from http://www.iimahd.ernet.in/egov/ifip/aug2002/article4.htm

Badgaiyan, B. D. (2002). On sustainability of a developmental website: Is it viable?*Information Technology in Developing Countries, 12*(3). Retrieved February 9, 2011, from http://www.iimahd.ernet.in/egov/ifip/dec2002/article3.htm

Bahl, M. (2008). *ICT in agricultural marketing*. India Science and Technology Report: 2008. New Delhi, India: NISTADS. Retrieved February 6, 2011, from http://www.nistads.res.in/indiasnt2008/India-S&T-2008-Full.pdf

Balakrishnan, P., Golait, R., & Kumar, P. (2008). *Agriculture growth in India since 1991*. Study number 27, DEAP, Reserve Bank of India, Mumbai. Retrieved February 7, 2011, from http://rbidocs.rbi.org.in/rdocs/content/pdfs/85240.pdf

Bhatnagar, S., Dewan, A., Torres, M. M., & Kanungo, P. (2004). *E–Choupal: ITC's rural networking project*. Retrieved February 6, 2011, from http://www.siteresources.worldbank.org/INTEMPOWERMENT/Resources/14647_E-Choupal_web.pdf

Boehlje, M., Akridge, J., & Downey, D. (1995). Restructuring agribusiness for 21st century. *Agribusiness*, *11*(6), 493–500. doi:10.1002/1520-6297(199511/12)11:6<493::AID-AGR2720110602>3.0.CO;2-G

Bowonder, B., Gupta, V., & Singh, A. (2004). *Developing a rural market e-hub: The case study of e-Choupal experience of ITC*. Retrieved February 6, 2011, from http://www.planningcommission.nic.in/reports/sreports/ser/stdy_ict/4_e-choupal.pdf

Bowonder, B., & Yadav, Y. (2003). *Developing ICT platform for enhancing agricultural productivity: Case study of EID Parry*. Retrieved February 06, 2011, from http://www.anningcommission.nic.in/reports/sreports/ser/stdy_ict/6_eid.pdf

Chakracarty, R. (2000). *IT at milk collection centers in cooperative dairies: The National Dairy Development Board Experience*.

Chand, R. (Ed.). (2004). *India's national agricultural policy: A critique. India's agricultural challenges: Reflections on policy, technology and other issues*. New Delhi, India: Centre for Trade and Development. Retrieved February 9, 2011, from www.iegindia.org/dis_rc_85.pdf

Chawala, R., & Bhatnagar, S. (2004). *Online delivery of land titles to rural farmers in Karnataka, India.* Presented at Scaling up Poverty Reduction: A Global Learning Process and Conference, May 25 – 27, 2004, Shanghai. Retrieved February 6, 2011, from http://www.siteresources.worldbank.org/

Dangi, N., & Singh, H. (2010). e-Choupal: Hope or hype? *American Journal of Economics and Business Administration, 2*(2), 179–184. doi:10.3844/ajebasp.2010.179.184

Dey, S. (2010, June 12). M-Choupal. *Business Outlook.* Retrieved February 16, 2011, from http://business.outlookindia.com/article.aspx?265712

Gupta, R. K. (2003). *E–Choupal so far and ahead: A case study in a village in M.P.* Theme paper, fieldwork segment, PRM 2002 – 2C IRMA, Anand, Gujrat, India, February.

IICD. (2006). *ICTs for agricultural livelihoods: Impact and lessons learned from IICD supported activities.* Retrieved February 9, 2011, from http://www.iicd.org/files/IICD-agri-impact-2006.pdf

Jha, P., & Negre, M. (2007). *Indian economy in the era of contemporary globalization: Some core elements of the balance sheet.* Centre for Economic Studies and Planning, Jawaharlal Nehru University, New Delhi, India. Retrieved February 7, 2011, from http://www.macroscan.org/anl/may07/pdf/IndianEconomy.pdf

Jhunjhunwala, A., Ramachandran, A., & Ramachander, S. (2006). Connecting rural India: Taking a step back for two forward. *Information Technology in Developing Countries, 16*(1). Retrieved on February 9, 2011, from http://www.iimahd.ernet.in/egov/ifip /feb2006/article1.htm

Kumar, R. (2004). E-Choupal: A study on the financial sustainability of village Internet centers in rural Madhya Pradesh. *Information Technology and International Development, 2*(1), 45–73. doi:10.1162/1544752043971161

Maru, A. (2003). *Information and communication technology use in agriculture extension in India.* CTA. Retrieved February 6, 2011, from http://www.cta.int/en/content/download/2811/..../Case_Study_india.pdf

Maru, A. (2004). *ICT enabled information systems for agricultural development in India: Status, issues and future directions.* Presented at IAITA Second National Workshop on ICT in Agriculture and Rural Development, Dhirubhai Ambani Institute for Information and Communications Technology (DA-IICT), Gandhinagar.

Mehrotra, N. (2010). *Emerging patterns in share of small farms in production and credit implications for policy formulation.* Presented at 12th Annual Conference on Money and Finance in the Indian Economy, Indira Gandhi Institute of Development Research, Mumbai, India. Retrieved February 7, 2011, from http://www.igidr.ac.in/~money/Emerging%20Patterns%20in%20Share%20of%20Small%20Farms%20in%20Production%20and%20Credit_%20Implications%20for%20Policy%20Formulation.pdf

Misra, D. C., & Vijayaditya, N. (2006). Development informatics - Reaching the rural India role of NIC. *Information Technology in Developing Countries, 16*(1). Retrieved February 9, 2011, from http://www.iimahd.ernet.in/egov/Feb2006/article2.Html

Mohan, C. P. (2007). *Overview of agriculture and rural sector.* RB channel, Knowledge Bank, College of Agriculture Banking, Reserve Bank of India, Pune, India. Retrieved February 6, 2011, from http://www.cab.org.in/Lists/KnowledgeBank/Attachments/4 /Overview AgriRur Sector_2_17_2007.pdf

Moni, M. (2004). *Impact of economic reforms on Indian agriculture sector: Application of geomantic technology to reduce marginalization and vulnerability of small farmers in India*. Retrieved February 9, 2011, from http://www. gisdevelopment.net/application/agriculture/production/agric0003pf.htm

Narayanan, A., & Chakraverty, G. (2009). Running ICT telecenters in rural India. *Information Technology in Developing Countries, 19*(1). Retrieved on February 9, 2011, from http://www.iimahd.ernet.in/egov/ifip/feb2009/arvindd-narayanan.htm

Okyere, K. A., Davis, K., & Aredo, D. (2008). *Advancing agriculture in developing countries through knowledge and innovation –Synopsis of an international conference*. ISBN: 10-digit: 0-89629-780-2 13-digit: 9788-0-89629-780-7

Paul, P. (2003, January 06). The Mandi is now on the Net. *Deccan Herald*. Retrieved from http://www.itcpotal.com

Prahlad, C. K. (2004). *The EID Parry story. The fortune at the bottom of the pyramid*. Retrieved February 6, 2011, from http://www.eidparry.com/casestudy.asp

Sanganeria, S. (2003, February 13). ITC e-Choupal in wheat buying spree. *Business Standard*, p. 12.

Sharma, A., & Yadav, A. (2003). *What works: AKASHGANGA'S IT tools for Indian Dairy industry –Using IT to increase efficiency in rural dairy co-operatives*. Retrieved February 17, 2011, from http://www.nextbillion.net/archive/files/AKASHGANGA.pdf

Sharma, D. (2007). ICT and rural livelihoods: Whose livelihoods are we talking about? *Information Technology in Developing Countries, 17*(2). Retrieved February 9, 2011, from http://www.iimahd.ernet.in/egov/ifip/aug2007/devinder-sharma.htm

Shelke, R. D. (2009). Role of information technology on agriculture and its scope in India. In Ahmed, R. (Ed.), *Globalization and economic development–Part–1* (pp. 207–222). New Delhi, India: Mittal Publications.

Shenoy, M. (2002). Transforming agribusiness the e-way. *Business India, 634*, 102–104.

Shunmugam, V., & Bahl, S. (2009). ICTs as enablers of process change in agriculture. *Information Technology in Developing Countries, 19*(1). Retrieved February 9, 2011, from http://www.iimahd.ernet.in/egov/ifip/feb2009/shunmugam.htm

Singh, A. (2005). *Case study: Community based information system, India EID Parry Indiagriline*. Retrieved February 6, 2011, from http://www.fao.org/rdd/doc/EID Parry Indiagriline.pdf

Singh, A. (2005). *Case study: Community based information systems, India ITC e-Choupal*. Retrieved February 6, 2011, from http://www.fao.org/rdd/doc/ITC e-Choupal.pdf

Singh, A. P. (2005). Lokvani: An e-effort to empower citizens. *Newsletter of the IFIP Working Group, 15*(2). Retrieved February 9, 2011, from http://www.iimahd.ernet.in/egov/ifip/nov2005/article1.htm

Singh, S. (2004). Leveraging ICT for agricultural development: A case study of e–Choupal of ITC paradigm. *Journal of IMT, 3*(1).

Singh, S. (2008). Time for agriculture to metamorphose in agribusiness. *GBPUAT Newsletter, 6*(3), 4-5. Retrieved from http://ssrn.com/abstract=1526051

Singh, S. (2010). *Potential of information technology as a tool for transforming Indian agriculture*. Retrieved from http://ssrn.com/abstract=1548882

Swaminathan, M. S. (2009). Agricultural renewal and agrarian prosperity. In Ahmed, R. (Ed.), *Globalisation and economic development –Part – 1* (pp. 3–20). New Delhi, India: Mittal Publications.

Swaminathan, M. S. (2009). Can science and technology feed the world in 2025? In Ahmed, R. (Ed.), *Globalization and economic development –Part – 1* (pp. 21–36). New Delhi, India: Mittal Publications.

Tewari, D. (2005). A field trip to rural tele-centers in Gujarat, information technology in developing countries. *A Newsletter of the IFIP Working Group, 15*(2). Retrieved February 9, 2011, from http://www.iimahd.ernet.in/egov /ifip/nov2005/article3.htm

Trinque, B. M. (1994). *Dynamism and enervation contrasts in India's Agrarian economy.* Discussion paper series No. 94-18, November 1994 at Department of Economics, Southern Illinois University, Carbondale, IL. Retrieved February 7, 2011, from http://129.3.20.41/eps/dev/papers/9505/9505001.pdf

Upadhyay, D. K. (2009). ePanchayats ICT for rural local government in India. *Information Technology in Developing Countries, 19*(3). Retrieved February 9, 2011, from http://www.iimahd.ernet.in/egov/ifip/oct2009/upadhyay.htm

Upton, R. M., & Fuller, V. A. (2004). *The ITC e-Choupal initiative*. Boston, MA: Harvard Business School Publishing.

Venniyoor, S., & Sah, G. (2007). Spreading the ICT revolution in rural India: Experiences and examples. *Information Technology in Developing Countries, 17*(2). Retrieved February 9, 2011, from http://www.iimahd.ernet.in/egov/ifip/ aug2007/sajan-venniyoor.htm

Chapter 13
ICT Based Pest Management System for Sustainable Pulse Production:
A Case Study

O P Sharma
National Centre for Integrated Pest Management, India

S Vennila
National Centre for Integrated Pest Management, India

Niranjan Singh
National Centre for Integrated Pest Management, India

Someshwar Bhagat
National Centre for Integrated Pest Management, India

Archana Bhardwaj
National Centre for Integrated Pest Management, India

Neelam Mehta
National Centre for Integrated Pest Management, India

Kamlesh Kumar
National Centre for Integrated Pest Management, India

ABSTRACT

"E-National Pest Reporting and Alert System" in pulse crops is a unique ICT based decision support system, which is very effective and easy to operate through a centralized server system at National Centre for Integrated Pest Management, New Delhi, connected with internet and mobile phones. This system has developed to cater to the needs of rural farmers of India, who grow pulse crops. The useful information is collected, stored, processed, and interpreted, and the appropriate advisories are sent to the registered farmers through centralized server system. They apply suitable corrective measures as per advisories at right time, and thereby, heavy loss caused by various pests can be checked/minimized below economic threshold level. Based on the past experiences and larger response of the stakeholders, Department of Agriculture and Cooperation, Govt. of India, suggested that this program should be extended and implemented in all pulse growing states. This system is quiet useful bottom to top level officials/policy makers, involved in E-Pest Surveillance programme.

DOI: 10.4018/978-1-4666-2655-3.ch013

INTRODUCTION

Pulses are the major source of dietary protein for majority of Indian population. India is the largest producer of pulses with 25% share in the global production but it is not sufficient to fulfil nutritional requirements. India relies on import of additional pulses (1.5-2.8 MT per year) from neighbouring countries. This un-wanted import had an economic brunt on the national exchequer to the tune of 1000 of crores per annum. Maharashtra (20%), Madhya Pradesh (17%), Rajasthan (11%), Uttar Pradesh (11%) and Andhra Pradesh (11%) together account for 70% of the total production of 14.76 MT from an area of 23.63 M ha (www.agricoop.nic.in, 2009-10). The reasons for shortage of pulses in India are technological barriers of high yielding varieties, erratic & inadequate rains and upsurge in pests hitherto minor. The impact of climate change has also resulted in resurgence of pests with varying intensity. Hence, there is urgent need to increase the production of pulses, by managing these pests causing significant yield loss. Pest surveillance or monitoring is a cornerstone of pest management rather than calendar-based treatments. The purpose of pest surveillance is to know whether pest in present in the fields and hence to apply the appropriate pest management practice. By continuous and systematic pest surveillance, epidemic situations of any pests can be avoided by detecting damage prior to firm establishment of a higher pest population. In fact, pest surveillance provides field-specific information on pest pressure and crop injury leading to take appropriate decision about selection and application of pest management procedures. Use of information and communication technology (ICT- Internet and mobile) in pest surveillance constitutes e-pest surveillance i.e. basically a web based Decision Support System (DSS) for timely collection of data with regard to pest population and preparation of reports which greatly facilitates the execution of pest management activity. Currently, Directorate of Plant Protection, Quarantine and Storage (DPPQ & S), Faridabad, is responsible to undertake pest surveillance with the help of countrywide located 32 Central Integrated Pest Management Centers (CIPMCs), which is being carried out through rapid roving surveys. Present set up of this organization do not provide real-time information about pest built up, as a result pest damage is occurred before action could be initiated. Whole operation is based on the manual processing of data required manpower as well as time to take up any emergency situation. Unfortunately, systems of compiling and distributing the data are not suitable in devising management strategies. Additionally, lack of capacity and technology within the infrastructure has undone previous investment in IPM and unintentionally gave way to promotion of use of chemical pesticide.

In view of the futuristic requirements as well as strategies to meet unforeseen circumstances, Department of Agriculture and Co-operation (DAC) and National Centre for Integrated Pest Management (NCIPM), New Delhi, took the initiative under the ambit of National Food Security Mission (NFSM) to increase the yield by reducing pest infestation, with the help of "e-Pest Surveillance" and implementation of Area-wide Integrated Pest Management System by involving State Agricultural Departments (SADs), KVK (Krishi Vigyan Kendra) officials and farmers by conveying the clear cut message to the farmers about appropriate and timely decisions to be taken for pest management and the response of farmers are highly significant and encouraging. In this programme, IPM nuclear model villages have been established in 5 major pulse growing states (Uttar Pradesh, Madhya Pradesh, Andhra Pradesh, Maharashtra and Karnataka) of India to implement plant protection measures, covering 76,000 ha in farmers' participatory mode in collaboration with State Agricultural Universities (SAUs). The Department of Agricultural and Co-operation (DAC) has provided separate support to State Agricultural Departments (SDAs) of 13 states for state-wide implementation of IPM under Rastiya Kisan Vikas Yojna (RKVY) and National Food Security Mission (NFSM).

Decision Support Systems in Pest Management (DSS)

Computer–based Decision Support Systems (DSS) integrates a user-friendly front end to often-complex models, knowledge bases, expert systems, and database technologies. DSS are important component of information technology in IPM. Web-based models and DSS are becoming popular because little or no client software is required, thus reducing software management and distribution costs. Several internet-based DSS have been developed for industry, medicine, business, meteorology, and agriculture. DSS have emerged as essential tools to bridge the gap between science-based technology and end-users that make day-to-day management decisions. An IPM Decision Support Systems should provide users all necessary information about pests, sampling, threshold levels of pests, pest/disease developmental models linked to weather networks, pest management methods, plus currently available pesticides, and their safety issues and environmental impacts. At international level several western countries have successfully used DSSs in improving the effectiveness and the quality of decision-making by processing voluminous data and providing alternative agro solutions. Paz et al., (2004) recently has explored multi-facet usefulness of internet using World Wide Web to provide the same services and named as WebGro. WebGro is a Web-Based Soybean Management Decision Support System (http://webgro.ae.iastate.edu) helping soybean producers to meet different stresses e.g., water, soybean cyst nematode (*Heterodera glycines*), herbicide injury, *Rhizoctonia* root rot disease (*Rhizoctonia solani*) and hail damage which limit the yield of crop in fields. This system can be run interactively by simulating the effects of one or more stresses at a time. The user has option to set up all necessary information, by selecting cultivar, planting date, plant population, soil type, and the nearest weather station using a web form. Different stress levels can also be entered, and the model can be run interactively by simulating the effects of one or more stresses at a time. WebHADSS is software designed by Bennet et al., (2003) primarily for extension education and for weed managers. WebHADSS is a fully functioning decision aid that makes recommendations for POST and PDS treatments which can be accessed at www.cropsci.ncsu.edu/webhadss. Similarly USDA-ARS Agricultural Systems Research Unit (ASRU), in collaboration with Colorado State University (CSU), has developed Great Plains Framework for Agricultural Resource Management (GPFARM) to serve as a whole-farm/ranch for evaluation of strategic agro ecosystem sustainability. Ntaliani et al., (2008) proposes a framework for putting traditional public services online by combining organization, management and technology advancements. It identifies stakeholders who will create or use the public service, describes and models the service, implements the e-government service and evaluates it. Workflow patterns are adopted in order to model the service in a well defined, flexible and reusable graphical presentation, achieving organizational interoperability of all involved stakeholders. Web services are used for implementing the e-government service as a cost-effective and technical interoperable solution. The framework has been tested for the case of an agricultural public service of the Greek Ministry of Agriculture. A web-based decision support system for pest management recommendations has been developed for green bug management in Wheat (Elliot et al., 2004). This DSS considers the highly variable abiotic, biotic and agronomic conditions under which winter wheat is grown. The weather database in this system includes data for

Oklahoma, Kansas, eastern Colorado and northern Texas. This system is maintained on-line by Oklahoma State University. To access this system use a web browser to go to the Entomology and Plant Pathology at Oklahoma State University homepage then select "Agricultural Models" and "Cereal Aphids Pest Management," or visit the mirror site hosted by the USDA-ARS. The system has modules to help in insecticide selection, cereal aphid identification, and natural enemy identification. Another DSS developed for tree fruit growers at Washington that integrates environmental data [140 Washington State University (WSU) of all stations plus weather forecasts from NOAA], model predictions (ten insects, four diseases and a horticultural model), management recommendation and a pesticide database that provides information on non-target impacts on other pests and natural enemies. New web-based warning system called VIPS in Norway aims to give open access to all information needed for farmers to reduce their reliance on plant protection products. VIPS calculates warnings for 70 weather stations for several pests in selected fruits, vegetables and cereals. Registered users may adjust the climatic data used in the models and record field observations to get private warnings. They may also use the system to record farm practices. Pl@nteInfo is a unique Web based DSS system for Crop Production Management in Denmark (www.planteinfo.dk) and provides farmers and agricultural advisers with just-in-time information and decision support about risks for crop diseases and pests, irrigation etc., with the real time processing of user tailored information. Growth model-based DSS for crop management (GMDSSCM) was developed including process based models of 4 different crops, i.e. wheat, rice, rapeseed and cotton. This system aims in facilitating simulation and application of crop models for different purposes. These GMDSSCM systems have been used to predict dynamically crop growth and to make decisions regarding to management systems. The Crop Weather Program for South Texas is a Web-based decision support system designed to help farmers and consultants making more appropriate decisions while managing cotton crops. This decision support system is delivered through the Web site http://cwp.tamu.edu/ where crop managers can easily access historical and current weather data and a selection of suitable simulation tools for performing field-specific crop/weather calculations and obtain useful information to support crop management decisions. Antonopoulou et. al., (2010) developed a Web-based semi-structured decisions making software called as Major Field Crops (MAFIC-DSS), which includes several modules, covering collaboration, information and managerial skill of farmers. This module provide information about the cultivation process and marketing of the supported crops, simulation and forecast models of general interest and expert systems for pest control and selection of cultivation practices. Moreover, MAFIC-DSS has been designed to offer services through mobile phones as it is accessible to large number of peaoples, even in population groups like farmers not having access to computers. Potential users of the system are mainly farmers or farm advisers acting on behalf of them. Moreover, policy makers, land owners and agronomists can benefit from the system. Likewise another web-based application, the South Eastern Agro forestry Decision Support System (SEADSS) was developed to assist landowners and extension agents in the Southeast United States to evaluate potential sites and suitable tree and shrub species for agro forestry planning. SEADSS offers on-line access to county-level spatial information, such as topography, hydrology, soils and land use, which are essential in evaluating potential agro forestry sites and suitable tree species. SEADSS links a geographical information system component with a subtropical tree/shrub database, which enables the user to query for biophysically, economically and managerially suitable trees and shrubs (Ellis et. al., 2005).

National Initiatives

In India, Government has taken up a number of initiatives to encourage the ICT based technologies like "Kissan call centers" (www.kissancallcenter.net) for sustainable increase of crop production. The call centre based extension service delivers knowledge and information exactly as per the query of the farming community. The Kisan Call Centre scheme has been functioning from year 2004 and can be accessed by Toll Free Number 1551 across the country. Similarly "e-choupal" (www.echoupal.com/) was launched in June 2000, linking rural farmers directly for the procurement of agricultural / aquaculture produce like soyabean, coffee, wheat, rice, pulses, shrimp etc. comprising of more than 40,000 villages through 6500 kiosks across ten states (Madhya Pradesh, Haryana, Uttarakhand, Karnataka, Andhra Pradesh, Uttar Pradesh, Rajasthan, Maharashtra, Kerala and Tamil Nadu). Reddy and Ankaiah, (2005) created "e-Sagu" (www.esagu.in/) (*Sagu* means *farmer* in Telugu language) to improve crop productivity and reduce costs of cultivation. It provides advice on weekly basis from sowing to harvesting by storing and processing farmer's data which in turn helpful in correct decision support system. The system of e-Sagu is fully adapted to available network, wether farmers are infrastructure supported and familiar or not with the information technology. aAQUA (almost All Questions Answered), (www.aaqua.org/) is a database of what farmers want to know and provides information in the context of rationale questions from farming communities. Agrocom Software (ASHA (www.assamagricbisiness.nic.in) and "Kissan kerala" (www.kissankerala.net) provides an online interactive platform for the farmers to interact with the scientists and agricultural extension officers to get correct solution of their problems. These portals also provides several dynamic advisory services like market information, weather and crop advisory, expert system on fertilizer recommendation, etc.. The mobile-based services like SMS, voice based Agri advisory services, videos through the mobile etc. and "e-krishi" (www.e-krishi.org/web/main/) in Kerala State addresses the existing gap in agriculture information flow and transaction management. One of private sector (Nagarjuna Group) developed agricultural Web site "Ikisan" (http://www.ikisan.com/) which provides online information on knowledge and business requirements for workers and traders in the agricultural sector (http://www.nagarjuna.com). Ikisan has been developed as a comprehensive agri-portal to address the information, knowledge and business requirements of various players in the agriculture including farmers, trade channel, partners and agriculture input/output companies. The portal also serves as an e-Market place where trade activities is being done online. This portal provides all kinds of agri-information, crop diagnostics, local information at the sub-district level and other useful information for the use of farmers. "e-Agriculture" is a global forum, where people from all over the world exchange information, ideas, and resources useful for information and communication technologies (ICT) for sustainable agriculture and rural development (www.e-agriculture.org). More than 7,000 members from 150 countries and territories, the e-Agriculture Community is made up of individual stakeholders such as information and communication specialists, researchers, farmers, students, policy makers, business people, development practitioners, and others. Some of the Popular and Successful ICT Projects in Rural Development in India are as follows:

- WARANA project in Maharashtra
- A wired village at WARANA
- NaiDisha-Haryana (naidisha.nic.in/)
- Bhoomi (Karnataka) (bhoomi.karnataka.gov.in/)
- Community Information Centre (http://www.cic.nic.in/)
- Gyandoot Project in DHAR (M.P) (http://gyandoot.nic.in/)

- Lokmitra (H.P) (himachal.nic.in/lokmitra.htm)
- Gram Darpan, an approach to e-gov ernance (http://rural.nic.in/panchayatIT/gramdarpan/main.htm)
- Janmitra Rajasthan (http://emitra.gov.in/)
- Mahiti Shakti - Gujarat (http://www.mahitishakti.net)
- Sustainable Access in Rural India (http://edev.media.mit.edu/SARI/sari-pilot-new-2.htm)
- E-Seva (A.P.) (http://www.esevaonline.com/)
- Tarahaat (http://www.tarahaat.com/)
- Dairy Information System Kiosk –NDDB (www.iimahd.ernet.in/~egov)
- Rural e-seva (west godavari) (www.westgodavari.org/)
- Lokmitra (rajasthan) (http://emitra.gov.in/egovConstruction.jsp)
- Drishtee (www.drishtee.com/)
- Vijay Vahini (www.newsreporter.in/tag/vijay-vahini)
- Chips (http://www.chips.gov.in/)
- Akshaya (http://www.akshaya.kerala.gov.in/)

Review of above web portal/DSSs have demonstrated the important role in Information and Communication Technologies (ICT) which play a significant role in rural development leading to several success stories. NCIPM has forayed in this technology and developed "Pesticide Advisor" (Singh et al., 2006); online Discussion Forum and "Pest Management Information System" (Sharma et al., 2004) which are being widely used by field workers. Realizing the proactive role played by "e-Pest surveillance", NCIPM and Maharashtra Govt. initiated this activity in the year 2009 and developed a system for Cotton and Soybean crops (http://www.ncipm.org.in/cropsap2009). This led to minimizing yield losses in Soybean and Cotton in Maharashtra. Based on these experiences, NCIPM and DAC, Govt. of India have initiated an e-National Pest Reporting and Alert System for Pulses to sustainable increase of pulse production under the ambit of National Food Security Mission.

E-NATIONAL PEST REPORTING AND ALERT SYSTEM FOR PULSES: THE DECISION SUPPORT SYSTEM (www.ncipm.org.in/A3P/UI/HOME/Login.aspx)

Methodology

With the assistance of DAC, Govt. of India, using inputs from crop based Indian Council Agricultural Research (ICAR) and State Agricultural Universities (SAUs), a real time pest surveillance system was designed to improve methods of pest data collection, management and sharing it with all stakeholders through web. The technical barriers in "e-National Pest Surveillance" in various states were identified and resolved with workshop and deliberations. In DAC workshops and meetings, State Agricultural Departments have agreed to utilize the same by training their district level master trainers and the training of 2nd level district officials for the use of system and generate a valid interpretations from the outputs for decisive actions. Various reports were designed for describing the real-time pest scenario based on systematic surveillance. Potential ambiguities were identified and the formats of data collection, timing and sampling procedures were standardized. National recommendations from All India Coordinated Projects (AICRP) were used for the improvement of sampling protocols and statistical analysis.

This broad initiative of pest surveillance is to:

1. Detect pest build-up and to monitor their progress through web-based query interface at Village/Taluk/District level.

2. Facilitate the electronic transfer of appropriate information from National Centre for Integrated Pest Management (NCIPM) to State Agricultural departments for initiating timely action.
3. To forewarn farmers through Short Messaging System (SMS) to ready themselves with preventive as well as curative pest management tools.
4. Recording of real time data.
5. To educate farmers to become decision makers of their own fields.
6. Reduction in pest incidence/intensity to increase pulse production.
7. Protection of the environment.
8. Savings in terms of money due to reduction in use chemical pesticides.
9. More coverage of IPM technologies for farmers (will result in healthy environment for human being, for pest defenders as well as for restoring soil health, which in long run reduce dependence on chemical nitrogenous fertilizers).
10. This real time data will be helpful for disaster management group.
11. Soil health will improved in terms of conservation of soil inhabiting beneficial flora and fauna.

Currently the pest surveillance aspect is being operated and maintained by NCIPM and has been implemented in 5 major pulses growing states of India. DAC has a plan to implement and expand the computerised system of "e-Pest Surveillance" for pulses crops, to all states. Despite several limitations, participating states have reconfirmed their commitment to implement at the earliest. The goal is to create a common national surveillance platform to identify trends and make a thumb rules, and target pest information collected at 'hotspots' for pest upsurges, to store data for future reference and policy decision.

Architecture of the Software

The "e-National Pest reporting & alert system" can be accessed at http://www.ncipm.org.in/A3P/UI/HOME/Login.aspx (See Figure 1). The structure of the system was decided on the basis of review and capabilities of available Decision Support Systems (DSS) in agricultural, web and mobile communication technologies. The above system uses 2 tier architecture (See Figure 2) consisting centralized database and with online application for pest data entry, reporting and issuing advisories to farmers through SMSs in local languages. This is a web based system developed using JavaScript, ASP.Net 3.5 and SQL server 2005 technologies. This is a platform independent and can be accessed from any computer having web browser and connected to the internet. The three layers of the system are:

1. **Presentation Layer:** This layer is implemented using HTML and JavaScript. This layer consists of forms for accepting information from the user and validating those forms using JavaScript.
2. **Application Layer:** Server side application layer is implemented using ASP.net 2.0.
3. **Storage Layer:** This layer is implemented using SQL server 2005. The relational approach has been used to design the database. This layer contains the database and its management systems to store and process data. It stores data for each one of the modules concerning the Registration/Approval and Export/Reporting module.

This system has four (6) modules. It is worth noticing that the selected modules are either functioning independently or in cooperation with other modules when the decision requires it. These four modules are described as follows:

Figure 1. The "e-National pest reporting and alert system"

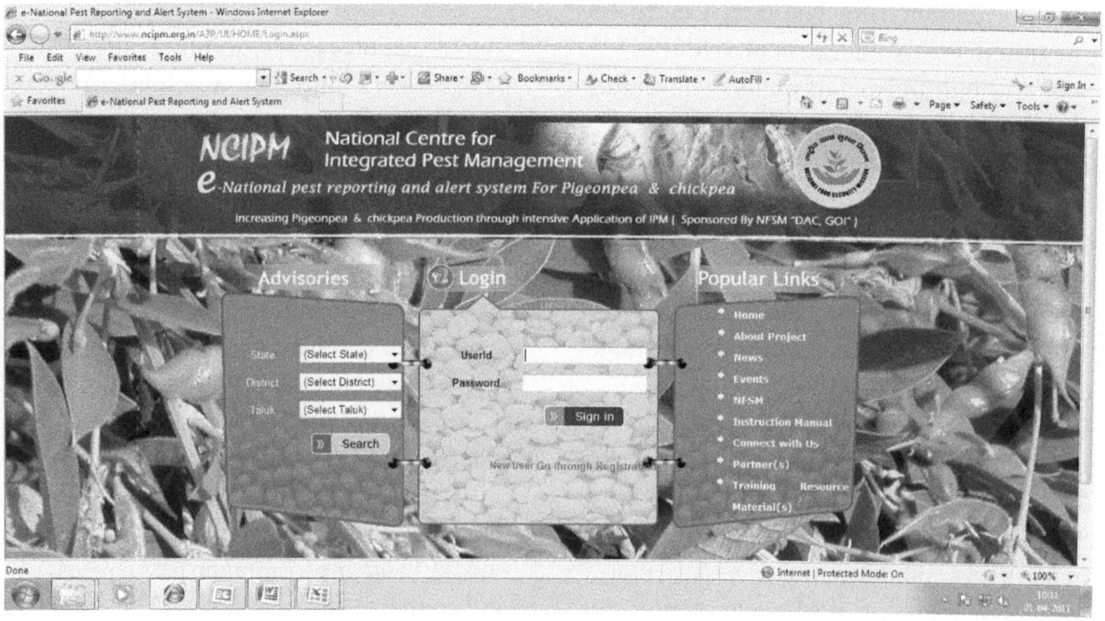

1. **System Admin Module:** The purpose of this module is to administer the whole application by creating centres; users such as Data entry operator, Approval officer and Pest experts; assigning access rights to the users; and maintaining database master files.
2. **Registration Module:** The purpose of this module is to register surveillance fields and location. The module consists 4 sub modules:
   a. **Register Scout Field:** Through the module, data entry user enter the information, the database pertaining to location, name of the District/Taluk/Block/Village latitude & longitude of villages, field Khasra No., details of owner of the field (Name, Address, Mobile No., Phone No., etc.), field type, irrigated or rainfed condition and informations about crop such as name, variety, etc.
   b. **Scout Survey Information:** This module captures the insect pest, disease and useful information collected by pest scouts from different field surveyed in standard data format (e.g., number of insects counted on 3 branches per plant, no of *Helicoverpa* eggs/larvae/plant, number of armyworm / cutworm larvae/no. of root grubs, *Coccinellids,* etc. per plant and for diseases e.g., no. of infected plants/spot for soft rot, botrytis, dry root rot, black rot, blight, and stunt disease, etc.). Other important information such as number of damaged pods, no. of pod fly immature stages, pheromone trap catches of *Helicoverpa* moths, inter culture operations, spray details, fertilizer details, weeding methods, etc., was also included.
   c. **Monitor Survey Information:** This module captures the details of monitor surveyed the fields randomly and pest information similarly as in case of scout survey.
   d. **Edit Scout & Monitor Survey Information:** Here the information entered into scout & monitor survey modules can be edited and updated before transfer to the centralised database.

Figure 2. The three tier architecture of "e-National Pest reporting & alert system"

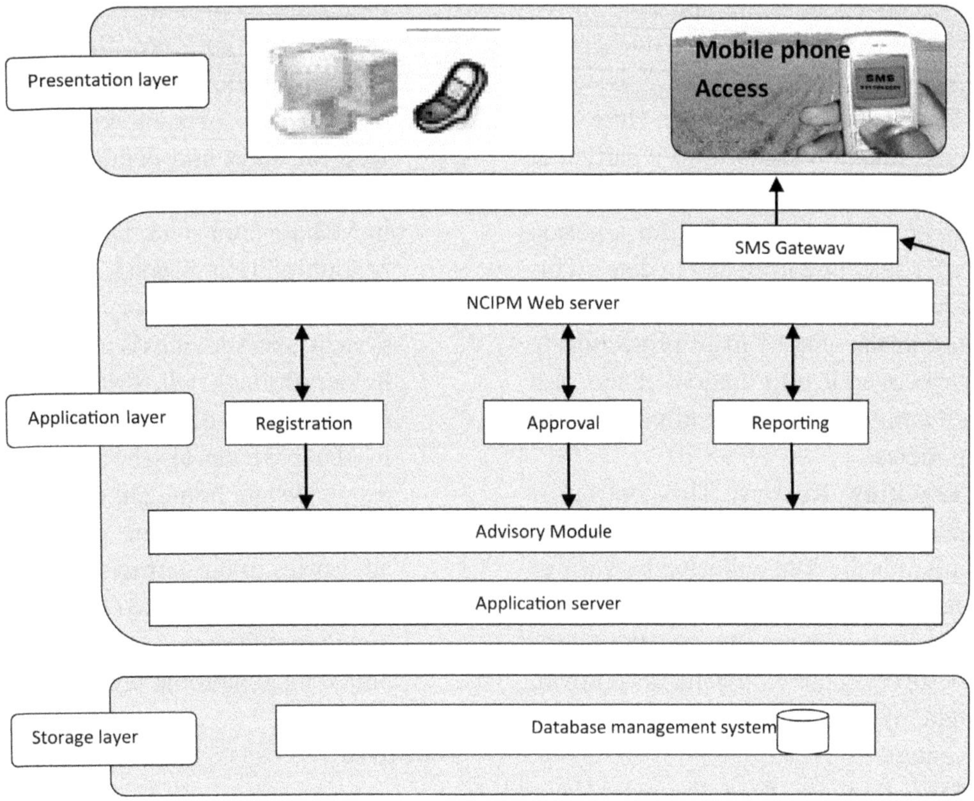

3. **Approval Module**: The purpose of this module was to verify the data feeded in the system and approval of concerned State Agricultural Department Authorities for the compiled pest details. This has provision for surveys pending for approval as well as surveys approved during particular period.
4. **Reporting Module:** The purpose of pest reporting was to communicate immediate or potential danger arising from the occurrence, outbreak or spread of a pest since accurate information on pest status facilitates technical justification of suitable plant protection measures and helps to minimize losses due to pest incidence. Thus, the necessary information passes on about pest situation to the concerned agencies without undue delay through web technology. The provision of reliable and prompt pest reports confirms the operation of effective and viable surveillance. This module produces reports in different formats for current as well as temporal pest reports for particular location. The report can be viewed village wise, taluka wise as well as district wise in tabular and graphical formats. These reports are available to registered policy makers, state agencies involved in pest surveillance and progressive growers on the click of the mouse.
   a. **Current Pest Report:** This module shows the real-time current pest situation/scenario in particular location and reflects the information in tabular as well as graphical formats along with date of observation, maximum, minimum and average values of all the pests. If the averages value of the particular

pest crosses the ETL level then the incidence values are appeared in red colour (font) and beneficial insects in green colour (font).

b. **Temporal Pest Report:** Here also pest situation/scenario of a particular location is produced in tabular as well as graphical formats but for a period after selecting from and to dates. This enables different govt agencies working in the area of plant protection to take overall information of the pest situation and to make effective draft policies.

c. **Tracking Report:** This option is meant for regulators and supervisors to monitor the data collected by various scouts and monitors as per the schedule. Herein system monitors see the details of surveys done with status of approval, date of observation, etc. for different centres.

d. **Miscellaneous/Trap Report:** These reports are to summarise the other relevant information such as details of pesticide spray, fertiliser details, pest trap data collected, etc. recorded with pest surveillance.

e. **The Map Based Reports:** This system Google maps are in use for geographical objects and spatial data manipulation. The expert module has been implemented using Java script in conjunction with suitable SQL queries. Herein the pest data for a particular district or taluka and all necessary spatial and ancillary informations are combined to report information about pest incidence on Google map and by colouring those areas with different colours as per pest intensity/ severity. Third option is good for research manages to forewarn the likelihood of pest spread in adjoining area.

5. **Advisory Module:** The purpose of this module is to provide pest alerts to all persons/groups who are directly involved in crop cultivation. Based on the pest situation/scenario of a taluka the pest experts feed the short and detailed advisories into the system in different local languages i.e. in Maharashtra state, the advisory is fed in Marathi and Hindi; and Kanada in Karnataka state. Once this advisory is submitted to the system, short advices is send to all the stakeholders through bulk SMS service module of the system and detailed advisory is stored in the database, can also be available for public use in system home page. State based A3P co-operators are using it to convey these advisories to the farmers through SMSs in their own languages to (13245 SMSs) 4650 member of farmers in 592 villages covering 36000 ha at national level.

Results

Implementation of A3P programme in general has proved very successful and increased 3 million tonnes of pulses in comparison to previous years, despite failures in Anantapur (AP) as well as Chindwara & Narsingpur (MP). Adverse climatic condition has resulted in resurgence of lesser known pest in these areas. The e-pest surveillance system has also enabled to identify the potential areas wherein yield can be increased by minimizing losses caused by proliferating pests (Maruca web, Pod bugs & Pod fly) as well as identification of endemic areas (Sterility mosaic virus & leaf spot diseases).

The "e-National Pest Surveillance" system in pulses crops (chickpea and pigeonpea) is functioning in five major pulse growing states of India and more than 13245 SMSs as IPM advisories have sent to farmers. The system "e-Pest Surveillance" has been evaluated for functionality and the user friendliness by stakeholders (farmers, agronomists, researchers' & pest advisors). This

Table 1. Adopted area & no. of farmers under National Electronic Pest Surveillance (2010-2012)

State / Districts	Pigeonpea		Chickpea	
	No of villages	No of Farmers	No of villages	No of Farmers
Karnataka (Gulbarga)	34	3301	23	1529
Maharashtra (Badnapur, Parbhani, Osmanabad, Nanded)	26	5017	20	3587
Andhra Pradesh (Anantapur)	25	482	12	1000
Madhya Pradesh (Chindwara & Narsimpur)	316	2967	55	1708
Uttar Pradesh (Hamirpur & Banda)	69	3000	12	557
Total	470	14767	122	8381

has benefited 3545 farmers, 10 agronomists/entomologists and 6 researchers in various terms. Post evaluation of this system suggests that it has fulfilled their basic requirements in terms of preparing themselves to manage pest incidences in more effective effectively. The results shown that about 60% of the stakeholders realised that the system contains sufficient information for pest diagnosis, selection of appropriate pesticide and advisories for sprays at right time. The use of mobile services seems to be very attractive since almost all the stakeholders can afford mobile and it has multifacet use (See Table 1).

Transmission of forewarning messages through SMS has resulted into reduced use of pesticides starting with eco-friendly and their number has been reduced to 3 from 6 in Gulbarga district. Use of this technology (e-pest surveillance) has benefited the farmers in terms of lesser pest incidence vis-a-vis conservation of beneficial insects because of timely action (See Figure 3) with eco-friendly management practices.

DISCUSSION

The existing extension systems is one to one from extension personnel's to farmers. However, due to shortage of extension personnel in the state departments of agriculture and Krishi Vigyan Kendras (KVKs), the agricultural technology developed under National Agricultural Research System (NARS) and State Agriculture Universities (SAUs) are unable reach the farming community. Therefore there is need to use the Information and Communication Technology (ICT) to transfer agricultural technology quickly to the farmers. This also facilitates in brining the farmer and scientists along with the subject matter specialists together in sharing the knowledge which was not possible in traditional extension serviceThe National Electronic Pest Surveillance is an initiative that promotes the use of data and information system standards to advance the development of efficient, integrated, and inter-operable surveillance systems at National, State and Local levels. This system aims to bring convergence in the effort of State Govt agencies and State Agricultural Universities (SAUs) to have reliable and timely information on the real time pest status and send useful information (advisory) in regional languages, based on suitability of agro-ecological principles of IPM to begin with pigeon pea and chickpea. e-Pest Surveillance Systems has enabled collection and monitoring of real-time pest data for trend interpretation and/or outbreaks leading to action in initial stage itself. This has happened in Kamasin Block of Banda (UP) wherein pod borer epidemics was managed with Emammectin benzoate. Similarly, pod borer infestation in Gulbarga (Karnataka) was managed

Figure 3. The data sequence diagram

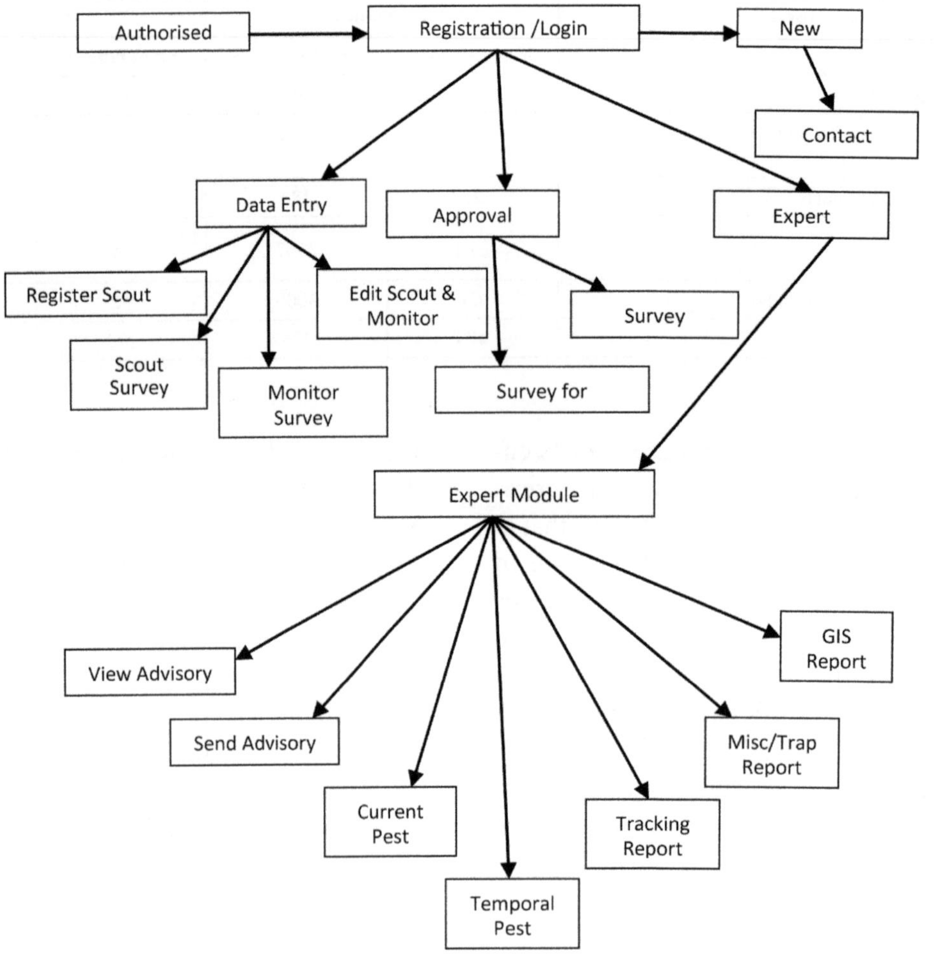

with Renaxypyr (See Table 2). In Nanded (Maharashtra), excessive flower drop was checked by timely convey of message to spray NAA.

ICT based technologies are helpful in making quick decision to apply appropriate plant protection measures as per level of pest infestation.. "E-National Pest Reporting and Alert System" for pulses crops in India is a major breakthrough in ICT based plant protection technology, initiated by NCIPM in collaboration with DAC, Govt. of India, has certainly benefited the farmers of five states (U.P., M.P., Maharshtra, Andhra Pradesh and Karnataka) in the form of short and detail advisories in relation to pest infestation. Several ICT based technologies have been successfully used in India and abroad. A web-based decision support system for pest management recommendations has been developed for greenbug management. This guide provides information about identity, pest management, and biological thresholds for this important pest (Elliot et al., 2004). Similarly, Jones et al., developed a decision support system for Washington tree fruit growers that integrates environmental data and pest and disease incidence to make a prediction model. Warana Wired Village Project covering 70 villages in

Table 2. The impact of "e-National Pest Reporting & Alert System" on pesticide use (2010-11)

A3P Farmers		**Non-A3P farmers**	
Total sprays used (4)		**Total sprays used (6)**	
Ovicide	Profenophos	Organo chloro compounds	Endosulphan
Bio-insecticide	Neem (Azadirictin)	Organo phospho compounds	Monocrotophos, Chlorpyriphos DDVP, Profenophos, Quinalphos, Acephate
Microbial insecticide	HaNPV	Carbamates	Methomyl, Thiodicarb, Carbaryl
Green labeled insecticide	Rynaxypyr	Pyrithroids	Alphamethrin, Fenvalrate, Cypermethrin
		Dusts	Malathion, Quinalphos, Fenvalrate, Endosulphan
		New molecules	Indoxacarb, Spinosad, Emamectin benzoate

Maharashtra the existing cooperative structure has been used with state of the art infrastructure to allow Internet access to existing cooperative societies. The aim is to provide information to villagers by establishing networked booths in the villages. The villages in this sugarcane-growing region have computers that are linked to a central network that provides farmers access to essential pieces of information such as the ideal time for planting and harvesting sugarcane, the current market rates of their produce, and payments made by the factories. Apart from Agriculture, aAQUA, Bhav Puchiye and the Digital Library can be configured and customized for expert advice and content in education, healthcare, e-Governance esp. by organizations who are working in connecting laymen with experts. ITC's e-choupal is a unique example of using ITC's for agricultural development; e-choupal has already become the largest initiative among all internet based interventions in rural India. e-Choupal link rural farmers directly for the procurement of agricultural / aquaculture produce like soya, coffee, prawns etc. eliminating the role of the middleman. The principle of the e-Choupals is to inform, empower and compete. The mobile-based agricultural information service is a giant leap of KISSAN towards becoming an epitome of complete Agri information systems. This new venture, which offers a plethora of services, could very well be positioned as a right tool for information dissemination that leverages on the modern technology. The information regarding various mobile-based services like SMS based information services, voice based Agri advisory services, and videos though the mobile etc. Real-time information and customised knowledge provided by 'e-Choupal' enhance the ability of farmers to take decisions and align their farm output with market demand and secure quality & productivity. The aggregation of the demand for farm inputs from individual farmers gives them access to high quality inputs from established and reputed manufacturers at fair prices. As a direct marketing channel, virtually linked to the 'mandi' system for price discovery, 'e-Choupal' eliminates wasteful intermediation and multiple handling. Thereby it significantly reduces transaction costs. The pathologists / entomologists / agronomists and researchers expressed confidence in the "e-Pest Surveillance" system and is suitably address the problems of farming communities and this system can be extended to other crops and leftover states with similar needs/ pest problems.

Major thrust in Agricultural Policy at national level has been the promotion and implementation of Integrated Pest Management (IPM) strategies to reduce pesticide use and support the development of environmentally benign tools. The success of IPM depends upon correct and timely identification of pest so that they can be managed with lesser toxic molecules and their spread can be suppressed before they reach to epidemic level. The overall IPM framework has provided the platform on which the output of NFSM project been able to build confidence and strengthen the pest and disease surveillance system. Advance warning about pests and diseases will allow farmers and support the workers in time to implement pest management tools promoted through the field schools. Thus, the output will facilitate efforts to conserve native natural enemies and to reduce use of chemical pesticides.

CONCLUSION

With the encouraging response of farmers of all 13 states of India, the Govt. of India has taken a decision that this program will be extended to other states and every state will select farmers representing specific villages, which in turn will be trained by officials of State Department of Agriculture (SDA) and will be actively involved in e-pest surveillance programme. Their operation will be to cover adjoining villages other than whole cultivated area of their village and will also re-enforce learning of IPM. Nominated individuals, "contract farmers" will be able to collect data on specific endemic pests and submit the same to NCIPM trained SDA, who in turn will feed the same in national "e-Pest Surveillance System". Once "e-Pest Surveillance System" is implemented across the country enormous yield losses can be saved with minimal use of pesticides and resources will be judiciously utilised. Establishment of this system will also help in storing the pest related data in a standard format at a central place, which can be utilised for drawing inferences in the time of eventualities. By using centralised database, fertilizer, chemical and water usage can be optimized to reduce production costs and increase crop production. It will benefit resource poor farmers apart from reducing agricultural waste and environmental pollution.

REFERENCES

Antonopoulou, E., Karetsos, S. T., Maliappis, M., & Sideridis, A. B. (2010). Web and mobile technologies in a prototype DSS for major field crops. *Computers and Electronics in Agriculture*, *70*(2), 292–301. doi:10.1016/j.compag.2009.07.024doi:10.1016/j.compag.2009.07.024

Ascough, J. C., Dunn, G. H., McMaster, G. S., Ahuja, L. R., & Andales, A. A. (2005). Producers, decision support systems and GPFARM: Lessons learned from a decade of development in Zerger. *Proceedings of International Congress on Modelling and Simulation 2005*, (pp. 170–176).

Bennet, A. C., Price, A. J., Sturgill, M. C., Buol, G. S., & Wilkerson, G. G. (2003). HADSS, pocket HERB, and WebHADSS: Decision aids for field crops. *Weed Technology*, *17*(2), 412–420. doi:10.1614/0890-037X(2003)017[0412:HPHAWD]2.0.CO;2doi:10.1614/0890-037X(2003)017[0412:HPHAWD]2.0.CO;2

Brodal, G., & Rafoss, T. (2006). Forecasting pests and diseases in Norway. *Proceedings 2006 Bioforsk Plantehelse - Ås Plant Health and Plant Protection.*

Cao, W., Tang, L., Zhu, Y., Pan, J., Li, W., & Chen, B. (2006). Development of growth model-based decision support system for crop management. *Frontiers of Agriculture in China*, *1*(3).

Elliott, N. C., Royer, T. A., Giles, K. L., Kindler, S. D., Porter, D. R., Elliott, D. T., & Waits, D. A. (2004). *A web-based decision support system for managing greenbugs in wheat*. Crop Management. doi:10.1094/CM-2004-1006-01-MGdoi:10.1094/CM-2004-1006-01-MG

Ellis, E. A., Nair, P. K. R., & Jeswani, D. S. (2005). Development of a web-based application for agroforestry planning and tree selection. *Computers and Electronics in Agriculture*, *49*(1), 129–141. doi:10.1016/j.compag.2005.02.008doi:10.1016/j.compag.2005.02.008

Jensen, A. L., Thysen, I., Boll, P. S., Hansen, J. G., Secher, B. J. M., & Juhl, O. (1997). *Pl@nteInfo – Using the internet for custom tailored crop information*. Retrieved from www.planteinfo.dk/information/publikationer/efita97/

Jones, V. P., Brunner, J. F., Grove, G. G., Petit, B., Tangren, G. V., & Jones, W. E. (2010). A web-based decision support system to enhance IPM programs in Washington tree fruit. *Pest Management Science*, *66*(6), 587–595. PubMed

Niranjan, S., Dhandapani, A., Sharma, O. P., Garg, D. K., & Verma, P. V. (2007). Pest management information system decision support tool for pest management in Basmati rice. *ICAR News: A Science and Technology Newsletter, 13*(2), 15-16.

Ntaliani, M., Costopoulou, C., & Karetsos, S. (2008). Mobile government: A challenge for agriculture. *Government Information Quarterly*, *25*(4), 699–716. doi:10.1016/j.giq.2007.04.010doi:10.1016/j.giq.2007.04.010

Paz, J., Batchelor, W., & Pedersen, P. (2004). WebGro: A web-based soybean management decision support system. *Agronomy Journal*, *96*, 1771–1779. doi:10.2134/agronj2004.1771doi:10.2134/agronj2004.1771

Reddy, K., & Ankaiah, R. (2005). A framework of information technology-based agriculture information dissemination system to improve crop productivity. *Current Science*, *88*(12), 1905–1913.

Sharma, O. P., Dhandapani, A., & Singh, N. (2004). Computer based decision support system for integrated pest management. In A. Singh, H. R. Sardana, & N. Sabir (Eds.), *Validated IPM technologies for selected crops* (pp. 181–190). New Delhi, India: National Centre for Integrated Pest Management.

Singh, A., Sharma, O. P., Dhandapani, A., Singh, N., Nagarajan, S., Satya, K., & Mehta, N. (2006). Pesticide advisor (Version 2005 1.0): An expert system for judicious use of pesticides for management of pests. *ICAR News: A Science and Technology Newsletter, 12*(4), 17-19.

Section 4

E-Agriculture and Rural Development in Europe and North America

Chapter 14
ICTs in Agribusinesses:
Opportunities for Extranet Services to Provide Information to Farmers

Pádraig Wims
University College Dublin, Ireland

ABSTRACT

This chapter critiques the level of adoption of ICTs among the farming community in Ireland and compares this with adoption by the general population. It was found that although computer access and internet use among Irish households have been increasing rapidly in recent years, farm households have been lagging behind in adoption of these technologies. Broadband access has been particularly problematic in rural areas, where farm households are located. The development of a novel extranet service by a large agribusiness is also evaluated in this chapter; this service allows dairy farmer clients to access all information relevant to their accounts with this agribusiness through a secure website. A structured survey of clients was carried out, and respondents were stratified on the basis of their usage of this extranet service. It was found that three factors are constraining adoption and use of ICTs among Irish farmers: low levels of computer skills; lack of awareness of the potential of ICTs to contribute to the farm business; and thirdly, access to Broadband in rural areas. These are fundamental problems that constrain adoption of ICT.

INTRODUCTION

Ireland is a small, trade-dependent economy in the extreme West of the European Union. It experienced rapid economic expansion from 1995 to 2007, when annual GDP growth averaged 6%, but economic activity dropped sharply in 2008 and Ireland entered into a recession for the first time in well over a decade with the onset of the world financial crisis and subsequent severe slowdown in property and construction markets. Agriculture, once the most important sector of the Irish economy, is now dwarfed by industry and services. Nonetheless, agriculture remains very important

DOI: 10.4018/978-1-4666-2655-3.ch014

to the Irish economy; the agri-food sector is one of Ireland's most important indigenous industries and is particularly central to the economic and social vitality of rural communities.

There is increasing acceptance that information and knowledge are central to socio-economic development. However, there is a perception that rural areas, and farming families in particular, are lagging behind in adopting and using ICTs. The objectives of this chapter are to critique the level of adoption of ICTs among the farming community in Ireland and compare this with adoption by the general population. It also reviews and evaluates the experience of one large agribusiness that has developed a novel extranet service. This is a dedicated secure site whose main purpose is to allow farmers have access to all information relevant to their business with this company. The objective of this research was to explore the effectiveness of the extranet project in providing information to its farmers and to determine how this could be improved.

Agriculture in the Irish Economy

The agri-food sector in Ireland accounts for over 6% of Gross Value Added, almost 8% of total employment, and 10% of exports (Department of Agriculture, Fisheries and Food, 2010a). While agriculture is of major importance to economic welfare and development in Ireland it is particularly central to the economic and social vitality of rural communities because of the dispersed nature and composition of the agri-food sector in Ireland.

The land area of Ireland is 6.9 million hectares of which 4.2 million hectares (60.9%) is used for agriculture and a further 737,000 hectares (10.7%) for forestry (Department of Agriculture, Fisheries and Food, 2010b). About 80% of the agricultural area in Ireland is devoted to grassland, 11% to rough grazing (mainly hill and mountain pastures) and 10% to crop production (ibid). Beef and milk production currently account for 66% of agricultural output at producer prices (CSO, 2010), both of these are grassland based farm enterprises. Data from CSO's Farm Structures Survey show that in 2007 (the most recent year for which data are available) there were 128,200 individual farm holdings, with an average farm size of 32.3 hectares (Department of Agriculture, Fisheries and Food, 2010a).

Information and Communication Technology Policy in Ireland

It is now recognised that information and knowledge are at the very heart of socio-economic development (O'Donnell et al, 2003). ICT and broadband are enabling tools and infrastructures for accessing, developing, using and sharing information and knowledge (ibid). A study conducted by Chambers Ireland (2007), involving Irish businesses, showed that Information and Communication Technology (ICT) tools and e-business solutions had become crucial for increasing efficiencies across all sectors of business in Ireland.

Both the EU and Irish Government policy has focused on promoting adoption and use of ICT in the home, in education and in the workplace. In March 2000, in Lisbon, the European Union (EU) set itself the ambitious target of becoming the world's *"most competitive and dynamic knowledge based economy"* within ten years (European Commission, 2007). It recognised that attaining this goal depended on making the best possible use of ICT. The Lisbon Strategy placed greater emphasis on the knowledge-based society within existing policy processes and launched the eEurope 2002 Action Plan as a roadmap to modernise the European economy (ibid). The i2010 strategy was launched by the European Commission in June 2005 and is in place until 2010. It builds on the eEurope initiative, which came to an end in 2005 (ibid).

Irish Government policy on ICT was built around two Action Plans. In January 1999, the Government launched the first plan for implementing the information society in Ireland. This set out a series of initiatives which included development of telecommunications infrastructure which included national availability of broadband, new legislation and other measures to enable both public and private businesses to operate on-line and a range of eGovernment initiatives and projects (Government of Ireland, 1999).

A study by O'Donnell et al (2003) showed there had been significant increases in levels of engagement with ICT in recent years. The dramatic increase in adoption of ICT was further supported by the second Action Plan of Government (New Connections, 2002) which accepted that it was timely to assess how we had progressed, to review our priorities, and to put in place a new strategic framework to take the country forward.

The Government Ministry of Agriculture and Food (Department of Agriculture and Food, 1999) recognised that the country was embracing ICT at a rapid rate and this had major implications for rural areas in particular. However it did caution that without the necessary infrastructural investment, rural areas could be disadvantaged. Since then much emphasis has been placed on engagement with business and the individual in relation to ICT. However it was recognised from an early stage that some groups were not engaging with ICT and as a result were being excluded from the benefits accruing. These included agricultural workers (O'Donnell et al, 2003). The application of eGovernment to agriculture had the potential to totally transform the way farmers engage with the Department of Agriculture. The Comptroller and Auditor General (2008) concluded that:

"There has been mixed progress in eGovernment projects that had a high transactional objective. There were a number of outstandingly successful projects including motor tax, ROS, On-line Area Aid, and animal disease eradication." (p.38)

However, the Department of Agriculture was held up as an example of how successful eGovernment could be. Today the Department of Agriculture and Food, together with Teagasc (Irish Agricultural Development Authority, responsible for research, extension and education) have a number of excellent projects operating under the eGovernment initiative. These initiatives indicate a very definite commitment on behalf of the Department of Agriculture and Teagasc to ICT and eGovernment initiatives.

ADOPTION OF ICTs BY THE GENERAL POPULATION AND BY THE FARMING COMMUNITY IN IRELAND

The adoption of ICT by the Irish population is a multifaceted area made up of a number of subsections which will be presented and discussed:

- The level of computer availability in Irish households;
- The level of access to the internet;
- The level of broadband penetration; and
- Adoption of ICTs by Irish farmers.

Level of Computer Availability

When computer access in households was measured during the last three months of 2009 it was found that 75% of Irish households had either a personal computer or a laptop (Millward Brown IMS, 2010a). However, further analysis revealed that only 53% of farming households had either a personal computer or a laptop indicating that farmers are lagging behind in computer use (ibid). Only the retired and the unemployed socio-demographic groups lagged behind the farming households, 54% of these households (i.e. only 1% higher than farming households) did not have either a personal computer or laptop.

Level of Access to the Internet

Internet use among the Irish population is increasing rapidly. By mid 2010 some 77% of the population used the internet on a regular basis and 95% of these did so from their own homes (Millward Brown IMS, 2010b); this had increased significantly from 58% just two years previously (Millward Brown IMS, 2008). Internet use among businesses was considerably higher; research (Millward Brown IMS, 2010c) indicated that internet access among Small and Medium size Enterprises (SMEs, defined as companies with less than 100 employees) has remained steady at 92%, while corporate access (i.e. companies employing 100 or more employees) was almost universal. Millward Brown IMS went on to report that smaller companies (i.e. employing less than 10 staff) were least likely to have internet access with 84% connected to the internet. However farming households were among the lowest internet users at 54% (Millward Brown IMS, 2010b). This was only surpassed by the oldest age category (52% of the over 65 age group did not have internet access). These figures indicate, once again, that farming households are not embracing ICT.

Comreg (2010a) reported that there were just under 1.6 million active internet subscriptions in Ireland at the end of June 2010. This represented a 0.2% increase on the previous three months and a very significant 8.1% increase on the same period the previous year. While the total number of subscribers is increasing rapidly the data in Table 1 show that dial-up internet access is declining rapidly while broadband is increasing.

According to Millward Brown IMS (2010a) the main uses for the internet included sending and receiving e-mail with nine out of every ten using this application. Research purposes were also considered important, together with transactional type activities (banking, shopping, booking tickets, etc). Social networking sites were also prominent driven by the younger age groups.

Level of Broadband Penetration

Broadband subscriptions are increasing at a rapid rate. Comreg (2010a) reported that by the end of June, 2010, there were 1,518,349 broadband subscriptions in Ireland. This represented an increase of 16.9% in the number of subscriptions since the previous year. Mobile broadband represents 33.4% of all broadband subscriptions in Ireland. Based on these data, the broadband per capita penetration rate at the end of June 2010 was 34.0% compared with 33.4% in the previous quarter and 29.4% the previous year. When mobile broadband is excluded, the penetration rate was 22.6%. The broadband household penetration rate was 63.1% at the end of June, 2010, up slightly from 62.4% in the previous quarter.

Comreg (2010b) compared broadband penetration rates in Ireland with other EU countries. It reported that Ireland's household broadband penetration rate ranked 14th among the EU 27 countries and was above the EU27 average of 56.7%. The Netherlands and Denmark had the highest household broadband penetration rates at 86% and 82% respectively. This situation was confirmed when fixed broadband per capita penetration rates for EU countries were compared. Comreg (2010b) reported that Ireland's broadband penetration at 22.6% placed Ireland below the EU average of 24.8% and again in 14th place in the EU 27.

Amarach Consulting (2007) surveyed subscribers' attitude to broadband. This study found that internet users outside of Dublin, the capital city, had far more problems with the poor service they were getting relative to Dublin both in terms of availability and competing choice. Meanwhile a national survey conducted by the Irish Farmers' Association (Smith, 2008) calculated that 75% of rural households did not have access to broadband in 2007. It was also stated that infrastructure to provide broadband to rural areas was totally inadequate (*ibid*).

Adoption of ICTs by the Farming Community in Ireland

There has been a long-standing conviction in Ireland that farmers and farm families are among the slowest adopters of ICT. Morrow *et al* (2003) for instance reported that Irish farm families were falling behind other sectors of Irish society with regard to home internet access and cited evidence

Table 1. Total number of active Internet subscriptions in Ireland

Subscription Type	Quarter (Q) 2, 2010 Subscriptions	Quarterly Growth Q 1 2010 - Q 2 2010	Year-on-Year Growth Q 2 2009 - Q 2 2010
Total Narrowband	77,243	-26.5%	-56.5%
DSL Broadband	729,892	+0.8%	+6.1%
Mobile Broadband	508,620	+3.8%	+40.4%
Cable Broadband	173,146	+5.9%	+39.3%
Other Broadband	106,691	-4.2%	-15.1%
Total Internet Subscriptions	**1,595,592**	**+0.2%**	**+8.1%**

Source: Comreg, 2010a

which showed that 21% of farmers had internet in their homes compared with the national average of 35% at that time. They also commented that this compared unfavourably with the farming communities of other developed economies. The author conducted research to measure the level of computer and internet usage among Irish farmers, to analyse the purpose to which these were put and to examine the factors which influenced the adoption of ICTs. The results were compared with the level of adoption among the general population in Ireland and with farm families in other comparable countries (Wims, 2007). It was found that a slightly higher proportion of farmers than the general population had home PCs and internet connections. However, when international comparisons were made it was found that the Scandinavian countries were forging ahead, with significantly higher proportions of farmers in Finland, Norway and Sweden having a home computer and internet access. Other countries with high levels of adoption were Switzerland, New Zealand, UK and USA (Ross and Waksman, 2004). From the figures of ICT adoption established in this study, it emerged that Irish farmers were approaching a level of adoption that was similar to the UK, USA and New Zealand.

It was found that Irish farmers who were married and with dependent children, particularly older dependent children, were much more likely to have adopted ICT within their homes; it is evident that these children have greater demand for ICT for educational purposes and for recreational purposes and are a considerable influence in encouraging adoption of ICT at household level. In addition, there were trends whereby farmers and/or their spouses who had off-farm employment were more likely to have adopted ICT than full-time farmers. This is likely to occur for two reasons, one being the increased level of disposable income available in these households and the other being exposure to sources of information about ICT external to their immediate social system. Farmers engaged in specialist enterprises were more likely to have adopted ICT; however, the proportions engaged in less profitable enterprises that have adopted ICT was surprisingly high but it is suggested that this can be attributed to the fact that these farmers or their spouses were more likely to have off-farm employment. Finally, it emerged that those with medium-sizes farms were most likely to have adopted ICT; *apriori* expectations were that those farming larger areas would have greater requirements for ICT within their farming systems, it was discovered that this latter category were least likely to have off-farm employment, thus reducing their disposable income and their exposure to ICT in the workplace.

DEPLOYMENT OF EXTRANET BY AGRIBUSINESSES

In this section, a case study of an extranet website is presented. The term "extranet", according to Vlosky et al. (2000) is defined as a network that links business partners to one another over the internet. They explain that this linkage is usually accomplished by companies allowing their partners to access certain areas of their intranet. There has been considerable interest in researching the role of extranets in the business environment in recent years. Most of this research has focussed on exploring the benefits accruing to business from adopting extranets and also analysing the factors affecting their adoption. It has been found that the use of extranets can facilitate the sharing of information, decrease operating costs, save time and resources, improve customer service and generally improve business-to-business relationships (Anderson (1998), cited by Vlosky et al., 2000). In spite of these benefits, Windrum et al (2003) found that the adoption of extranets involved major organisational innovation and altered communication between the firm and its clients and suppliers.

However, very little research has been conducted among agribusinesses and especially among farm businesses on their adoption of extranets. This present case study helps to address this information deficit by analysing the experience of one extranet service. The name of this extranet site is "AgriLink" and it was developed by Glanbia, an international cheese and nutrition company, whose headquarters are in Ireland. Glanbia was formed in 1997 from the merger of two previously long-established dairy food companies. The group has 4,500 farmer milk suppliers and so is one of the larger milk processing companies in Ireland. Glanbia set up an extranet site (www.agrilink.ie) in 1999 on a pilot basis and launched it among its suppliers in 2000. It was established to provide a link between the Milk Division and Agribusiness section of Glanbia and the farmer customers. It is a dedicated secure site whose purpose is to allow farmers have access to all information relevant to their business with Glanbia. Milk suppliers have access to a range of information regarding their milk account. This includes the latest milk test results, individual milk collection details and copies of milk statements. A superlevy statement is updated after every collection and help pages are provided with advice on non-conforming test results. Trading account holders can view the current balance on their accounts. Grain suppliers can select by grain type or intake branch to confirm the latest test results, weight, price paid and value of grain supplied. Copies of grain statements are also provided. AgriLink also provides a comprehensive news section. This includes updates on the Glanbia Superlevy position, technical information, links to other agri-websites and the latest Glanbia promotions. Thus, AgriLink has an important extension element. Registering for access to the AgriLink website has always been free to customers of Glanbia. In order to access AgriLink milk suppliers require a Glanbia account number and access to the internet. This extranet site is not unique in Ireland; in fact almost all of the large dairy cooperatives have established similar services in recent years.

The main benefits for clients of using AgriLink include:

1. 24 hour per day access to account information through AgriLink seven days a week.
2. AgriLink is updated daily, therefore the most current information regarding accounts is provided.
3. AgriLink also provides historical information since 1999; this allows the farmer to compare one month against the same month of another year.
4. Copy statements for milk, trading and grain accounts are provided.
5. Supplier's superlevy position updated after every milk collection.
6. Branch transactions updated daily after close of business.
7. Trading Statement download. This service allows a customer to select any month and a file containing a copy of the customer's

trading or milk account statements will be forwarded by email to the customer. This file can then be downloaded into a farm management accounts package. The benefit is the time saved as there is no need to key-in branch dockets into the accounts package.

The features on the AgriLink website that were most accessed during 2007 are presented in Figure 1.

Research was recently conducted by the author to evaluate the AgriLink website. The overall objective of this research was to explore the effectiveness of the AgriLink project in providing information to its farmers and to determine how this could be improved. The specific objectives included, *inter alia*, to describe the history and evolution of the website; to describe the present content and how it is presented; to determine how many of the target users were actually using it; to analyse/measure the success of the website; and to examine what constraints were impeding farmers from using it.

RESEARCH METHODOLOGY

Data were obtained from the Glanbia website development team, Agribusiness Division and Farm Development Section and from dairy farmers who supplied milk to Glanbia Co-Op who provided primary data.

The quantitative research took the form of a postal questionnaire to a sample of Glanbia dairy farmers in order to establish farmer views on their adoption and use of ICT and AgriLink. As the questionnaire was being posted to farmers, care was taken to ensure that all questions

Figure 1. Hits on AgriLink during 2007 (N=857 suppliers who used AgriLink) Source: AgriLink Development Team (personal communication)

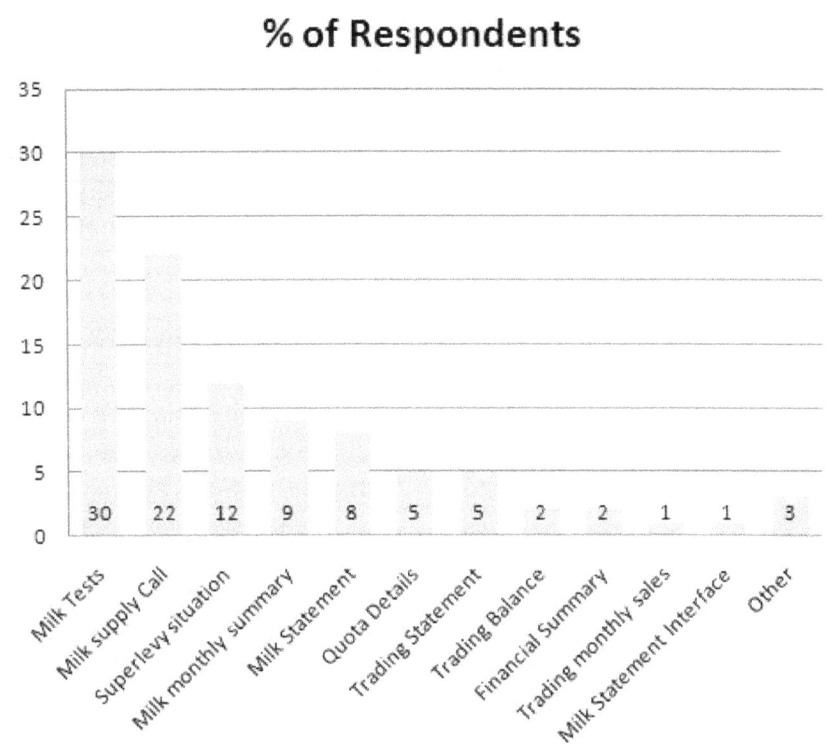

were clear and unambiguous. The questionnaire was presented in A5 format in booklet form. The questionnaire was initially piloted with a sample of five dairy farmers and final adjustments were made to the questionnaire arising from this exercise. A letter was prepared explaining the purpose of the research and Glanbia also provided a cover letter of support.

Sample selection was on the basis of stratified random sample, stratified on the basis of milk supplier usage of the AgriLink service. Of the 4,500 milk suppliers to Glanbia some 2,200 were registered to use AgriLink but only 857 of these used the site during 2007 (Source: Glanbia IT section, personal communication). Glanbia staff compiled a list of all suppliers with their corresponding usage of AgriLink, where usage was measured by the number of times each milk supplier logged on to the site. Hits on the website were recorded each time milk suppliers logged on and sought information regarding their accounts. The range of hits on the website varied from 1 to 2,500 during 2007. For this research, a sample of 165 dairy farmers in three categories was selected: these consisted of 55 frequent users of AgriLink, 55 occasional users and 55 non users of the website. The most frequent 55 users for the research were all those farmers who recorded hits on the AgriLink site of more than 460 times during the year. All 55 milk suppliers in this group (top 6% of users) were selected to participate in this study. This group was designated as the frequent users group.

The middle user group were identified as those who had between 192 to 460 hits during 2007. Some 101 milk suppliers (12% of all users) were in this category and 55 were randomly selected to participate in the study and were designated as the occasional users group.

In order to derive a non user group the complete Glanbia supplier database was accessed and all AgriLink registered users were removed. It was observed that none of those suppliers already selected had less than 21,997 gallons (100,000 litres) of milk quota. In order to maintain a comparable sample in terms of enterprise scale, all suppliers in this category were also removed and then 55 suppliers were selected by random sample. This group was designated as the non-user group.

Questionnaires and cover letters were posted to all three groups. A stamped self addressed envelope and a return date were also included with the questionnaires. Eight days later a text message reminder was sent to all non respondents. Some 113 fully completed questionnaires were returned. This represented an excellent response rate of 68% (113 out of 165). The breakdown of the three groups represented 43 frequent users (78% of sample), 41 occasional users (75% of sample), and 29 non users (53% of sample).

The data were entered into the Statistical Package for the Social Sciences (SPSS 11.0® for Windows) and analysed using this data manipulation programme. Descriptive statistics (frequency distribution, percentages, measures of central tendency and variability) were used to describe the data. Crosstabulations were used to examine relationships between variables. This methodology is typical in research of this nature which analyses adoption and use of ICTs (see, for instance, articles in well recognised peer-reviewed journals by Alam and Noor (2009), Bayo-Moriones and Lera-Lopez (2007) and Fabiani et al. (2005)).

RESEARCH FINDINGS

Analysis of the data revealed that the profile of the typical AgriLink user was male, engaged in full time farming, either young or middle aged (i.e. 35-50 years of age), married with dependent children, and owner or joint owner of the farm business. He was likely to be a Teagasc client, well educated and with a good level of agricultural education. Non users of AgriLink were also likely to be male, full-time farmers but were older (in age range 35 to 65 years), less likely to be a Teagasc client and also had a lower level of education, including agricultural education.

When home PC ownership was examined it was found that 94% of respondents had a home PC; this was expected as all AgriLink users had a home PC. However 76% of respondents who did not use AgriLink also had a home PC. This was considerably higher than the national average of 53%. Connectivity to the internet was examined and it was established that 49% of respondents used broadband, 2% used ISDN and the remaining 48% used dial up. This half and half split between broadband and dial up was poorer than the rate of broadband penetration nationally. The distribution of respondents by their means of accessing the internet is presented in Table 2.

When the three user groups were examined, the level of internet access was 100% among users while access was at 65% among non users. The use of broadband was also much lower among non users at 24% whereas over half of the frequent users had broadband.

Adoption and Use of PC and Internet Among Respondents

When PC usage was examined it was found that everybody in the respondents' households could operate the computer in 55% of cases while the respondent alone operated the computer in 34% of households. The remaining 11% of respondents indicated that their spouses (4%) or their children (7%) could operate the PC. When the results were analyzed by AgriLink user groups a different picture emerges and the results are presented in Table 3.

Table 3 indicates that respondents themselves were operating home PCs in almost all cases where they used AgriLink whereas in non-users' households only 60% of respondents could use a computer.

Computer use by respondents was examined to ascertain for what purposes the PC was used. It emerged that respondents who used computers applied them to farm business use. The four options selected most often were use of AgriLink (by 79% of respondents), maintaining herd records (75% of respondents), maintaining miscellaneous farm records (68% of respondents), and Entertainment (by 60% of respondents).

AgriLink Usage Among Respondents

Respondents who used AgriLink were asked how often they did so and their replies are summarized in Figure 2.

From Figure 2 it emerged that the majority of respondents used AgriLink regularly; a significant proportion (32%) used the facility daily and 77% of respondents used it at least weekly.

Respondents were also asked for what purpose they used AgriLink and their replies are summarized in Figure 3.

Table 2. Type of Internet connection among respondents by AgriLink user group (n = 113)

Internet connection type	Frequent Users (n=43)		Occasional Users (n=41)		Non Users (n=29)		Total
	No	%	No	%	No	%	No
Dial-up	16	37%	22	54%	12	41%	**50**
Broadband	25	58%	18	44%	7	24%	**50**
ISDN	2	5%	1	2%	0	0	**3**
None	0	0	0	0	10	35%	**10**
Total	**43**	**100%**	**41**	**100%**	**29**	**100%**	**113**

Figure 2. Distribution of respondents according to frequency of use of AgriLink (n=88)

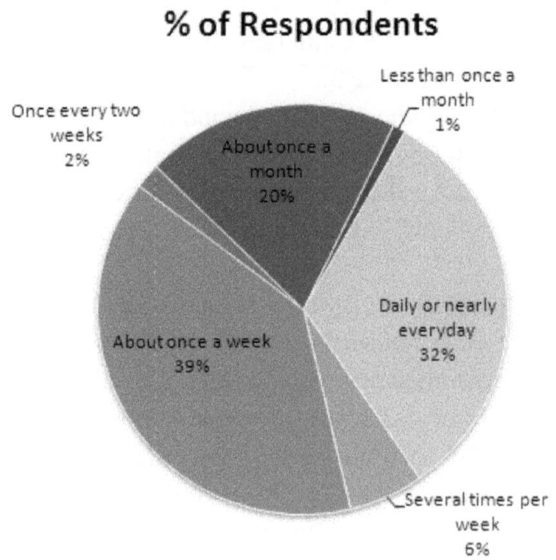

Respondents were asked to rate eleven features on AgriLink. Each feature was evaluated on the basis of being very useful, of some use or of no use to the respondents. All eleven features were easily accessed on AgriLink and were strongly promoted by the website developers. The distribution of respondents by their responses to these features is presented in Table 4.

From Table 4 it is clear that respondents who used AgriLink were very familiar with all milk related features on the website. However Milk Statements and Annual Summaries were not as popular as all other milk related features; despite this most respondents still found them useful or of some use. On the other hand all agribusiness and online business features were much less popular.

The bulk feed ordering feature was used by 26 respondents but only nine of these reported it to be of any use (i.e. very useful or of some use). The General News section was also regarded as poor with only ten respondents finding it very useful. The Statement Download facility was only useful to those who had a farm software package to utilize it. While some 60% of respondents did indicate that they had a farm management software package it is clear that not all of them were incorporating AgriLink into it and obtaining the full benefits of it.

Factors Which Influenced PC and Internet Adoption and Non-Use of AgriLink

In this study PC and internet adoption were found to be universal for users of AgriLink. For non users it was necessary to examine the reasons for slower uptake of ICT. While 76% of respondents who did not use AgriLink had access to computers at home, it was considered unusual that they had not engaged with this resource and other factors, other than access to a home PC, were clearly preventing them from realising the benefits of AgriLink.

It was found that non users of AgriLink were much less likely to have completed computer

Table 3. Distribution of respondent by who in household is capable of operating PC (n = 106)

	Frequent Users (n=43)		Occasional Users (n=41)		Non Users (n=22)		Total
PC operator	**No**	**%**	**No**	**%**	**No**	**%**	**No**
Respondent	19	44%	13	32%	4	18%	**36**
Respondent's spouse	1	2%	2	5%	2	9%	**5**
Respondent's children	0	%	0	0%	7	32%	**7**
Everyone	23	54%	26	63%	9	41%	**58**
Total	**43**	**100%**	**41**	**100%**	**22**	**100%**	**106**

Figure 3. Distribution of respondents according to purpose of AgriLink use (n=84); Multiple responses allowed

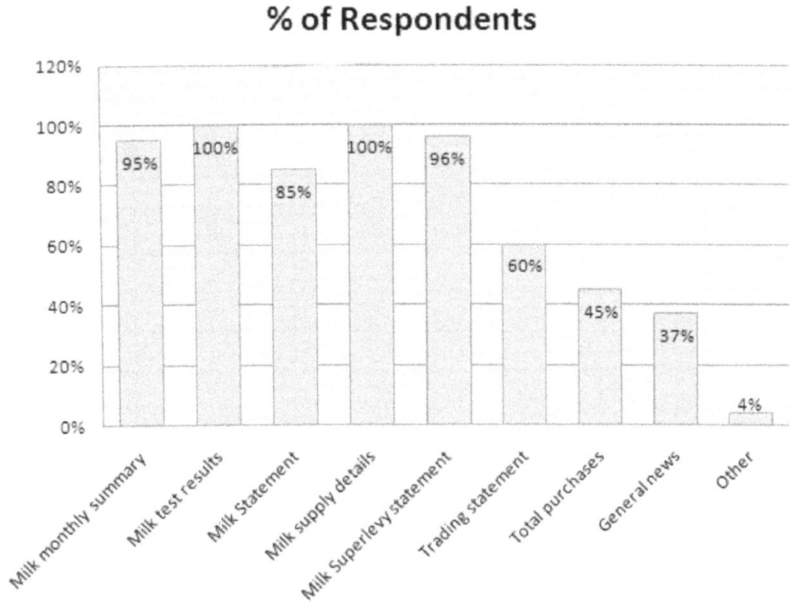

training than either frequent users or occasional users. The distribution of respondents by their level of computer training is presented in Table 5.

From Table 5 it is notable that 56% of AgriLink users had engaged with computer training whereas only 33% of non users had taken computer training. It may be concluded that this low level of training among non users has influenced the adoption of ICT and the use of AgriLink. Analysis revealed that 75% of non AgriLink users had a home PC and 66% were connected to the internet. Figure 4 indicates what would encourage Non User Respondents to use AgriLink.

It is clear that a lack of computer training was a major impediment to increased computer use and increased AgriLink use. Infrastructure was also an issue as more than 20% of respondents felt that improved internet access would encourage them. A number of respondents also were totally unaware that an AgriLink type service was available as they felt that access to Trading and Milk Account details would encourage them to use AgriLink. This indicates that an awareness campaign by Glanbia, the service provider, might encourage increased use.

Conclusions to AgriLink Study

This study examined the Glanbia AgriLink extranet site and the respondents to the survey could not be taken as representative of how Irish farmers are adopting ICT on their farms. However the study did discover a number of issues which may be slowing the rate of adoption of ICT. The sample of farmers who did not use AgriLink have also embraced ICT but not to its full potential. Some 75% of those farmers have a home PC and 60% of them can operate it. Therefore there is a huge onus on IT support teams both in Glanbia and in training bodies such as Teagasc to provide bespoke training to help those farmers in the years ahead to embrace ICT more fully and so reap the benefits accruing.

Table 4. Distribution of respondents by how useful they rated AgriLink features (n=88)

AgriLink Feature	Very Useful	Of some Use	Of no Use	Have not tried it	Total Responses
Milk Monthly Summary	68	13	3	0	**84**
Milk Supply Details	77	6	1	0	**84**
Milk Test Results	79	4	1	0	**84**
Milk Statement	54	26	4	0	**84**
Super Levy Statement	73	6	3	2	**84**
Annual Summary	37	31	6	6	**80**
Trading Statement	39	22	8	12	**81**
Total Purchases	28	26	9	13	**76**
General News	10	29	26	13	**78**
Bulk Feed Ordering	5	4	17	47	**73**
Statement Download	19	16	10	30	**75**

CONCLUSION

This chapter has analysed the penetration of ICTs among the general population and among farmers in Ireland. It is apparent from this chapter that considerable strides have been made in ICT and eGovernment initiatives in Ireland in recent years. Unfortunately, however, without the necessary infrastructural developments which include broadband, Irish farmers could continue to be marginalised when it comes to ICT adoption.

Constraints that Impede Farmers from Using ICT

It emerged throughout this chapter that older farmers are less likely to use ICTs than farmers under 50 years of age. Older farmers have less understanding of the benefits of ICT adoption. This is partly due to their lower education levels. According to ISITA (2003) computer training is an issue for farmers in the adoption of ICT and research findings presented in this chapter also found that the lack of computer skills is an issue. Less than one third of non users of AgriLink, for instance, had completed any formal computer training while none had completed more advanced ECDL training compared with 57% of users who had training. This lack of training will continue to be a barrier to increased uptake of ICT among non users.

While the skills necessary to engage with ICT continue to limit uptake, the lack of infrastructure necessary to use ICT to its full potential continues to be a barrier. The lack of availability of broadband in rural areas is limiting the impact of ICT among Irish farmers. While rates of broadband use among AgriLink users, for example, was high at 51%, it was much lower among non users at 24% indicating that lack of availability and lack of awareness of the potential of broadband was evident among non users. It may be recalled that recent research by Smith (2008) indicates that 75% of rural areas in Ireland do not have an adequate broadband supply. This is significant since 20% of non users of AgriLink did indicate that improved internet access would encourage them to use it.

RECOMMENDATIONS

The adoption of computer technology to enhance the flow of information to farmers is seen as a vital tool to allow farmers remain competitive on

Figure 4. Distribution of non-using Respondents by Main Factor that would Encourage them to use AgriLink (n=29); Multiple responses allowed

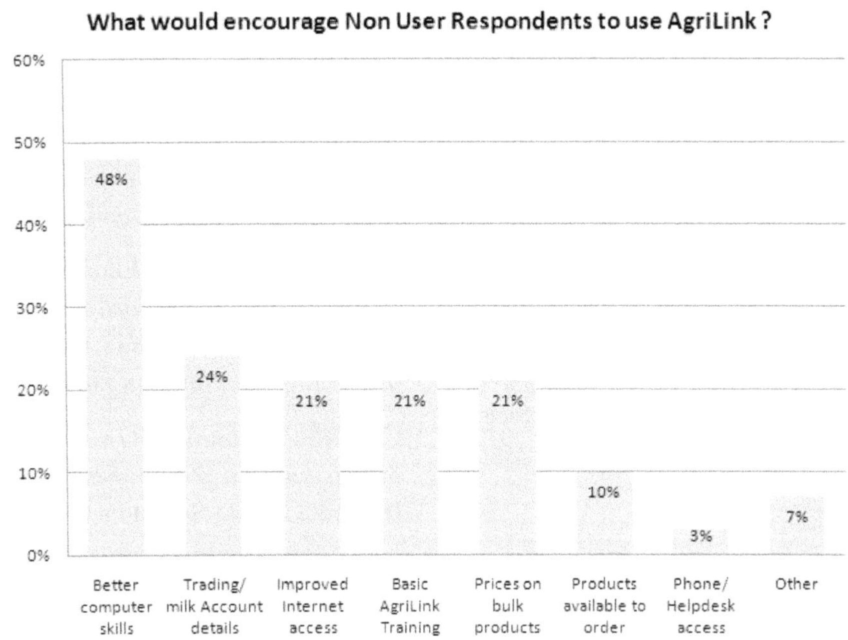

a global stage. What is clear from this chapter is that a core group of farmers have embraced ICT and are benefiting from its adoption. However, it is essential that barriers to the adoption of ICT by farmers who are not using it are overcome. Efforts must be made to focus all players within the agricultural industry to make ICTs more relevant to farmers.

The transformation which has taken place in Irish agriculture due to eGovernment has been remarkable with the facilities to interact with the Ministry of Agriculture on a number of schemes for farmers. This development is welcome and must continue. However, one area where Government has not been so successful is in broadband provision to rural areas, where the farming community are concentrated. Today, there is a major imbalance in the provision of broadband to urban and rural areas with many rural areas trailing in ICT penetration. The commitment given by Government to extend the benefits of the Information Society to all parts of the country including farm-

Table 5. Level of computer training completed by respondents according to AgriLink user group (n=110)

Computer training	Frequent Users (n=42)		Occasional Users (n=41)		Non Users (n=27)		Total
	No	%	No	%	No	%	No
Beginners course	16	38%	18	44%	9	33%	**43**
Advanced course (ECDL)	8	19%	5	12%	0	0	**13**
No Training	18	43%	18	44%	18	67%	**54**
Total	**42**	**100%**	**41**	**100%**	**27**	**100%**	**110**

ing communities and the rural population must be delivered. Broadband is now accepted as vital to rural sustainable development and without it farmers cannot make full use of ICT.

Finally, it is evident that there is a desire among farmers to embrace ICT but without the necessary skills to do so, this will not happen. If farmers are to adopt ICT on a grand scale it will only happen if they are taught the skills which they can implement immediately in relation to their farming businesses. While agribusinesses themselves may not want to get involved directly in training, it is important that they are supportive of organisations such as Teagasc or other local organisations in delivering the type of training required by farmers.

REFERENCES

Alam, S., & Noor, M. (2009). ICT adoption in small and medium enterprises: an empirical evidence of service sectors in Malaysia. *International Journal of Business and Management*, *4*(2), 112–125.

Amarach Consulting. (2007). *A research de-brief for Comreg.* Prepared by Amárach Consulting. Retrieved from http://www.comreg.ie/_fileupload/publications/ComReg0708.pdf

Bayo-Moriones, A., & Lera-Lopez, F. (2007). A firm-level analysis of determinants of ICT adoption in Spain. *Technovation: The International Journal of Technological Innovation. Entrepreneurship and Technology Management*, *27*, 352–366.

Chambers Ireland. (2007). *E-business survey 2007,* Vol. 1. Co Partnered Chambers Ireland and Comreg Dublin. Retrieved from http://www.chambers.ie/ktmlpro/files/uploads/eBusiness-Survey%2007.pdf

Comptroller & Auditor General. (2008). *Special report: eGovernment (2007).* Government of Ireland. Retrieved from http://www.audgen.gov.ie/documents/vfmreports/58_eGovernment.pdf

Comreg. (2010a). *Irish communications market: Quarterly key data report; Data as of second quarter 2010.* Document no 10/73R. Commission for Communication Regulation, Dublin.

Comreg. (2010b). *Irish communications market: Quarterly key data report; Data as of first quarter 2010.* Document no 10/43. Commission for Communication Regulation, Dublin.

CSO. (2010, June). *CSO output, input and income in agriculture, 2009: Final estimate.* Dublin, Ireland: Central Statistics Office.

Department of Agriculture. Fisheries and Food. (2010a). *Annual review and outlook for agriculture, fisheries and food 2009/2010.* Dublin, Ireland: Author.

Department of Agriculture. Fisheries and Food. (2010b). *Factsheet on Irish agriculture – August 2010.*

Department of Agriculture and Food. (1999). *Ensuring the future- A strategy for rural development: A white paper on rural development in Ireland.* Dublin, Ireland: Author.

Economics and Planning Division, Dublin, Ireland: Author European Commission. (2007). *Europa: Activities of the European Union, summary legislation.* Retrieved from: http://europa.eu/legislation_summaries/information_society/index_en.htm

Fabiani, S., Schivardi, F., & Trento, S. (2005). ICT adoption in Italian manufacturing: Firm-level evidence. *Journal of Industrial and Corporate Change*, *14*(2), 225–249. doi:10.1093/icc/dth050

Government of Ireland. (1999). *Implementing the information society in Ireland: An action plan.* (PN6727) Government of Ireland. Retrieved from http://www.taoiseach.gov.ie/attached_files/Pdf%20files/InfoSocietyActionPlanJan99.pdf on 9 December, 2010

ISITA. (2003). *Identification of the obstacles to ICT uptake on farms and the strategic way forward.* Annual Technical Meeting of the Irish Society for Information Technology in Agriculture (ISITA), 23 October, 2003, Newbridge, Ireland. Retrieved from http://www.iol.ie/~harkin/isita.htm

Millward Brown, I. M. S. (2008). *Consumer ICT survey: Quarter 2, 2008, A review of findings.* Millward Brown IMS. Retrieved from http://www.odtr.ie/_fileupload/publications/ComReg0849.pdf

Millward Brown, I. M. S. (2010a). *Consumer ICT survey: Quarter 4, 2009, A review of findings.* Millward Brown IMS. Retrieved from http://www.odtr.ie/_fileupload/publications/ComReg1022.pdf

Millward Brown, I. M. S. (2010b). *Consumer ICT survey: Quarter 2, 2010, A review of findings.* Millward Brown IMS. Retrieved from http://www.odtr.ie/_fileupload/publications/ComReg_1062r.pdf

Millward Brown, I. M. S. (2010c). *SME and corporate ICT research; First half, 2010.* Millward Brown IMS. Retrieved from http://www.odtr.ie/_fileupload/publications/ComReg1033.pdf

Morrow, L., Kelly, T., & Kirley, T. (2003). *ICT – Its potential as a channel for enhanced extension services.* Paper presented at the AIAEE conference, University College Dublin, June 2003.

New Connections. (2002). *A strategy to realise the potential of the Information Society.* Second Government Action Plan, Government of Ireland. Retrieved from http://www.taoiseach.ie/attached_files/Pdf%20files/NewConnections-March2002.pdf

O'Donnell, S., McQuillan, H., & Malina, A. (2003). *E-inclusion: Expanding the Information Society in Ireland.* Report to the Information Society Commission, Dublin.

Ross, B., & Waksman, G. (2004). *Internet use in agriculture: Progress driven by demand, not by supply.* Retrieved from http://www.acta-informatique.fr/telechar/NewAg/progress_driven_by_demand.pdf

Smith, P. (2008, February 23). No Broadband in 75% of rural homes in Ireland. A national survey of Irish Farmers' Association 945 branches conducted in 2007. *Irish Farmers Journal*, 8.

Vlosky, R. P., Fontenot, R., & Blalock, L. (2000). Extranets: Impacts on business practices and relationships. *Journal of Business and Industrial Marketing*, *15*(6), 438–457. doi:10.1108/08858620010349510

Wims, P. (2007). Analysis of adoption and use of ICTs among Irish farm families. *Journal of Extension Systems*, *23*(1), 14–28.

Windrum P., & de Berranger, P. (2003). *Factors affecting the adoption of intranets and extranets by SMEs: A UK study.* Maastricht Economic Research Institute on Innovation and Technology, report no 023s.

Chapter 15
Implementation of E-Learning in Forest Workers' Training for Sustainable Forest Management

Petros A. Tsioras
Aristotle University of Thessaloniki, Greece

ABSTRACT

Forest workers training is recognized as a necessary precondition for sustainable forest management. As sustainable forest management sets higher levels of standards, the adoption of new information and communication technologies (ICT) in forest workers' training and extension systems is a necessity in the Information Age, which will help the current and future workforce to better fit in the increasingly digitalized and demanding world of forest work. However, limitations on the use of ICTs do exist and should be seriously taken into consideration. Digital divide is evident both in developed and developing countries, making the implementation of ICT applications difficult, especially in the case of African and Asian countries. Furthermore, the development of e-learning systems for forest professionals must adhere to very specific design standards and educational theories. In this context, research should focus on the needs of the end-users at regional or national level. However, all efforts will be ineffective unless they are incorporated and facilitated by a stable framework of political and financial support. The combination of long-term planning, clear target setting, and cost effectiveness is critical for the successful promotion of sustainable forest management through e-learning.

INTRODUCTION

New information and communication technologies have changed the way we live and work. Recent developments in these technologies find their way with increasing frequency in everyday life and are recognized as important factors for social and economic development. As Keegan stated in 2000, "most of the new business opportunities in the next few decades will either be the creation of digital goods or the application of digital and internet techniques to old economy

DOI: 10.4018/978-1-4666-2655-3.ch015

products". Although such statements may sound very optimistic, or far-fetched, the trend of implementing new technologies is evident not only in the production systems but to practically all aspects of life.

The increase of knowledge puts higher standards in terms of skills and competencies. Good knowledge of ICTs is crucial for today's workers around the globe (Zarini, Varis, & Mar, 2009). Information Age requires different knowledge and skills and the very nature of employment has changed. The role of technical and vocational education and training (TVET) is of primary importance and complicated because it has to prepare the students of vocational schools for the new working environment. The investment on ICTs is nowadays regarded by TVET systems as a necessity, which will increase efficiency of the educational and training processes.

Forest workers' training systems (FWTSs) have emerged from the need for a better forestry workforce with certified qualifications, a very important precondition towards sustainable forest management. Since the majority of forest workers work and live in rural mountainous areas, new flexible forms of training should be used. Therefore, the geographical dispersion of the specific target group is an important factor for the implementation of ICTs in specialized vocational training programs with respect to forestry. In this context, a large number of websites and multimedia software has been developed as training tools for forest workers.

Aim of this book chapter is to present the special characteristics and needs that the use of ICTs in forest workers' training systems and forestry extension services must address. The role of Digital Divide and limitations of e-Learning in vocational training systems are also discussed. Successful case studies from all over the world are presented and finally research priorities are described. Despite the problems described, the implementation of ICTs in vocational training and extension services in forestry should be further encouraged because it is combined with multifold advantages: it can enhance the professional and entrepreneurial capacity of forest workers and forest owners, promote sustainable forest management and actively support rural and regional development.

THE GROWING ROLE OF ICTS IN TVET

Technical Vocational Education and Training (TVET) is concerned with the acquisition of knowledge and skills for the world of work (International Centre for Technical and Vocational Education and Training, 2010). Various terms have been used to describe elements of the field that are now conceived as comprising TVET, such as Apprenticeship Training, Vocational Education, Technical Education and Vocational Education and Training (VET) among others. In the Revised Recommendation concerning Technical and Vocational Education, United Nations Educational Scientific and Cultural Organization / International Labour Organization (2002) offer the following definition of TVET:

Technical and vocational education is used as a comprehensive term referring to those aspects of the educational process involving, in addition to general education, the study of technologies and related sciences, and the acquisition of practical skills, attitudes, understanding and knowledge relating to occupations in various sectors of economic and social life. Technical and vocational education is further understood to be:

1. An integral part of general education;
2. A means of preparing for occupational fields and for effective participation in the world of work;

3. An aspect of lifelong learning and a preparation for responsible citizenship;
4. An instrument for promoting environmentally sound sustainable development;
5. A method of facilitating poverty alleviation.

The paradigm shift as described by International Labour Organization (2003) sets new targets for the modern TVET systems. Modern training should be demand-driven and, at the same time, focused on the trainee. Lifelong training is introduced and multiple directions and choices should be available. In order all this to take place, extended use of ICTs in TVET systems is expected to facilitate the change from the old to the new paradigm of TVET.

ICTs were used in order to facilitate the development and strengthening of TVET around the world by enhancing networking and knowledge-sharing opportunities. The first electronic mailing lists dedicated to TVET emerged in the mid 1990s, such as VOCNET and 'Technology-edu', developed by the former National Center for Research on Vocational Education and Training at the University of California (Berkeley) and the Council on Technology Teacher Education respectively. Since then, the Internet has been extensively used as a communication and knowledge-sharing tool, and on-line communities in the field of vocational training have been developed. The European Training Village (ETV) is an interactive platform developed by the Centre for the Development of Vocational Training (CEDEFOP) accessible to anyone interested or involved in TVET. The ETV offers detailed up-to-date information on TVET in Europe, by means of its digital library of documents and databases, and facilitates networking through its forums.

A large number of terms has been used in order to describe the use of ICTs in TVET such as Remote Teaching, Computer Assisted Instruction, Computer Based Learning and many others. However, the term e-Learning is a wider definition which refers to learning that is supported by information and communication technologies (European Training Thesaurus 2010) and could offer flexible forms of training (European Centre for the Development of Vocational Training, 2002a). That is possible through the development of information delivery and exchange mechanisms unconstrained by location, time and distance, which will be based on multiple or hybrid methodologies, such as the use of internet, online learning and other electronic or interactive means (European Centre for the Development of Vocational Training, 2002b).

E-learning has been recognized as one of the major contributors in the change of Distance Learning, by covering needs and weaknesses of the "traditional" training systems (Herremans, 1995; Porter, 1997). New forms of training have been introduced and TVET today is not confined to the educational facility, but to the access to ICTs. These developments set new demands on all types of educational systems, which according to Haddad and Draxler (2002), were developed in the context of the Industrial Age, but now fail to meet the needs of Information Age (Butcher, 2003). To overcome this problem, new technologies must be introduced, that will enhance the flexibility of the training process and lift restrictions on time and place at reduced cost.

FOREST WORKERS' TRAINING AND SUSTAINABLE FOREST MANAGEMENT IN THE INFORMATION AGE

Forest work presents significant differences in many aspects compared to work in other production sectors. There is a broad variety of changing parameters with regard to the object of work. In wood harvesting for example, tree dimensions and forms can vary greatly, as well as other tree characteristics such as the number of branches, faults and tree health, all of which having a distinct impact on the productivity of the forest worker (Gröger, & Lewark, 2002). Another difference is the changing workplace. Forest stands can differ in many ways either referring to management actions (thinning, clearcutting, protection from erosion or even fire fighting in some countries) or stand conditions (slope, area, vegetation type, accessibility). Finally, everyone working in the forest is exposed to the weather conditions all year long.

The tasks a trained forest worker can complete is quite impressive; Efthymiou (2001) has estimated this number of works to exceed 70, including various works for forest protection, silviculture, torrent control and construction of forest roads among others. The majority of these tasks are very demanding in terms of energy expenditure (Apud, 1989). The increased difficulty in forest professions is also responsible for high accident rates and prevalence of profession-related diseases. An extensive study in Germany has shown, that one third of forest (33%) workers with at least ten work experience suffered from back pain, a ratio that was doubled (66%) at the case of very experienced forest workers (Poschen, 1993). Training is highly efficient in upgrading the skills of the forest workers, with subsequent results in the promotion of safety and health during work (Tsioras & Efthymiou, 2007).

Many studies have shown that the training of forest workers can result in increased productivity. In some cases the findings are impressive. In Germany during the period 1974-1975 the average productivity for wood harvesting (felling and skidding) was found to be $0.65m^3/h$ compared to $0.32m^3/h$ in Greece. According to Efthymiou (1989) this difference could be mainly attributed to the better level of training of the German forest workers. In another study, the combination of a new planting tool and training resulted in a productivity increase of 23%, reduction of the work cost by 11% (including the cost for training) and reduction of the heart rate, a significant factor measuring the work intensity, from 110 to 101 (Johansson & Strehlke, 1996).

Productivity, quality, efficiency, health and safety in forest operations need a well trained and motivated workforce. The first forest workers' training systems (FWTSs) have emerged from these needs at the end of 1920's in Germany (Gröger & Lewark, 2002). Since then, FWTSs are in a continuous evolutionary process, in order to offer training programs to a number of target groups, the most of which are hard to reach. The fact that the majority of forest workers works and lives in rural mountainous areas underlines the need for new flexible forms of training. The geographical dispersion of the specific target group is an important factor for the implementation of ICTs in specialized vocational training programs with respect to forestry. For this reason, a large number of websites and multimedia software has been developed to be used as training tools for forest workers.

There has been a growing awareness of the multiple use of forests since the 1980's. Forests can provide a broad variety of products and services, such as the production of wood and non-wood products, water production, carbon sequestration and tourism. The proper utilization of the forests directly constitutes one of the main pillars of sustainable development, under the precondition that it is implemented in a sustainable way. Sustainable forest management is considered as one of the most important contributions which the forestry sector can make to the sustainable development objectives of any nation, particularly those richly endowed with forest (FAO 2010b).

The UNCED, held in Rio de Janeiro in 1992, set the guiding principles for sustainable forest management as a contribution towards sustainable development. The "Forest Principles" and Chapter

11 of Agenda 21, adopted during the conference, include the commitments made by countries in the forestry domain in order to achieve sustainable development in all types of forests on the planet. In many paragraphs of Chapter 11 it is made clear, that the development of human resources through training represents a prerequisite towards sustainable forest management. International certification systems (FSC, PEFC, CSA) also recognize the importance of forest workers' training with regard to sustainable forest management. The reason is very simple: sustainability and multiple use of the forest resources is a utopia, unless the local populations and all the people employed in forestry are well trained and informed. Trained forest workers can make possible the design and implementation of modern harvesting systems, which can guarantee higher productivity, increased safety during work and reduced impacts to the forest stand (Efthymiou 2003). Such developments can bring forestry closer to the demands of the society, with increased economic and ecologic efficiency.

Forest management has become significantly more difficult in the last decades as a result of the often conflicting forest products and values, especially in the context of multiple purpose forestry objectives over a range of spatial and temporal scales. The use of ICTs in FWTSs can provide new exciting ways and increased efficiency in teaching the complexity of sustainable forest management goals and help the future forest workers acquire a sound knowledge background on these subjects.

Forestry in the Information Age is changing to Digital Forestry. According to Zhao et al. (2005), Digital Forestry is the science, technology, and art of systematically acquiring, integrating, analyzing and applying digital information to support sustainable forests. Although many forest scientists and forestry professionals are already applying a wide variety of digital technologies in forestry, the gap between contemporary forestry and the concepts embodied in digital forestry is significant (Shao & Reynolds, 2006). However, a large variety of forest machinery and forestry equipment have been adopted in a wide range of forest operations in the last decades. This complicated machinery integrates ICTs with increasing frequency. Modern harvesters and forwarders are nowadays equipped with a modified PC unit, while other technologies like GIS and GPS are increasingly used on portable devices. These developments have added new possibilities and increased productivity, but also impose increased demands on behalf of the forest workers in terms of their ICT skills and competences. For this reason, the curricula of forest worker training schools have been enriched with training modules which aim to increase the computer skills of the trainees. Also ICTs have been introduced in forest workers training schools as a means to improve the teaching of theoretical subjects. Such changes comply with the modern definition of forest worker, according to which he/she is not limited to the tasks of the traditional logger, but is regarded as a versatile collaborator in forestry, a self-sufficient skilled worker, who is resourceful at the practical level (Evju, 1989).

It is obvious that FWTSs have to adapt to the new demands which arise from the modern working environment in forestry. The use of ICTs is necessary due to the ever increasing use of digital technologies in forest machinery, equipment and production methods in forestry. Findings in studies confirm a relationship between the perception of usefulness of computers to perform job tasks and the level of usage (Koohang, 1989). However, the social aspect should be also stressed: Forest professions suffer from low occupational status and professional isolation (Egan & Taggart, 2009). This situation sometimes leads to displacement of forest workers in other fields of production (Helvoigt, Adams, & Ayre, 2003). The creation or participation in digital communities for forestry professionals can be used as a means to strengthen the bondages between the community members. Such actions can contribute to improvement of the public image of the forestry professions and can weaken the feeling of professional isolation, thus contributing to maintain the number of forest workers, a target very important in terms of rural development (Tsioras, 2004).

THE ROLE OF FOREST EXTENSION IN SUSTAINABLE FOREST MANAGEMENT

If the role of FWTSs is to prepare for working life, then forest extension services can be regarded as the advisory services aiming at the solving of problems during working life. Extension in forestry is "a process of integrating knowledge, attitudes and skills to determine what is needed, how it can be done, what local cooperation and resources can be mobilized, and what additional assistance is available and may be necessary to overcome particular obstacles in achieving forestry objectives" (International Union of Forest Research Organizations, 2002). It is a targeted educational process which involves forest owners or forest personnel, which can be used as an individual instrument or in support of other instruments, such as obligations, prohibitions or material incentives (Wild-Eck, Zimmerman, & Schmithüsen, 2006).

Forest extension services are provided by both international and national organizations, universities and private companies and differ from country to country according to the special needs and conditions. Despite the variety of target groups, forest extension refers to the promotion of problem solving, as well as participatory and multi-stakeholder approaches to enhance the contribution of trees and forests to sustainable land use (Food and Agriculture Organization, 2010). Extension services are recognized as one important instrument used to promote sustainable forest management (Wild-Eck et al., 2006) and sustainable development through capacity building and the strengthening of local structures and knowledge (Organization for the Economic Cooperation and Developent, 1995; Rifkin, 1999; Wild-Eck et al., 2006).

Forest extension is provided to a large number of beneficiaries such as forest workers, forest owners, forest contractors and small companies and/or local populations. The efficiency of forest extension programs is often under question, due to the fact that the requested programs are delivered to only a small portion of the respective target group (Boyd et al., 1988). Another trend is the privatization of public extension services, a debate point because of its possible implications on resource poor target groups (Oladele & James, 2008). The use of ICTs can overcome these difficulties by enabling the delivery of information to all target groups unrestricted of time and place limitations. In this context, websites represent a constant information point for every one interested in forest management, expanding their reach to stakeholder groups often difficult to be exactly defined and communicated, such as small forest owners living in urban centers. Information and training material has also been developed in various digital forms, such as of text, image video files or web courses covering specific local needs. Email lists and web conferencing can also facilitate communication between extension services' agents and interested parts (Morford & Hollstedt, 2007). Evolving communication and information systems, including the Internet and village telecentres, are potentially powerful tools which can facilitate two-way communication with marginalized rural and forest dependent communities (Food and Agriculture Organization, 2010a).

ICT advances and innovations in the past 20 years have enabled significant changes in the practice of forest management and the respective tools (Reynolds et al., 2005). A wide variety of software applications is available to support decision making in forest management, including databases, growth and yield models, wildlife models, silvicultural expert systems, financial models, geographical information systems (GIS), and visualization tools (Schuster, Leefers, & Thompson, 1993). Special emphasis should be given to the advance of Decision Support Systems (DSSs), which have been especially developed in order to improve the consultation process in forest extension systems (Lexer, Vacik, Hönninger, & Unegg, 2005). DSSs are computer-based frameworks for integrating data and expert opinion with analytical and operational research models, by means of graphic display and tabular reports, to reveal the basis of support for alternative decisions (Fischer, Scholten, & Unwin, 1996).

The complexity of goals, as they are set by modern forest management, requires extension services of good quality. Forest workers and forest owners have to keep up with the latest developments, in order to remain competitive and secure their income. As forestry has been enriched, during the last decades, with a broad variety of products and services, such as the non-wood forest products, flow of up-to-date information to all interested members is now, more than ever, a necessity in times of continuous changes, like ours. For this reason, extension services have adopted ICTs, in their educational and advisory functions, in order to provide cost-effective mix of seminars, distance delivery and web based extension (Alberta Forest Extension Network, 2009).

LIMITATIONS AND SPECIAL CONSIDERATIONS ON THE USE OF ICT IN FWTS

Digital Divide

Investment on ICTs represents a big chance for the development of developing countries (International Labour Organization, 2003) as well as for the poor areas in developed countries (Leiken, 2002). It is a necessary move to integrate technologies that change the global economy for everyone who wants to remain informed and competitive. The adoption of ICTs in rural areas for agricultural purposes is a unique challenge and has long been of specific public concern with regional, national and international strategic significance (Gelb & Offer, 2006). However, some members of society or areas of the world are left behind others who have access to new information and communication technologies due to the "digital divide". Digital divide is evident in many parts of the world in a variety of forms (Reding, 2001):

- Between the United States and the European Community;
- At European level between the digitally literate and those who don't possess computer skills;
- Between genders (the majority of internet users is male); and
- Between age groups, as younger people use the internet more than older people.

Two world regions heavily affected by the digital divide are Africa and Asia. According to the latest data on internet usage (See Table 1) these areas have the lowest penetration rate in their populations. Internet use is globally expanding, albeit at a slower pace than in the former years. In 2009, an estimated 26 per cent of the world's population were using the Internet (International Telecommunication Union, 2010), while in 2010 this percentage has increased to 28.7 (Internetworldstats 2011). In developed countries, the percentage remains much higher than in the developing world where four out of five people are still excluded from the benefits of being online (International Telecommunication Union, 2010). However, digital divide is also present between regions within the same country, as a result of differences in levels of the necessary infrastructure (Sern, 2001). Rural areas are negatively affected by the uneven distribution of networks, because newer technologies are, at most cases, made available by the communication companies in the densely populated urban areas.

Table 1. World internet usage and population statistics (Source: internetworldstats 2011)

World Regions	Population (2010 Est.)	Internet Users		Penetration (% Population)	Growth 2000-2010	Users % of Table
		Dec. 31, 2000	Latest Data			
Africa	1,013,779,050	4,514,400	110,931,700	10.9%	2,357.3%	5.6%
Asia	3,834,792,852	114,304,000	825,094,396	21.5%	621.8%	42.0%
Europe	813,319,511	105,096,093	475,069,448	58.4%	352.0%	24.2%
Middle East	212,336,924	3,284,800	63,240,946	29.8%	1,825.3%	3.2%
North America	344,124,450	108,096,800	266,224,500	77.4%	146.3%	13.5%
Latin America/ Caribbean	592,556,972	18,068,919	204,689,836	34.5%	1,032.8%	10.4%
Oceania / Australia	34,700,201	7,620,480	21,263,990	61.3%	179.0%	1.1%
World total	6,845,609,960	360,985,492	1,966,514,816	28.7%	444.8%	100.0%

Still, the lack of possession of computers and network connections is only one part of the problem. Equally significant are the lack of digital skills, as well as the lack of meaningful usage opportunities (Besser, 2001; Van Dijk & Hacker, 2003; Fuchs and Horak, 2008). Despite the fact that the gap between developed and developing countries is closing in terms of physical access to ICTs, it is widening with regard to the digital skills required and the usage opportunities (Van Dijk & Hacker, 2003). Digital divide is a multidimensional phenomenon (Norris 2001) and bridging it requires equal treatment of all its components.

The above mentioned dimensions of digital divide are evident in the forest sector of both developed and developing countries. Leary and Berge (2006) revealed that even in the advanced economies of Europe and America, e-learning adoption in agriculture falls significantly behind the current adoption rates seen in non-agricultural sectors. In the case of forestry, the majority of forest workers and forest owners live in remote areas, which is a restricting factor with regard to broadband internet. Furthermore, forest workers perceive the priority of training in the use of computers as low (Egan, 2005).

The advance of new technologies has influenced the curricula of FWTSs. In this new era, forest workers' training schools replace traditional teaching methods with ICT-supported ones and add computer in their curricula. This is a necessary development, because digital literacy is regarded as a prerequisite, in order to make better use of the existing technologies. At the case of FWTSs, limited knowledge of ICTs inhibits the integration of new technologies in the educational procedure, as well as possible future use in the form of extension services. Young people in developed countries nowadays have the necessary background to make use of the provided material, which does not happen with older forest workers, forest owners with limited computer skills or forest workers in developing countries. For the latter, training or extension agents, playing the role of intermediaries, can facilitate the diffusion of knowledge in digital form to them during seminars.

Limitations on the Use of E-Learning in TVET and FWTSs

The ever increasing potential of ICTs should not be treated as panaceas for all kinds and activities in TVET. In fact, in many areas of TVET such as the FWTS, where hands-on experience and manual skills are required, ICTs and virtual information cannot replace the need for fixed education and training facilities with specialized equipment and trained specialist instructors (Zarini et al. 2009).

Similarly, e-learning has been reported to be a limiting factor in the learning process of trainees, who learn better after using their own physical activity (Zirkle, 2003).

In TVET areas where manual skills are crucial, such as the training of forest workers, ICTs can only substitute for a small portion of hands-on, practical work. Blended learning, which refers to a combination of e-learning and traditional learning methods, can reduce the need for fixed education and training facilities, especially in the case of some theoretical training subjects, by offering at the same time, more individualized learning to the students (Herremans, 1995; Planas, 2001; International Labour Organization, 2003). This combination of methods allows for face-to-face interaction between teachers and students and among students, which should not be underestimated. Furthermore, the provision of material and instruction through "free instructors" on the internet can facilitate extra help to the trainees who wish to further personalize their training process (Zarini et al. 2009).

Teachers Delivering Digital Educational Material in TVET

The role of the teacher in vocational training systems is crucial; he must be able to use the new technologies, but his primary focus must always remain his students and the training procedure (Shumer, 2001). The new teaching methods and tools should be used in order to motivate the students, so that they will be actively involved and not passively attend the instructor (Rathenberg & Mielk, 1986). As a result, the advisory, supportive and encouraging role of the instructor is substantially upgraded (Lionarakis, 2001).

The teacher using the new technologies doesn't have to be a computer specialist or technician knowing everything on computers and communication technologies (Porter, 1997). However, he must know how to use the educational software efficiently and be frequently informed on new developments (Herremans, 1995; International Labour Organization, 2003; Kotrlik, Harrisson, & Redman, 2000). For this reason, in the majority of European countries, in-service education in ICT is part of a national programme to initiate, develop and improve the use of ICT by teachers (Delhaxhe, 2004).

Such initiatives and programs have had a very positive impact on the teachers' ICT skills (Zarini et al., 2009). However, 'their poor ICT competence, low motivation and lack of confidence in using new technologies in teaching are significant determinants of their levels of engagement in ICT. These are directly related to the quality and quantity of teacher training programmes' (Balanskat, Blamire, & Kefala, 2006). It is also very difficult for teachers on their own to develop and implement pedagogical use of digital media in the classroom (Lieberg & Morgan, 1997). An issue that requires considerable attention is the preparation of teachers for ICTs in TVET. This problem can be overcome through the definition of nationally accepted standards of content or skill, that teachers are expected to possess in order to use new technologies effectively in the classroom (Lieberg & Morgan, 1997).

Cost of Equipment Acquisition and Maintenance

The implementation of e-learning requires a minimum technological platform: necessary hardware, adequate telecommunication capabilities, current browser versions, access to software, etc. In an effort to not create distrust and frustration in employees, the organization who offers or organizes e-learning courses should ensure appropriate technological capability (Minton, 2000). The system should be completely tested and user problems should also be anticipated and addressed. Once these problems are addressed, possible problems concerning the end users' equipment should also be taken into account. E-learning courses will not provide cross-platform flexibility, unless thorough consideration is given to all the possible permutations of equipment that may be used to complete on-line coursework (Minton, 2000).

Finding hardware and software at affordable prices is another challenge. The rapid obsolescence of quite expensive workshop or laboratory equipment, are now amplified by the even shorter life-span of computer equipment, as the economic life of ICT is measured in months before a new or improved model becomes available (Zarini et al., 2009). Third world schools often purchase used equipment or rely on donated hardware that may be of limited functionality (Tiene, 2002). A positive development however, is the declining cost of computer hardware as a result of evolution in the manufacturing technology of semiconductors (Flamm, 2003). Last but not least, the lack of technical expertise is another problem to be addressed. The ability to find the technical support to set up systems and repair equipment that malfunctions can be extremely problematic. School administrators and teachers with little technology experience need some guidance as to how to work with the equipment and how to integrate its use into the curriculum (Castro, 1998).

This implies that the policy-makers have to balance investment cost considerations with considerations of technological obsolescence and the pedagogical necessity to have ICTs available for instruction (Zarini et al. 2009). The complexity of issues involved in these decisions certainly does not make easy the selection among alternatives. The use of e-learning should be implemented in order to enhance the educational process only when the training organization has sufficient funding and trained personnel to cope with possible technical problems.

Preparation of E-Learning Material

The introduction of new technologies in vocational education is successful when the combination of technologies and training material satisfies both the target set by the respective curricula and the needs of the trainees. Before new training methods or material is introduced in a vocational training system the following methodology must be applied (Lipsmeier, 1991):

- Analysis of needs and identification of problems.
- Analysis of the educational material and the trainees.
- Choice and preparation of alternative educational material.
- Recognition and description of limitations regarding implementation.
- Assessment of the alternative training material in terms of efficiency and applicability.
- Design and implementation of training means and material.
- Implementation of the new system under control.
- Assessment of the results - improvements.

The strategy to use e-learning and traditional teaching methods is not competitive, but rather the two methods are complementary in nature (Shumer, 2001). However, making use of the new technologies is subject to very specific design standards and methodologies, which aim at improving the efficiency of the educational process. In this context, the language used must be simple, understandable and, should be interesting in order to gain the trainee's attention (Avouris et al. 2001). The training material should be divided in smaller chapters (sections), which in turn will be accompanied by short introductory texts. Also, the use of practical examples, case studies and self-assessment exercises can facilitate better understanding of the training material. Finally, if the course is offered through a website, it should look attractive and must be updated regularly. This is necessary because the learners need changes, which keep the course current, interesting and accurate (Porter, 1997).

One very important aspect of training software is the design of user interface, also known as Human-Computer Interaction or HCI. User interface is also subject to specific standards, such as the ISO 9126 standard (International Organization for Standardization, 2001), in which quality factors such as usability, functionality, efficiency, reliability and maintainability are defined. User interfaces are considered to be a prime ingredient of computer user satisfaction (Mavromoustakos, Papanikolaou, Andreou, & Papadopoulos, 2005). Whether Graphical User Interface (GUI) or Web-based User Interface (WUI) it is important that user interface fulfills its target to facilitate increased educational efficiency through a well designed and easy to navigate software. Optimized User Interface Design requires a systematic approach. Especially in the case of e-learning design process, usability testing is of vital importance, due to the remote nature of the e-learning activities (Wong, Nguyen, Chang, & Jayaratna, 2003).

CASE STUDIES OF ICT IMPLEMENTATION IN FWTS AND FOREST EXTENSION

There have always been efforts to make use of modern media, tools and methods in vocational training systems for forest workers. In this context, educational material for forest workers was developed in the form of videotapes (Nugent, 1982) while today there is a large variety of such material in the form of videotapes or DVDs. Also, the expansion of radio and television makes these media attractive solutions for reaching rural populations (Food and Agriculture Organization, 2010). Television shows aiming at small forest owners have been very successful in Canada and USA (Blumenstock, 1988). Nowadays, the ICTs have further increased the possible communication options, making use of new options such as of the internet and the organization of village telecenters. Such solutions can facilitate two-way communication with marginalized rural and forest dependent communities and help bridging the digital divide for the benefit of sustainable forest management (Food and Agriculture Organization, 2010).

The potential of Informatics and its positive uses in terms of forest management impact has been recognized and utilized in forestry since their very appearance. As a result, ICTs have been used in the vocational training of forest workers, since the 1990s with increasing frequency and ways, replacing or supporting the traditional teaching methods. However, as it has already been mentioned, the dexterities and skills demand practical training which cannot be replaced by e-learning. For this reason, there is no offer for complete online courses leading to vocational certification of forest workers.

The potential of new technologies upgraded their role as supportive means of teaching. Through simulation environments, it is easy to visualize changes that will happen in forest ecosystems when management parameters are changed (Gerding, 2001). At the forest workers' training school of Neuheim-Hüsten in Germany, wood harvesting machine simulators have been used for the training of forest machine operators (Wagner, 1997). In the forest workers' training school of Münchenhof software has been developed, aiming at the teaching of crane loaders in the fields of safety, ergonomics, protection of the environment, work organization, legislation, wood harvesting and basic knowledge on machinery maintenance and repair (Morat, 1997).

Some other efforts concentrate on the development of multimedia titles for forest workers. Alberta Forest Products Association (AFPA) created the multimedia CD "Leadership in Health and Safety (LH&S) Training". This CD is an alternative interactive version to the three day LH&S Training in a classroom setting, and is designed to be user friendly for learners at all level (Alberta Forest Products Association, 2010). This training material is also provided, along with four other training sessions for online training for the Association members (www.albertaforestproducts.ca/resources/training.aspx).

The multimedia CD "Forest work in Europe" (original title in German: Die Waldarbeit in Europa) resulted from the cooperation of the forest workers training schools of Ort (Austria), Königsbronn (Germany) Abertamy (Czech Republic) in the framework of Leonardo da Vinci program, in 2001. It is primarily focused on the safe use of chainsaw, starting from the basic knowledge demanded, and extends progressively to the analysis of specific techniques. Main focus of this multimedia title is the promotion of safety and health during work.

The multimedia title "Flexiform" was also developed in the framework of the Leonardo da Vinci Program with the aim to facilitate the continuing vocational education training of self employed people with limited access to such programs. It consists of three CDs and provides a modular training solution in the field of sustainable forest management (Descombes, 1998). The product was also designed to be used as a training tool for forest workers' training schools and centres. It is important to stress that this product can help its users carry out self-assessment before starting a training activity (European Commission, 2002).

The Institute of Intercultural Didactics (University of Göttingen, Germany) developed the DVD-ROM Ecosystem Forest (Ökosystem Wald). The educational material is accompanied by 400 photos, 50 videos and many panoramic images 360° and handles subjects like the atmosphere, human interventions and biodiversity. This multimedia title is used by the forest workers' training school of Weiburg (Germany) (Gerding, 2001).

The development of training material is not restricted in multimedia titles. Webpages have been also designed in order to facilitate the training and inform the forest workers in many countries. The forest training center of Neheim-Hüsten (www.forstliches-bildungszentrum.nrw.de) offers courses on work safety and ergonomics. The training environment contains useful information on work safety and ergonomics, noise, vibration, gases, accident protection and work-related diseases, as well as the possibility to measure their level of knowledge on these subjects. Web courses are also provided by the Norwegian Forestry Extension Institute (http://www.skogkurs.no/nettkurs.asp), with a more general focus on instructions for forest owners.

The website www.eduforest.org contains very useful information about the forest professions and contact informations with forest workers' training schools and centers in most European countries.

The website of the German Board of Trustees for Forest Work and Forest Technology - KWF (www.kwf-online.de) provides plenty of information to everyone interested in forest operations. It contains updated information on the curricula of forest workers' training programs in the federal states of Germany, and useful material, including publications, reports and results from the testing of new forest machines. Similar content is provided on the website www.codoc.ch by the Swiss Authority of Coordination and Documentation for Forest Training.

The information and communication platform "waldwissen.net" (http://www.waldwissen.net/index) is the result of the cooperation of six German, Austrian and Swiss forest institutes and provides an impressive amount of material. It is a very interesting source of information with a large variety of thematic fields, ranging from ecology and forest protection to entrepreneurship in forestry and issues of forest technique. Finally, Extension Service Southern Regional (http://www.sref.info/) has created a website with lots of information including, among others, a forest encyclopedia, online publications freely available, list of events and podcasts on selected forest issues.

International organizations have a strong internet presence. The website of International Tropical Timber Organization (www.itto.org) includes a lot of information and interesting statistical data on tropical forests. Among the material presented there are also brochures and manuals in different languages on the sustainable management of tropical forests. Similar material is included in the website of the Center of International Forestry Research (http://www.cifor.cgiar.org). A large number of manuals in different languages on capacity building in forestry is available at the website of the Food and Agriculture Organization as part of their Education for Rural People Program (ERP) program http://www.fao.org/sd/erp/toolkit). Finally, the World Agro-forestry Centre (http://worldagroforestry.org) offers a large number of manuals in different languages with special focus on agro-forestry.

Forestweb originated as a two-year (2007-2009) Leonardo-project with the common goal to create a European network of knowledge on forestry and providing high quality education in sustainable forestry through the internet (INVERDE, 2010). Nine partners of eight European countries (Sweden, Austria, the Netherlands, the Czech Republic, Germany, Spain, Scotland and Belgium) participated and created a pilot formula with different course modules that can be used by other teachers/trainers as course material or as an example to develop new e-course modules. Forest workers and vocational students in forestry were the two target groups of the project. Following an analysis of needs of the target groups, a total of 18 different modules were developed, which had been available online through the website www.forestweb.eu (currently not available).

FUTURE RESEARCH NEEDS AND PERSPECTIVES

The potential of harnessing ICT in forest workers' training and extension services is at least promising. Now it is easier than ever to train the forest professionals and keep them informed about the principles of sustainable forest management. However, the shift from traditional to modern methods incorporates significant challenges; ingrained professional and cultural habits and attitudes will have to be overcome; resources need to be invested in new learner-based techniques of education and training; and the ICT infrastructure, including in schools and training institutions, will need strengthening. In this framework, some future research priorities should be set.

Human resources development is a difficult and very demanding process. Prior to the introduction of modern training methods and equipment, research efforts must determine the specific attitudes and perceptions of the target groups. Such actions will enable better understanding of their needs. Some e-learning systems failed to satisfy the learners' needs and requirements due to the absence of key human and technological issues in the engineering process. The successful development of E-Learning systems requires the identification of human issues such as, social and cultural factors (Mavromoustakos et al., 2005).

Detailed research is needed in order to find the optimal ratio between the use of ICT and human presence, taking into consideration the educational theories (Karoulis & Pombortsis, 2003). Pedagogy requirements should not be overlooked, and cannot be replaced by top-notch equipment. However, in some multimedia titles more attention has been given in the usage of new technological innovations rather than on any educational theory.

E-learning in forestry and forest extension has still many challenges ahead. The development of educational software and digital educational resources in small nations is an area for governmental attention and economic support due to the small size of their commercial market (Lieberg & Morgan, 1997). Furthermore, some countries lag, with regard to the implementation of the new training methods. Research projects, such the Leonardo da Vinci Programme represent an interesting opportunity, for the development, evaluation and delivery of internationally available training material, overcoming basic problems, such as funding issues. Also, technology transfer is facilitated through such projects from forestry schools and centers with large experience to the benefit of others with limited experience.

Much more research is necessary in the application of computer-aided decision making in the next several years (Reynolds & Schmoldt, 2006). Given the wide range of objectives of forest extension services, efforts should focus on the development and implementation of new tools. Evaluation and assessment studies will provide useful information regarding the qualitative and quantitative impact of these tools on the target groups. This knowledge is crucial for the policy makers to decide further adoption or cancelation of ICT-based extension software projects.

The first step towards the successful design and implementation of e-learning in FWTSs and forest extension systems is the setting of distinct educational and advisory targets. Analysis of the social, economic, cultural and forest management context will enable the precise determination of the local needs. If the targets set are not clear, success of the proposed methods and tools is at risk. However, it should be stressed that the greatest challenge of all is providing basic education to all people, as a prerequisite for access to the information and knowledge society. As the *World Employment Report 2001* strongly argues, "digital literacy is essential, but there remains an order of priorities in which literacy and access to a basic education of high quality are most fundamental" (International Labour Organization, 2001).

CONCLUSION

The use of ICTs in specialized vocational training systems can facilitate the provision of large amounts of information regardless of time and space. In the case of forest professionals, lifting these two limitations is very important, mainly due to their residence in remote mountainous areas. ICTs can facilitate better and more effective learning in forest workers' training, as research projects and implementation efforts have shown. The role of forest workers training systems is crucial in the integration of new technologies, and can act as the starting point for some forest professionals to make use of existing training, educational and extension material in digital form. Modern forest workers' training systems combined with extension services can significantly upgrade the skills, dexterities and even the entrepreneurial spirit of forest workers and owners, transforming the sustainable forest management from a buzz word to reality.

Bridging the digital divide is a key point for the effective use of ICTs towards sustainable forest management. It is evident both in developed and developing countries, and the gap is, unfortunately, widening. It is the role of policy makers and researchers to provide effective solutions. In this context, ICT projects and initiatives should consist of elements of long-term national or regional development policies and not standalone applications. These policies should address, in a successful way, all aspects of digital divide, with special focus on the development of the necessary digital skills of the target group or population and the effective use of the new technologies.

The combination of policy measures and financial support will create the necessary framework, in which research can flourish. Research should aim at cost-effective, user-oriented and not technology-driven applications (Mansell & Wehn, 1998). ICT applications are expected to fail or deliver poor results unless are driven of the specific demands of each region under study. This fact underlines the significance of international cooperation. Developing countries can benefit from the expertise of developed countries, as well as from financial and material support. Furthermore, the dissemination of information at national and international level will facilitate benchmarking the results of past and on-going projects, which, in turn, will enable better decision making in the future.

Despite the limitations discussed, the presented examples of ICT applications show a very serious potential with regard to the development of human resources in forestry. As neither capacity building nor forest production are fast processes, long-term planning and concerted efforts are required in order to deliver rewarding results. Well implemented and supported ICT applications can convince the end users of their usefulness and make them invest time or resources in them. Therefore, building knowledge partnerships and engaging the private sector and technology drivers in the pursuit of rural development goals is paramount if ICTs are to have a role in the promotion of sustainable forest management worldwide.

REFERENCES

Alberta Forest Extension. (2009). *An introduction to the Alberta Forest Extension Network*. Retrieved from www.afex.ca/images/stories/docs/afex_powerpoint_for_website_june_2009.pdf

Alberta Forest Products Association. (2010). *Training*. Retrieved from http://www.albertaforestproducts.ca/resources/training.aspx

Apud, E. (1989). Human biological methods for ergonomics research in forestry. In Apud, E., Bostrand, L., Mobbs, I. D., & Strehlke, B. (Eds.), *Guide-lines on ergonomic study in forestry*. Geneva, Switzerland: International Labour Organization.

Avouris, N., Solomos, K., & Tselios, N. (2001). Internet as a tool for open and remote teaching: Alternative mechanisms of quality control. In *Proceedings of the First Panhellenic Conference on Open and Remote Teaching*. Retrieved from http://www.eap.gr/news/EXAGGELIA_SYNEDRIOU/synedrio/html/sect5/22.htm

Balanskat, A., Blamire, R., & Kefala, S. (2006). *The ICT Impact Report: a review of studies of ICT impact on schools in Europe*. Brussels, Belgium: European Communities.

Besser, H. (2001). *The next digital divides*. Retrieved from http://www.tcla.gseis.ucla.edu/divide/politics/besser.html

Blumenstock, R. (1987). Reaching the woodlot owners with the use of mass media. In *Proceedings of the Seminar on Extension activities for owners of small woodlands* (pp. 98-109). Fredericton, Canada: Canadian Forestry Service.

Boyd, G. D., Daniels, B. J., Fallon, R., & Hyde, W. F. (1988). Measuring the effectiveness of public forestry assistance programs. *Forest Ecology and Management, 23*(4), 297–309. doi:10.1016/0378-1127(88)90060-6

Butcher, N. (2003). *Technological infrastructure and use of ICT in education in Africa: An overview*. Paris, France: Association for the Development of Africa.

Castro, C. (Ed.). (1998). *Education in the information age*. Washington, DC: Inter-American Development Bank.

Delhaxhe, A. (2004). *Key data on information and communication technology in schools in Europe*. Brussels, Belgium: EURYDICE.

Descombes, P. (1998). Flexiform – Self training tools suited to the profession. In *Forestry training for target groups that are hard to reach* (pp. 327-344). La Bastide-des-Jourdans, France: Centre Forestier de la Région Provence Alpes Côte d'Azur.

Efthymiou, P. N. (1989). Ergonomic problems and accidents in the mountain logging of Greece. In *Seminar on the mechanization of Harvesting Operations in mountainous terrain*. Antalya/Turkey 20-24 Nov. 1989.

Efthymiou, P. N. (2001). Forest protection and forest work. In Konstantinidis, P. N. (Ed.), *Development and protection of forests – Forest work*. Thessaloniki, Greece: Ministry of Agriculture, Forest Research Institute.

Efthymiou, P. N. (2003). Efficiency problems in harvesting small dimensioned wood. In Arzberger, U., & Grimoldi, M. (Eds.), *New trends in wood harvesting with cable systems for sustainable forest management in the mountains* (pp. 173–178). Rome, Italy: Food and Agriculture Organization, United Nations Economic Commission for Europe, International Labour Organization, International Union of Forest Research Organizations.

Egan, A. (2005). Training preferences and attitudes among loggers in northern New England. *Forest Products Journal*, *55*(3), 19–26.

Egan, A., & Taggart, D. (2009). Public perception of the logging profession in Maine and implications for logger recruitment. *Northern Journal of Applied Forestry*, *26*(3), 93–98.

European Centre for the Development of Vocational Training. (2002a). *Common vocational certificates project – Methodology for construction – Basic methodology*. Retrieved from www.trainingvillage.gr

European Centre for the Development of Vocational Training. (2002b). *Elearning Europe 2002: Survey*. Retrieved from www.trainingvillage.gr

European Commission. (2002). *Leonardo da Vinci series good practices: Distance learning. Luxemburg*. Office of the Official Publications of the European Communities.

European Training Thesaurus. (2010). Retrieved from http://libserver.cedefop.europa.eu/ett/en

Evju, O. (1989). Further training of professional forest workers. In *Proceedings of the Joint FAO/ECE/ILO Committee Seminar "The Training of Professional Forest Workers"* (pp. 129-142). Helsinki, Finland: The Finnish Government Printing Centre.

Fischer, M. M., Scholten, H. J., & Unwin, D. (1996). Geographic information systems, spatial data analysis and spatial modeling: An introduction. In Fischer, M. M., Scholten, H. J., & Unwin, D. (Eds.), *Spatial analytical perspectives on GIS*. Taylor and Francis.

Flamm, K. (2003). Moore's law and the economics of semiconductor price trends. *International Journal of Technology. Policy and Management*, *3*(2), 127–141.

Food and Agriculture Organization. (2010a). *Equitable partnerships, markets and forest extension*. Retrieved from http://www.fao.org/forestry/partnerships/en/

Food and Agriculture Organization. (2010b). *Criteria and indicators for sustainable forest management*. Retrieved from http://www.fao.org/forestry/ci/en/

Fuchs, C., & Horak, E. (2008). Africa and the digital divide. *Telematics and Informatics*, *25*, 99–116. doi:10.1016/j.tele.2006.06.004

Gelb, E., & Offer, A. (2006). *ICT in agriculture: Perspectives of technological innovation*. Hebrew University of Jerusalem, Centre for Agricultural Economic Research.

Gerding, V. (2001). Die Learntec 2001 in Karlsruhe – Vom vernetzten Denken zu multimedialer Didaktik. *Forsttechnische Informationen*, *8*, 84–87.

Gilbert, L., Sim, Y. W., & Wang, C. (2005). An e-learning systems engineering methodology. In *The 5th IEEE International Conference on Advanced Learning Technologies*, Kaohsiung, Taiwan (pp. 150-154).

Gröger, V., & Lewark, S. (2002). *Der arbeitende Mensch im Wald*. Dortmund, Germany: Bundesanstalt für Arbeitsschutz und Arbeitsmedizin.

Haddad, W. D., & Draxler, A. (2002). *Technologies for education. Paris, France: UNESCO*. Washington, DC: Academy for Educational Development.

Helvoigt, T. L., Adams, D. M., & Ayre, A. L. (2003). Employment transitions in Oregon's wood products sector during the 1990s. *Journal of Forestry*, *4*(1), 42–46.

Herremans, A. (1995). *New training technologies*. Turin, Italy: International Training Center of the ILO.

Hetemäki, L., & Nilsson, S. (2005). Introduction. In Hetemäki, L., & Nilsson, S. (Eds.), *Information technology and the forest sector* (*Vol. 18*, pp. 1–7). Vienna, Austria: IUFRO.

International Centre for Technical and Vocational Education and Training. (2010). *What is TVET?* Retrieved on March 5, 2010, from http://www.unevoc.unesco.org/2.0.html?tx_drwiki_pi1[keyword]=What%20is%20TVET

International Labour Organization. (2001). *World employment report 2001: Life at work in the information economy*. Geneva, Switzerland: International Labour Organization.

International Labour Organization. (2003). *Learning and training for work in the knowledge society.* Report 91 IV. Retrieved January 3, 2010, from http://www.ilo.org/public/libdoc/ilo/2002/102B09_22_engl.pdf

International Organization for Standardization. (2001). *IEC 9126-1, Software engineering – Product quality – Part 1: Quality model.*

International Telecommunication Union. (2010). *Measuring the information society 2010.* Retrieved March 8, 2010, from http://www.itu.int/dms_pub/itu-d/opb/ind/D-IND-ICTOI-2010-SUM-PDF-E.pdf

International Union of Forest Research Organizations. (2002). *Terminology project – New definitions*. Vienna, Austria: IUFRO.

InternetWorldStats. (2000). *World internet usage and population statistics*. Retrieved from http://www.internetworldstats.com/stats.htm 2011

INVERDE. (2010). *Forestweb – E-learning about sustainable forestry*. Retrieved March 9, 2010, from http://www.inverde.be/en/forestweb_987

Johansson, K., & Strehlke, B. (1996). *Improving working conditions and increasing profits in forestry*. Geneva, Switzerland: ILO.

Joint FAO/ECE/ILO Committee on Forest Technology. Management and Training. (1997). *People, forests and sustainability: Social elements of sustainable forest management in Europe*. Geneva, Switzerland: International Labour Organization.

Karoulis, A., & Pombortsis, A. (2003). The cognitive transfer and the tutor's role in a CBL environment. [Vilnius, Ithuania: Institute of Mathematics and Informatics.]. *Informatics in Education*, *2*, 1–15.

Keegan, V. (2000, 14 December). Divide and rule out. *Guardian Unlimited Net News*.

Koohang, A. A. (1989). A study of attitudes toward computers: Anxiety, confidence, liking, and perception of usefulness. *Journal of Research on Computing in Education*, *20*, 137–151.

Kotrlik, J. W., Harrisson, B. C., & Redmann, D. H. (2000). A comparison of information technology training sources, value, knowledge and skills for Louisiana's secondary vocational teachers. *Journal of Vocational Education Research*, *25*(4). Retrieved from http://scholar.lib.vt.edu/ejournals/JVER/v25n4/ doi:10.5328/JVER25.4.396

Leary, J., & Berge, Z. (2006). Trends and challenges of elearning in national and international agricultural development. *International Journal of Education and Development using ICT, 2*(2), 51-59.

Leiken, S. (2002). *Extending the information revolution – Secondary and Post-secondary education.* Retrieved from www.digitaldividenetwork.org/content/stories/index.cfm?key=226

Levedys, A. (2004). *Trends and current status of the contribution of the forestry sector to national economies.* FAO working paper on financing sustainable forest management: FSFM/ACC/07.

Lexer, M. J., Vacik, H., Hönninger, K., & Unegg, F. (2000). Implementing a decision support system for silvicultural decision making in low-elevation Norway spruce forests. In E. Klimo, H. Hager, & J. Kulhavy, (Eds.), *EFI Proceedings on Spruce Monocultures in Central Europe: Problems and Prospects,* Vol. 33 (pp. 119–134).

Lieberg, S., & Morgan, J. (1997). *Information and communication technologies in the Danish school system: A state of art evaluation in 6 areas of government action.* Copenhagen, Denmark: Ministry of Education.

Lionarakis, A. (2001). What kind of distance education are we talking about? In *Proceedings of the First Panhellenic Conference on Open and Remote Teaching*. Retrieved from http://www.eap.gr/news/EXAGGELIA_SYNEDRIOU/synedrio/html/sect2/34.htm

Lipsmeier, A. (1991). *Wie konzipiert man Fernunterricht in der Beruflichen Bildung?* Mannheim, Germany: Zentralstelle für Gewerbliche Berufsförderung.

Mavromoustakos, S., Papanikolaou, K., Andreou, A. S., & Papadopoulos, G. A. (2005). Human and technological issues in the e-learning engineering process. In First International Conference on E-Business and E-learning (EBEL'05) (pp. 343-348). Amman, Jordan.

Minton, M. C. (2000). *Is your organization ready for e-Learning? Seven key questions that need to be answered.* Retrieved from http://www.comproj.com/Minton.htm

Morat, J. (1997). Weiterbildung für Fahrer von Krannvollernter in Deutschland. *Forsttechnische Informationen, 3*, 38–40.

Morford, S., & Hollstedt, C. (2007). *Revisiting a forest extension strategy for British Columbia: A survey of natural resource practitioners and information providers.* Victoria, Canada: B.C. Ministry of Forest Range, Resources.

Norris, P. (2001). *Digital divide. Civic engagement, information poverty, and the internet worldwide.* Cambridge, UK: Cambridge University Press. doi:10.1017/CBO9781139164887

Nugent, J. D. (1982). Videotape recording – A practical tool for training forest workers. In *Seminar on the Management of Forest Worker Training,* Vol. 2, (pp. 21-28).

OECD. (1995). *Developing environmental capacity. A framework for donor involvement.* Paris, France: OECD.

Oladele, O. I., & James, O. (2008). Perceived effect of privatization of extension services among researchers, extension agents and farmers in Oyo State, Nigeria. *Agricultural Journal, 3*(5), 409–413.

Planas, J. (2001). *Continuing training in Europe: A problem in common.* CEDEFOP Workshop on Continuing Training Policies in the European Context. Retrieved from studyvisits.cedefop.eu.int/assets/.../lifelonglearning_initialcontinuing_en.pdf

Porter, L. R. (1997). *Creating the virtual classroom: Distance learning with the internet.* New York, NY: Wiley.

Poschen, P. (1993). Forestry, a safe and healthy profession? *Unasylva, 172*(44), 3–12.

Poschen, P. (1997). Forestry and employment – Much more than meets the eye. In *Proceedings of the XI World Forestry Congress.* Retrieved from www.fao.org/montes/foda/wforcong/PUBLI/V5/T34E/Default.htm

Rathenberg, E., & Mielk, A. (1986). *Wie entwickelt man Medien für die Berufliche Bildung?* Mannheim, Germany: Deutsche Stiftung für Entwicklung.

Reding, V. (2001). *The role of the European Community in creating the knowledge and Information society.* Speech at the Zentrum für Europäischen Integrationsforschung in the University of Bonn on 7.3.2001. Retrieved from http://europa.eu/rapid/pressReleasesAction.do?reference=SPEECH/01/108&format=PDF&aged=1&language=EN&guiLanguage=en

Reynolds, K. M., Borges, J. M., Vacik, H., & Lexer, M. J. (2005). ICT in forest management and conservation. In Hetemäki, L., & Nilsson, S. (Eds.), *Information technology and the forest sector - IUFRO World Series* (*Vol. 18*, pp. 150–172). Vienna, Austria: IUFRO.

Reynolds, K. M., & Schmoldt, D. A. (2006). Computer aided decision making. In Shao, G., & Reynolds, K. M. (Eds.), *Computer applications in sustainable forest management including perspectives on collaboration and integration* (pp. 143–169). Springer.

Rifkin, S. B. (1999). Capacity-building: An approach to people-centred development. *The International Journal of Health Planning and Management, 14*(4), 336–338.

Schuster, E. G., Leefers, L. A., & Thompson, J. E. (1993). *A guide to computer-based analytical tools for implementing national forest plans.* Moscow, ID: U.S. Department of Agriculture, Forest Service, Intermountain Forest and Range Experiment Station.

Sern, G. (2001). The broadband divide: A widening chasm in higher education? *EDUCAUSE Review, Jan/Feb 2001* (pp. 60-61). Retrieved December from www.educause.edu/ir/library/pdf/erm0117.pdf

Shao, G., & Reynolds, K. M. (2006). Introduction. In Shao, G., & Reynolds, K. M. (Eds.), *Computer applications in sustainable forest management including perspectives on collaboration and integration* (pp. 3–16). Springer.

Shumer, R. (2001). *A new, old vision of learning, working, and living: Vocational education in the 21st century. Journal of Vocational Education Research, 25(2).* Virginia Polytechnic and State University.

Tienne, D. (2002). Addressing the global digital divide and its impact on educational opportunity. *Educational Media International, 39*(3/4), 211–222.

Tsioras, P. A. (2004). *Analysis and formation of modern forest workers' training systems for the Greek Forestry.* Unpublished doctoral dissertation, Aristotle University of Thessaloniki, Greece.

Tsioras, P. A., & Efthymiou, P. N. (2007). Need for forest workers' training in Greek forestry. [In Greek with abstract in English]. *Geotechnical Scientific Issues, 18*(2), 46–56.

UNESCO & International Labour Organization. (2002). *Revised recommendation concerning technical and vocational education. Paris, France: UNESCO.* Geneva, Switzerland: ILO.

U.S. Department of Commerce. (2004). *A nation online: Entering the Broadband age.* Retrieved from http://www.ntia.doc.gov/reports/anol/NationOnlineBroadband04.htm#_Toc78020939

Van Dijk, J., & Hacker, K. (2003). The digital divide as a complex, dynamic phenomenon. *The Information Society, 19*(4), 315–326. doi:10.1080/01972240309487

Wagner, T. (1997). Softwaregestützte Simulationstechnologie. *Forsttechnische Informationen, 7-8*, 96–99.

Wild- Eck, S., Zimmermann, W., & Schmithüsen, F. (2006). Extension for private forest owners: Insights from a representative opinion poll in Switzerland. *Small-Scale Forest Economics. Management and Policy*, *5*(2), 161–174.

Wilson, D. N. (2001). Reform of TVET for the changing world of work. *Prospects: Quarterly ReviewofComparative Education*, *31*(1), 21–37. doi:10.1007/BF03220047

Wong, S. K., Nguyen, T. T., Chang, E., & Jayaratna, N. (2003). Usability metrics for e-learning. *Lecture Notes in Computer Science*, *2889*, 235–252. doi:10.1007/978-3-540-39962-9_34

Zarini, M., Varis, T., & Mar, N. Y. (2009). Information and communication technologies (ICTs) in vocational education and training. In Maclean, R., & Wilson, D. (Eds.), *International handbook of education for the changing world of work*. Springer.

Zhao, G., Shao, G., Reynolds, K., Wimberly, M. C., Warner, T., & Moser, J. M. … Li, F. (2005). Digital forestry: A white paper. *Journal of Forestry*, *103*, 47-50.

Zirkle, C. (2003). Distance education and career and technical education: A review of the research literature. *Journal of Vocational Education Research*, *28*(2), 161–181. doi:10.5328/JVER28.2.161

KEY TERMS AND DEFINITIONS

Digital Divide: A term which refers to the gap between people with effective access to digital and information technology and those with limited or no access at all.

E-Learning: The use of Internet technologies to deliver a broad array of solutions that enhance knowledge and performance in all types of educational systems.

Forest Extension Service: A system under which technical knowledge available at a university or other educational or research institution is systematically communicated to practictioners (forest workers, forest cooperative members, forest owners) through instruction and practical demonstration by itinerant specialists (forest extension agents, extension workers).

Forest Worker: A person engaged in practical work associated with silviculture, harvesting of forest products and other forest professions.

Forest Workers' Training: Vocational training in forestry aims at providing the specialized skills required by forest workers, forest machine operators, and foremen.

ICT: Information and Communications Technologies is an umbrella term that includes any communication device or application as well as the various services and applications associated with them.

Sustainable Forest Management: The stewardship and use of forests and forest lands in a way, and at a rate, that maintains their biodiversity, productivity, regeneration capacity, vitality and their potential to fulfill, now and in the future, relevant ecological, economic and social functions, at local, national, and global levels, and that does not cause damage to other ecosystems.

Vocational Training: Training which aims to provide the knowledge, skills and attitudes which are required for effective and efficient performance within an occupation or group of occupations. It encompasses initial, refresher, up-dating and specialized job-related training. It may but does not necessarily include general educational subjects.

Chapter 16
Soft Computing Modeling of Wild Fire Risk Indices:
The Risk Profile of Peloponnesus Region in Greece

L. Iliadis
Democritus University of Thrace, Greece

T. Betsidou
Democritus University of Thrace, Greece

ABSTRACT

It is essential to find ways that can reduce the risk of devastating forest fires which have multiple negative ecological and financial consequences. This preliminary research effort focuses on the implementation of an intelligent rule based fuzzy inference system evaluating wild fire risk in the forest departments of Greece. The system uses soft computing techniques and was built in the Matlab integrated environment. The whole research is related to the wild fires in Greece during the period 1983-1997 with data coming from the general forest management service. It classifies all Greek forest departments (by assigning three labels) according to their forest fire risk due to distinct parameters. The estimation of the risk indices was done by using fuzzy triangular membership functions and Einstein fuzzy conjunction T-Norms. Moreover the system produces the profile of the forest departments located in the geographic area of "Peloponnesus." This is a region located in the southern part of the country and it has a vast number of annual forest fire breakouts. Meteorological, topographic, and historical (total burned area and intervention time) features were considered for the determination of the risk indices. The system has shown a good performance which can be improved further if more data is gathered and used. Its main advantage is that it offers an innovative and reliable model that can be employed in any part of the world as a basis for natural disasters' risk estimation.

DOI: 10.4018/978-1-4666-2655-3.ch016

1. INTRODUCTION

Forest fire risk estimation is a major issue. The necessity for more efficient allocation of fire fighting resources becomes more and more urgent, especially in the areas located around the Mediterranean basin (Portugal, Spain, France, Italy, Greece) (Dimitrakopoulos, 1994).

Wild fires are an interdisciplinary field of research and they constitute one of the most serious environmental threats. It is a fact that in our times, more and more environmental problems are caused by the climatic change which is mainly due to anthropogenic intervention. Through the international scientific community there is an increasing interest for the study of environmental disasters. Fuzzy logic has been widely recognized as a powerful tool for risk estimation and there are several such cases in the literature (Huang & Ruan, 2008).

In Greece during the last 40 years the number of forest fires has increased significantly. This is the result of human activities, such as farming, tourism, but also the result of financial development which raised the interest for building cottages in forest areas (Dimitrakopoulos, 1994). Wild fires considerably contribute in the climate change and the devalorisation of ground. The destruction of forests due to fires involves usually negative repercussions in the ground surface and also in the hydrological cycle, with the increase of surface flow, the reduction of evaporation, the increase of erosion, the appearance of extreme floods. Moreover the burning of high volumes of biomass contributes in the increase of greenhouse gasses. Thus it is well understood that there are several negative side effects.

Intelligent systems can have a major role in the protection of the environment and they offer powerful tools capable of estimating flexible risk indices even in cases with imprecise or inaccurate data. In several cases immediate action and planning is required whereas on the other hand it would take several years to gather proper data series. Approximation modeling is often required in order to carry out rational wild fire protection and prevention planning. Such systems are called to offer reliable and effective models of real world situations in the right timing. The use of fuzzy logic which is a universal approximator that can provide intelligent solutions by employing proper linguistics such as "*Risky,*" "*Not Risky,*" or "*Extremely Risky*" in order to avoid the use of irrational crisp boundaries.

1.1. Features Considered

This research effort considers the following factors related to forest fire risk:

- Intervention time which is the time required from the moment of spotting till the initiation of the fire fighting process. This parameter is very crucial, capable of determining the total burned area.
- Meteorological parameters such as temperature wind speed and relative humidity at the momentum and at the exact location of the fire incident.
- Moreover topographical data such as the average altitude and the average slope of the area are really important.
- Historical data like the annual total burned area of the forest department.

It cannot be claimed that all of the risk features were considered, however a wide range of them (having both structural and dynamic nature) was covered. It should be clarified that this is a pilot approach and thus its aim is to demonstrate the effective use of flexible soft computing modeling techniques in risk estimation.

The classification of the forest departments of Peloponnesus region is done based on the Partial Risk Indices (PRI) due to each of the above parameters independently. On the other hand an overall classification is performed based on the Unified Risk Indices (URI) which are produced when all parameters are considered collectively. This is achieved by performing fuzzy conjunction operations called T-Norms. All of the above tasks are executed by a novel fuzzy inference system developed in the integrated environment of Matlab software. The whole research effort has been carried out in the lab of forest informatics of the *Democritus University of Thrace*. The results were obtained by the postgraduate research team of the lab.

1.2. Literature Review

Several ICT applications have been reported towards forest fire risk modeling in the literature. However only lately the scientific community has began working on intelligent approaches. Brillinger (1986) developed an interesting model assessing the potential for forest fires in a particular place and time using a probabilistic methodology.

Another system for the estimation of forest fire risk was developed in 2003 by the Institute of Environmental and Sustainable land management, which is known as EFFIS (European Forest Fire Information System). It is a system estimating the daily and long-term forest fire risk using dynamic and static parameters. It is an ambitious and interesting attempt to create a common indicator of forest fire risk throughout Europe (Ayanz et al. 2003).

Deeming, Burgan, Cohen, in the USA have developed in 1978 a successful system estimating the daily wild fire risk for various American climate zones and it is named the National Fire Danger Rating System (NFDRS). This system was expanded in 1985.

Iliadis, 2005 has proposed a system based on a heuristic and fuzzy hybrid approach in order to estimate risk indices. Recently the same research team has developed a fuzzy inference system towards forest fire risk estimation based on three distinct Gaussian functions, with reliable results (Iliadis et al., 2010). Also (Iliadis and Zigkrika 2010)(Iliadis et al. 2004) have recently developed a system estimating forest fire risk indices by performing scenarios. These scenarios are based on the heuristic assignment of weights to various fuzzy risk indices related to distinct risk factors according to Box 1 function 1(Iliadis and Zigkrika, 2010) (Calvo et al., 2002). Where $f(\mu, w) = \mu^{\frac{1}{W}}$ and the Aggregation operator is any fuzzy conjunction T-Norm. Also Kaloudis (Kaloudis et al, 2005) and Huang (Huang et al. 2005) have worked with fuzzy logic for risk estimation.

2. MATERIALS AND METHODS

The data used in this research effort are related to each forest fire break out in each forest department of Greece for 15 years (1983-1997). After 1997 our research team has no detailed data on forest fires. Though from this point of view this is a limitation of this study, the period of fifteen years and the existing dozens of thousands of data vectors are quite supportive.

The Forest Fire Risk Fuzzy Inference System (FFIRIFIS) was designed and developed under the *Matlab* integrated environment. More specifically, the Inference Engine of the system is based on fuzzy logic and it uses three fuzzy sets that correspond to three properly designed "*Linguistics*" (*Low Risk, Moderate Risk, High Risk*) in order to characterize all Greek forest departments according to their wild fire risk (Chang et al. 1991) (Cheney, 1991). For this purpose Fuzzy Triangular Membership Functions (FTMF) were applied. FTMF are functions of the form: $\mu_A(x) \to [0,1]$ where $\mu_A(x)$ is the degree of membership of x to the Fuzzy Set A and it always takes real values in the closed interval [0,1] (Kecman, 2001) (Iliadis, 2007). The following functions 2 and 3 correspond to Triangular and Trapezoidal Membership Functions (MF) respectively. In this research the

Box 1.

$$\mu_{\underset{S}{\sim}}(x_i) = Agg\left(f\left(\mu_{\underset{\mathrm{A}}{\sim}}(x_i), w_1\right), f\left(\mu_{\underset{\mathrm{A}}{\sim}}(x_i), w_2\right), \ldots, f\left(\mu_{\underset{\mathrm{A}}{\sim}}(x_i), w_n\right)\right) \quad (1)$$

Triangular MF was chosen to be applied due to the fact that we did not require a tolerance interval and it was essential to have not many areas with a degree of membership equal to one (Cox,2005).

$$\mu_s(x) = \begin{cases} 0 \text{ if } X < a \\ (X\text{-}a) / (c\text{-}a) \text{ if } X \in [a,c] \\ (b\text{-}X) / (b\text{-}c) \text{ if } X \in [c,b] \\ 0 \text{ if } X \geq b \end{cases} \quad (2)$$

$$\mu_s(X) = \begin{cases} 0, \text{ if } X \leq a \\ (X\text{-}a) / (m\text{-}a), \text{ if } X \in (a,m) \\ 1, \text{ if } X \in [m,n] \\ (b\text{-}X) / (b\text{-}n), \text{ if } X \in (n,b) \\ 0, \text{ if } X \geq b \end{cases} \quad (3)$$

In this way a higher level of distinction between forest departments has been achieved. The values of the parameters (a,b,c) determine the three characteristic values of the Triangular membership function.

For each risk feature, a FTMF was build. The determination of the values of the characteristic parameters *a,b,c* for each risk factor was done automatically by the Fuzzy Toolbox of Matlab employing the *Mamdani* logic, after entering the highest and the lowest value ever recorded for each year.

Matlab is a software platform offering a high-performance language towards technical computing. It enables computation, visualization and programming in an integrated environment where problems and solutions are expressed in familiar mathematical notation. *Matlab* operations are performed on matrices and arrays and are related to data analysis, fuzzy inference systems, building of artificial neural networks, fuzzy clustering and others.

Before entering the data they were scaled according to the formula

$$Z_j = \frac{X_j - \mu_j}{\sigma_j} \quad (4)$$

where σ_j, μ_j are the standard deviation and the mean value of the jth parameter (Maris et al, 2010).

The FFIRIFIS performs the profile of the forest departments of the Peloponnesus region which is located in southern Greece and it always suffers by devastating forest fires. These profiles are determined by the unification of the produced fuzzy sets (each one corresponding to a PRI) in a URI.

The fuzzy unification operation can be performed by employing various types of conjunction operators called T-Norms. There are several types of T-Norms each one phasing the problem under evaluation under a different perspective. In this case the Einstein T-Norm was used because it offers an overall estimator considering each parameter on an equal basis (Iliadis, 2007). The Einstein T-Norm for the unification of two fuzzy sets can be seen in the following function 5 (Kecman, 2001) (Iliadis, 2007) (Leondes, 1998)

$$\mathrm{A} \cap \mathrm{B} = \frac{\mu\mathrm{A}(x)\mu\mathrm{B}(x)}{2 - [\mu\mathrm{A}(x) + \mu\mathrm{B}(x) - \mu\mathrm{A}(x)\mu\mathrm{B}(x)]} \quad (5)$$

3. THE DEVELOPMENT ENVIRONMENT

The fuzzy logic Toolbox is a collection of functions built in the *Matlab* numeric computing environment. It provides tools for the development and edition of Fuzzy Inference Systems (FIS). The user can also integrate fuzzy systems into simulations with *Simulink*, or even can build stand-alone C programs that can execute FIS. The fuzzy Toolbox relies heavily on Graphical User Interface (GUI) tools but the user can also work entirely from the command line for faster and more tailored-made approaches.

The toolbox provides three categories of tools:

- Command line functions
- Graphical, interactive tools
- Simulink blocks and examples

Many of these functions are *Matlab m-files* or series of *Matlab* statements that implement specialized fuzzy logic algorithms.

For the creation of a new FIS file, the user should type >>*fuzzy* in the command prompt of *Matlab*. The following window presented in Figure 1 corresponds to the Fuzzy Inference System editor. By employing the FIS editor, the input and output membership functions can be defined, the rule base can be built, and the fuzzy operators can be set.

As it has already been mentioned, the FFIRI-FIS has seven inputs and one output and it uses the *Mamdani* inference and aggregation method.

Figure 2, presents the actual construction environment of the FIS. The first priority is the determination and development of the proper MF. The following Figure 2 presents the definition of a FTMF. By default there are three membership functions, mf1, mf2, and mf3 which correspond to the linguistics "*low risk*" "*medium risk*" and "*high risk*" for each related feature respectively.

As it was mentioned before, three "linguistics" were defined for each risk parameter. In this way 3X7=21 "Linguistics" corresponding to 21 fuzzy sets were defined. In order for the system to produce the boundary values, the maximum and minimum ever recorded value for each parameter (after it has been normalized) was used in the range option of the system.

4. BUILDING THE FUZZY RULESET

After having defined all fuzzy sets for each factor respectively the fuzzy rule set was built. The rules had a form of the type (*IF* $Parameter_i$ *IS "*$Linguistic_i$*" THEN OUTPUT IS "*$Linguistic_j$*"*). In this way the rules have fuzzy antecedents and fuzzy consequents aiming in building a system as flexible as possible, capable of extracting as much as reasonably possible conclusions for each single risk index and also for the overall URI.

The number of features is quite high (seven) and combined with the three potential "*linguistics*" that were defined for each feature would result in a very high number of rules. Obviously, the problem of "*combinatorial explosion*" is present here and it always poses insuperable problems to

Figure 1. Screenshot from the development environment of the FIS

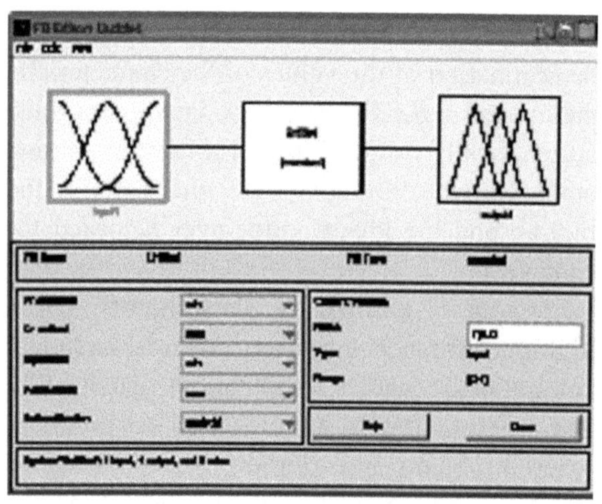

Figure 2. Screenshot from the design of the Triangular MF

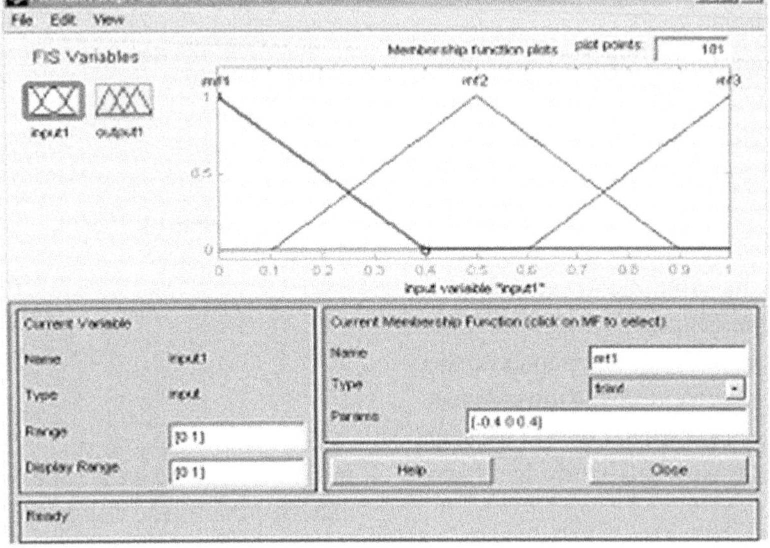

rule based systems developers. Factorials can be useful to facilitate such cases of combinations. For instance the number of *k*-permutations of *n* can be written as in the following function 6 (Biggs, 1979), (Borwein,1985) (Jackson, 1990).

$$n^{k} = \frac{n!}{(n-k)!}; \qquad (6)$$

A heuristic approach was used in order to develop proper decision tables and to solve this problem. The target was to avoid the use of a huge number of rules enabling the project to retain its simplicity. For this reason the factors were divided into smaller subgroups according to their thematic area. Proper decision tables were created in a rather heuristic manner and they were applied to each feature subgroup. In this way the number of rules was minimized significantly. The following Figure 3 is a small sample of the rule set. Actually the FIS developed used 41 rules in total.

The rule editor contains a friendly editable user interface for displaying and building rules. It also uses familiar landmarks similar to the ones used in the FIS editor and the MF editor, including the menu bar and the status line.

5. RESULTS AND DISCUSSION

The use of a properly designed FTMF under the *Matlab* environment produced the Degrees of Membership (DOM) (for each department and

Figure 3. A sample of the fuzzy ruleset

Meterological risk	LR	R	LR	R	ER	R	ER	ER	LR
Historical risk	LR	LR	R	R	R	ER	ER	LR	ER
OUTPUT									
	LR	LR	R	R	R	ER	ER	R	R
Sample of the Ruleset									

for each factor separately) to the three linguistics "*area of Low fire Risk*", "*area of Moderate fire Risk*", "*area of High fire Risk*".

On the other hand the use of the Einstein T-Norm produced the unified degree of membership to the three above *linguistics* corresponding to the URI based on the equal contribution of each feature.

It should be clarified that for the feature of humidity the higher its value the lower it degree of membership to the Linguistic "*area of High fire Risk*". Table 1 is used only to demonstrate the methodology that has been employed towards risk estimation due to each parameter and to present in a clear fashion the way this study has been conducted.

For 1983 "*Patra*" and "*Chania*" are the most risky cases due to the burned area, whereas all others except "*Peiraias*" (which is located close to the capital) have a very bad intervention time. "*Ioannina*" and "*Korinthos*" areas have a very high risk due to altitude, whereas "*Patra*", "*Chania*" and "*Korinthos*" have a high slope problem.

"*Thessaloniki*" and "*Volos*" are areas with a major problem regarding high temperatures and apart from them almost all other areas have a high problem with the wind (See Table 2). It should be clarified that these risk indices are related to the days with the forest fire breakouts and not to the whole of the year.

Figure 4. The areas of Peloponnesus under study

5.1. Detailed Results

The risk classification in Table 3 has been obtained for each forest fire risk parameter and for each forest department of Greece from 1983 to 1997. They clearly present the risk labels assigned to all forest departments based on various types of

Table 1. An indicative sample of the partial DOMs to the fuzzy set "area of moderate risk" for seven risk parameters for 1983

Forest Department 1983	DOM Burned area	DOM Intervention time	DOM Altitude	DOM Slope	DOM Humidity	DOM Temperature	DOM Wind
Alexandroupoli	0	0.3395	0.0872	0.4792	0.0068	0.1944	1
Didymoteixo	0	0.773	0	0.4318	0.4958	0.0758	0.8864
Thessaloniki	0	0.4002	0.2945	0.437	0.0204	0.8333	0.9022
Ioannina	0	0.3252	0.9576	0.375	0.3328	0.6667	1
Corfu	0	0.7256	0.0254	0.375	0.3195	0	1
Patra	0.044	0.6752	0.5657	1	0.2107	0.1667	1
Peiraias	0.0311	0.0502	0	0.375	0.1155	0.4167	1
Mesologgi	0	0.3252	0	0.375	0.2241	0.0833	0
Chania	0.0811	0.9499	0.1653	1	0.36	0.0833	1
Volos	0	0.7752	0.0975	0.375	0.5502	0.9167	0
Korinthos	0	0.7002	0.5064	1	0.2241	0.3333	1

Table 2. The frequency of appearance of each Forest Department in the fuzzy set "areas of high fire risk" for the "burned area" factor (1983- 1997)

Forest Department	Number of times been labelled as "High Risky" based on Burned area Ha	Forest Department	Number of times been labelled as "High Risky" based on Burned area Ha
Kavala	1	Istiaia	1
Thasos	2	Lesvos	1
Thessaloniki	1	Chios	3
Arnaia	1	Samos	1
Kozani	1	Dwdekanisa	3
Thesprwtia	1	Patra	2
Larisa	6	Olympias	1
Trikala	2	Kalamata	1
Lamia	1	Molaoi	1
Pentelh	1	Tripoli	1
Kapandriti	3	Korinthos	2
Peiraias	1	Mesologgi	1
Leibadiea	1	Chania	1
Chalkida	1	Lasithi	3

Table 3. Risk labels based on historical and topographic data

Risk related to the historical parameter "*Burned Area*"			Risk related to the topographic parameter "*Altitude*"		
The following forest departments are the ones that have the highest degree of membership to the corresponding fuzzy set more often (more than 50% of the years) 1983-1997			The following forest departments are the ones that have the highest degree of membership to the corresponding fuzzy set more often (more than 50% of the years) 1983-1997		
"Low Risky"	**"Moderate Risky"**	**"High Risky"**	**"Low Risky"**	**"Moderate Risky"**	**"High Risky"**
Rodopi	Thesprotia	Larisa	Alexandroupoli	Nevrokopi	Florina
Serres	Lamia	Hleia	Soufli	Kozani	Kastoria
Naousa	Xalkida	Kefallinia	Didymoteicho	Tsotyli	Konitsa
Metsovo	Samos	Kalamata	Thessaloniki	Elassona	Metsovo
Skopelos	Patra		Kassandra	Grevena	Karpenhsi
Karpenisi	Chania		Preveza	Kalamata	Drama
Leonidio	Lasithi		Skopelos	Sparti	Ioannina
			Lavrio	Kefallinia	Tripoli
			Peiraias	Leonidio	
			Dodekanhsa	Xylokastro	

risk factors. It is very important to clarify that the values of the risk parameters that were used as input to the system are the values that were recorded locally under the spatial and temporal conditions of the forest fire breakouts.

The period 1983-1997 is quite long and capable to offer reliable conclusions. Data were recorded by the central forest management authority of Greece (See Table 4 and Table 5). As soon as more data are gathered the research will be expanded.

5.2. Testing FFIRIFIS for 1995

The FFIRIFIS was tested for 1995 by comparing its output (based on data from 1983 to 1994) to the actually burned areas for 1995 (Iliadis et al. 2002).

Table 6 shows that the testing of FFIRIFIS is very encouraging and very promising.

6. THE RISK PROFILE OF PELOPONNESUS

"Peloponnesus" is an area with a severe forest fire problem. In 2007 Peloponnesus faced devastating forest fires which burned an area more or less equal to one million stemma whereas several people lost their lives. For this reason this research focuses mainly in this area of Greece. The following Figure 4 presents the areas under study.

The region of "*Kalamata*" is one of the most risky and most severely burned areas of "*Peloponnesus*". An effort was done to estimate the unified risk index due to several meteorological, topographic and historic factors. For each category of parameters a unified risk index was produced. More specifically the Einstein T-Norm was employed to perform fuzzy set conjunction for each category of parameters. Thus, a partial risk index was developed for each type of risk feature by the employment of fuzzy conjunction between all related parameters. For example, the

Table 4. Risk labels based on intervention time and meteorological data

Risk related to the historical parameter "*Intervention time*"			**Risk related to the meteorological parameter "*Humidity*" which has a negative effect on risk**		
The following forest departments have the highest degree of membership to the corresponding fuzzy set more often (more than 50% of the years) 1983-1997			The following forest departments have the highest degree of membership to the corresponding fuzzy set more often (more than 50% of the years) 1983-1997		
"Low Risky"	**"Moderate Risky"**	**"High Risky"**	**"Low Risky"**	**"Moderate Risky"**	**"High Risky"**
Thasos	Drama	None	Kavala	Thasos	Tripoli
Nevrokopi	Ioannina		Xanthi	Leivadia	Stavros
Stavros	Kerkyra		Stavroupoli	Lidoriki	Polygyros
Kassandra	Larisa		Rodopi	Aigio	Pendeli
Pendeli			Alexandroupoli	Tsotyli	Kilkis
Parnitha			Soufli	Kalavrita	Hgumenitsa
Lavrio			Didymoteicho	Nevrokopi	Florina
Aigaleo			Drama		
Lidoriki			Sidhrokastro		
Limni			Nigrita		

Table 5. Risk labels based on temperature

Risk Related to "Temperature". Forest departments with the highest DOM to the corresponding fuzzy set more frequently. More than 50% of the years. (1983-1997)					
"Low Risky"	**"Moderate Risky"**	**"High Risky"**	**"Low Risky"**	**"Moderate Risky"**	**"High Risky"**
Stavroupoli	Kavala	Thesprotia	Florina	Arnaia	Karditsa
Nevrokopi	Xanthi	Kerkyra		Kymh	Kalambaka
Skopelos	Thasos	Elassona		Sparth	Spercheiada
Kyparisia	Drama	Agia			Amfilochia

Table 6. Comparison of the FFIRIFIS output with actually burned areas

System's output. "*High Risky forest department*" due to various corresponding risk features for 1995					
"Burned area"	**"Intervention time"**	**"Humidity"**	**"Temperature"**	**"Wind Speed"**	**"Topography"**
Larisa PRI 0.999 **Kefallinia** 0.980	**Kavala & Thasos** 0.7523 **Chania** 0.9951	**Kavala &** Thasos 1 **Drama** 1 **Nevrokopi** 1 **Xanthi** 1	**Hleia** 0.375 **Aitoloakarnania** (Amfilochia 0.875 & Nafpaktos 0.375) **Lamia** (Sperxeiada 0.620) **Larisa** 0.6825 **Thesprotia** 0.625	**Stavroupoli Thessaloniki Polygyros Lamia Lesvos Samos Molaoi Arkadia** (Vytini) All with DOM 1	**Ioannina** (Konitsa 0.9606 Metsovo 0.981) **Florina** 0.7219
Actually most burned areas of Greece for 1995 CC means (correct classification)					
Kefallinia CC	Messinia	Attiki	**Hleia CC**	Evoia	**Larisa CC**
Irakleio	**Aitoloakarnania CC**	**Ioannina CC**	**Kavala & Thassos CC**	**Lamia CC**	**Arkadia CC**
Thesprotia CC	**Chania CC**	11 out of 15 were classified correctly **Correct RATE 73.3%**	4 out the 15 actually risky were not classified as risky though they were **Positive Error RATE 26.6%**	7 out of 20 were classified as risky and they were not in the top 15 burned areas **Negative Error RATE 35%**	

Table 7. The unified fire risk indices obtained by employing the Einstein T- Norms for the forest department of "Kalamata" located in Southern Peloponnesus

Unified Risk Index based on all Topographical features for "Kalamata" area		**Unified Risk Index based on all Meteorological features for "Kalamata" area**		**Unified Risk Index based on Historical features (Burned area & Intervention time) for "Kalamata" area**	
1983	0.6462	1983	0.302715815	1983	0.41177
1984	0.38958	1984	0.224058465	1984	0.1142
1985	0.5133	1985	0.046258359	1985	0.116321
1986	0.56003	1986	0	1986	0
1987	0.7179	1987	0.039753719	1987	0.258147
1988	0.17762	1988	0.000506615	1988	0
1989	0.20876	1989	0	1989	0.1318777
1990	0	1990	0.097706188	1990	0.2552686
1991	0.46491	1991	0.037117636	1991	0.4234
1992	0.0452	1992	0.137090871	1992	0.3236222
1993	0	1993	0.000901052	1993	0.0138225
1994	0	1994	0.020401728	1994	0.6148873
1995	0.04419	1995	0.305533652	1995	0.097979

Table 8. The unified fire risk indices obtained by employing the Einstein T- Norms for the forest department of "Gytheio" located in Southeastern Peloponnesus

Unified Risk Index based on all Topographical features for "Gytheio" area		Unified Risk Index based on all Meteorological features for the "Gytheio" area		Unified Risk Index based on Historical features (Burned area & Intervention time) for the "Gytheio" area	
1983	0.527500000	1983	0	1983	0.178713848
1984	0.501383768	1984	0.215425751	1984	0
1985	0.524300000	1985	0.027141194	1985	0.369621649
1986	0.140891315	1986	0.224669586	1986	0.031468932
1987	0.562900000	1987	0.011073457	1987	0.189581891
1988	0	1988	0.660756614	1988	0.320897028
1989	0.207629871	1989	0	1989	0.293239477
1990	0.587957067	1990	0.016292906	1990	0.249739568
1991	0.327280690	1991	0	1991	0.038966659
1992	0.268949272	1992	0.464500000	1992	0.395094357
1993	0.453933985	1993	0.018900782	1993	0.056697243
1994	0.119542026	1994	0.000656360	1994	0.404935084
1995	0.594108373	1995	0.520502322	1995	0

risk index due to the wind speed and the index due to the temperature were unified to a partial meteorological risk index.

The profiles are really useful for the civil protection authority in their effort to allocate properly and rationally the ground and air forces during the fire season.

From the unified topographic risk indices in Table 7, it is concluded that for the case of "*Kalamata*", in 8 out of 13 cases the forest fire breakouts occurred in areas with high slope or high altitude. Only in 3 cases the forest fire incidents took place in low altitude and not sloppy areas, whereas in 2 cases the fires occurred in moderate altitude/slope areas.

On the other hand from the unified meteorological risk indices it is concluded that "*Kalamata*" has a moderate risk as long as meteorological conditions are concerned. Only in three cases out of fifteen, the "meteorological risk" index is too low. Finally looking on the burned area/intervention time unified risk index, it is concluded that it is quite high in many cases.

It is remarkable that in 1986 when the risk due to meteorological features was equal to 0, and in 1988 when the meteorological risk was almost equal to 0 and the and the topographic risk was too low, the historical risk index was equal to zero. Moreover, by using the Einstein T-Norm the contribution of each parameter type to the total forest fire risk can be determined. The topographic unified risk index has the highest values in the case of "*Kalamata*" whereas the historical unified risk index comes second.

The area of "*Gytheio*" was analyzed second. It had a high historical risk index around five times and a very low one four times. The low unified historical risk index can be explained only once (in 1991) due to low meteorological risk.

Also for the case of 1994 where the unified historical index is quite high there cannot be given a reasonable explanation. All of the other cases appear quite reasonable results. In "*Gytheio*" area in nine (9/13) cases the forest fire breakouts occurred in areas with high average altitude or with high slope. On the other hand the risk due

Table 9. The unified fire risk indices obtained by employing the Einstein T- Norms for the forest department of "Aigio" located in Northern Peloponnesus

Unified Risk Index based on all Topographical features for "Aigio" area		Unified Risk Index based on all Meteorological features for "Aigio" area		Unified Risk Index based on Historical features (Burned area & Intervention time) for "Aigio" area	
1983	1	1983	0.04563976	1983	0.213940255
1984	0.532335634	1984	0.043601933	1984	0.518070394
1985	0.238033844	1985	0.617683236	1985	0.671871519
1986	0.574742928	1986	0.22921366	1986	0.753009803
1987	0.4279	1987	0.014410294	1987	0.76176863
1988	0.388559312	1988	0.9306	1988	0.629714652
1989	0.308146297	1989	0	1989	0.89771648
1990	0	1990	0.112191037	1990	0.564934796
1991	0.067543196	1991	0.4444	1991	0.888337631
1992	0.430669605	1992	0.25	1992	0
1993	0.633022734	1993	0.293606374	1993	0.792747905
1994	0.271959947	1994	0.054695181	1994	0.741630685
1995	0.04419	1995	0.305533652	1995	0.097979

Table 10. The unified fire risk indices obtained by employing the Einstein T- Norms for the forest department of "Xylokatro" located in Northeastern Peloponnesus

Unified Risk Index based on all Topographical features for "Xylokastro" area		Unified Risk Index based on all Meteorological features for "Xylokastro" area		Unified Risk Index based on Historical features (Burned area & Intervention time) for "Xylokastro" area	
1983	0.7246	1983	0	1983	0.039583811
1984	0.396811	1984	0.732836047	1984	0.200650981
1985	0.369486048	1985	0.6786	1985	0
1986	0.416628607	1986	0	1986	0
1987	0	1987	0.079258384	1987	0.576159902
1988	0.203709428	1988	0.573734348	1988	0.366988189
1989	0.4845587	1989	0.205705902	1989	0.619012907
1990	0	1990	0.25579192	1990	0.029797701
1991	0.476067852	1991	0.734261195	1991	0.406294548
1992	0.539885764	1992	0.125	1992	0
1993	0.574210604	1993	0	1993	0.611534495
1994	0.014263118	1994	0.042338482	1994	0.411650113
1995	0	1995	0	1995	0.506665084

to meteorological features was high only for three years, namely 1988 (the highest) 1992 and 1995. In 1988 the burned areas were high and the meteorological risk played a catalytic role in this, however in 1995 this did not seem to have any effect in the problem (See Table 8).

"*Aigio*" is a very interesting area, having an extremely high unified historical risk index in 10 out of the 13 cases whereas only twice the risk is very low and once it is moderate (See Table 9). In 1995 when the historical integrated risk is very low, also the topographical risk is low. In the other low overall historical risk there is no obvious explanation. The system has to consider also other factors (maybe qualitative) in order to explain it.

"*Xylokastro*" has a very high unified historical risk index due to topographic data only in 1 out of the 13 cases whereas in 4 cases the risk in moderate. On the other hand it has a high unified risk index based on meteorological data in three times. Finally it does not have high risk indices based on burned area and intervention time but it has moderate risk in four times. This can be seen in the following Table 10.

Results were obtained for several other forest departments of Peloponnesus region but it would be out of the scope of this paper to present them all due to lack of space. From the construction and from the analysis of the profile of each area it is obvious that their vulnerabilities are revealed. Of course more features and more data vectors should be used to obtain the optimal result.

7. CONCLUSION

Forest fires can occur due to a great number of factors. Often, many of these factors are extremely unpredictable and in some occasions the data inventories that we have in our disposal are not accurate. However civil protection authorities might require decisions to be taken in real time or in a very short period of time. These facts make the fire risk estimation a complicated problem that can be studied with the use of fuzzy logic.

FFIRIFIS uses a quite flexible way of thinking and it offers an innovative approach. The fact that it can use any kind of data available and that it can produce results as soon as the data is inserted, makes it a valuable tool for estimating which forest department is in danger and very often it is capable of determining the cause of the risk. In short the system can give degrees for each feature by using three "*Linguistics*" and one uniform degree which considers all factors together. The system really offered a clear estimation of the risk profile of the Peloponnesus sub-areas where the problem of forest fires is really intense. This will be very useful for the policy makers.

The system has shown that is really useful as an automated approach providing aid towards the proper distribution of fire fighting resources. Though it is a pilot research effort under evaluation in Greece, it has good perspectives to be employed by local authorities in cooperation with other neighbour countries in the near future and it has been used in the development of many research efforts around the globe. For example our approach has been referenced in the USA in a research effort to perform wild-land fire risk assessment of western military installations. This was a research project carried out by the *Colorado State University* (Beavers et al., 2007).

FFIRIFIS can surely improve its performance if more data is gathered and if more application time is invested. It should be mentioned that the input data vectors with the total burned areas include forests, rangelands and brush lands. In a future research effort these three types of burned areas should be distinguished and a more detailed analysis should be carried out for each type of area. Thus, there is a lot of future work remaining to be done but this pilot research offered a very fruitful outcome.

REFERENCES

Ayanz, J. S. M., Barbosa, P. M., Schmuck, G., & Liberta, G. (2003). *The European forest fire information system (EFFIS)*. European Commission. JIR Institute for Environment and sustainability. Retrieved from http://natural-hazards.jrc.it/documents/fires/2003- 591publications/EFFIS-earsel-paper.pdf

Beavers, A. M., Ruzycki, T. S., & Kunze, M. & CEMML. (2007). *Wildland fire risk assessment of western military installations*. Colorado State University, Department of Defense, Legacy Resource Management Program. (Project Number 04-230).

Biggs, N. L. (1979). The roots of combinatorics. *Historia Mathematica*, *6*, 109–136. doi:10.1016/0315-0860(79)90074-0

Borwein, P. (1985). On the complexity of calculating factorials. *Journal of Algorithms*, *6*, 376–380. doi:10.1016/0196-6774(85)90006-9

Brillinger, D. R., Preisler, H. K., & Benoit, J. W. (1986). *Risk assessment: A forest fire example*. University of California, Department of Statistics. Retrieved from http://stat-www.berkeley.edu/users/brill/Papers/speed23.pdf

Calvo, T., Mayor, G., & Mesira, R. (2002). *Aggregation operators: New trends and applications*. Heidelberg, Germany: Physica-Verlag Studies in Fuzziness and Soft Computing. doi:10.1007/978-3-7908-1787-4

Chang, T. C., Hasegawa, K., & Ibbs, C. W. (1991). The effects of membership function on fuzzy reasoning. *Fuzzy Sets and Systems*, *44*(2), 169–186. doi:10.1016/0165-0114(91)90001-7

Cheney, P. (1991). Models used for fire danger rating in Australia. In N. P. Cheney & A. M. Gill (Eds.), *Proceedings Conference on Bushfire Modeling and Fire Danger Rating Systems*, CSIRO Division of Forestry, Canberra, (pp. 119-126).

Cox, E. (2005). *Fuzzy modeling and genetic algorithms for data mining and exploration*. Elsevier Science, USA.

Deeming, J., Burgan, R., & Cohen, J. (1978). *The national fire danger rating system*. U.S. Department of Agriculture.

Dimitrakopoulos, A. (1994). *Analysis of forest fire causes in Greece*. Edition of MAI of Chania Greece

Huang, C., & Ruan, D. (2008). Fuzzy risks and an updating algorithm with new observations. *Risk Analysis-. International Journal (Toronto, Ont.)*, *28*(3), 681–694.

Huang, C. F. (2005). Risk assessment of natural disasters-Theory & practice. Beijing, Chian: Science Press.

Iliadis, L. (2005, May). A decision support system applying an integrated Fuzzy model for long - term forest fire risk estimation. *Journal Environmental Modelling and Software, 20*(5), 613-621. ISSN: 1364-8152

Iliadis, L. (2007). *Intelligent information systems and applications in risk estimation*. Thessaloniki, Greece: Stamouli Publishing.

Iliadis, L., Papastavrou, A., & Lefakis, P. (2002). A computer system that classifies areas of Greece in forest fire risk zones, using fuzzy sets. *Forest Policy and Economics*, *4*, 43–54. doi:10.1016/S1389-9341(01)00079-X

Iliadis, L., Skopianos, S., Tachos, S., & Spartalis, S. (2010). A fuzzy inference system using Gaussian Distribution Curves for forest fire risk estimation. *Proceedings of the 6th International Conference of Artificial Intelligence Applications and Innovations Springer AIAI 2010, IFIP AICT 339*, (pp. 376–386). IFIP (International Federation for Information Processing).

Iliadis, L., Spartalis, S., Maris, F., & Marinos, D. (2004). A decision support system unifying trapezoidal function membership values using t-norms. *Proceedings International Conference in Numerical Analysis and Applied Mathematics* (ICNAAM). Weinheim, Germany: J. Wiley-VCH Verlag GmbH Publishing Co.

Iliadis, L., & Zigkrika, N. (2010). Performing fuzzy multi-feature scenarios for the determination of forest fire risk Προφορική παρουσίαση. *Proceedings of the 3rd International Conference on Information and Communication Technologies in Agriculture, Food, Forestry and Environment* (ITAFFE 2010) (pp. 170-177).

Jackson, P. (1990). *Introduction to expert systems* (2nd ed.). Addison Wesley.

Kaloudis, S., Tocatlidou, A., Lorentzos, N., Sideridis, A., & Karteris, M. (2005). Assessing wildfire destruction danger: A decision support system incorporating uncertainty. *Ecological Modelling, 181*(10), 25–38. doi:10.1016/j.ecolmodel.2004.06.021

Kecman, V. (2001). *Learning and soft computing*. London, UK: MIT Press.

Leondes, C. (1998). *Fuzzy logic and expert systems applications*. USA: Academic Press.

Maris, F., Iliadis, L., Tachos, S., Loukas, A., Spartali, I., Vassileiou, A., & Pimenidis, E. (2010). Support vector machines-Kernel algorithms for the estimation of the water supply in Cyprus. *Proceedings of the 20th International Conference ICANN (International Conference of Artificial Neural Networks), Lecture Notes in Computer Science, Vol. 6353, Part* II, (pp. 21-29).

Chapter 17
Information and Communication Technology Uses in Agriculture:
Agribusiness Industry Opportunities and Future Challenges

Tizita Alemayehu Wasihun
West Virginia University, USA

Blessing Mukabeta Maumbe
West Virginia University, USA

ABSTRACT

The world has experienced an unprecedented growth in information and communication technologies (ICT) through the widespread use of personal computers, Internet, and mobile phones. The objectives of this chapter are to examine trends in ICT use in agriculture, identify key success factors for ICT utilization in agriculture, and investigate the implications of ICT-enabled value chains for the agribusiness industry. The chapter describes the strategic role of ICT in the development of both e-commerce and mobile commerce in agriculture globally. The chapter identifies the leading areas of ICT use in agriculture and agribusinesses as input procurement, production, marketing, food traceability, and financial service delivery. Producers are increasingly seeking ways to add value to their businesses by integrating ICT in the value chain. Similarly, consumers are becoming more knowledgeable about how they could use ICT to articulate their preferences. The chapter discusses key success factors for ICT applications affecting both the internal and external environment of agribusiness firms. The chapter concludes by drawing implications for ICT use in agriculture and agribusiness value chains.

DOI: 10.4018/978-1-4666-2655-3.ch017

1. INTRODUCTION

The world has experienced an unprecedented growth in information and communication technologies (ICT) through the widespread use of personal computers, Internet and mobile phones (Lio and Liu, 2006). Information and Communication Technologies (ICT) refers to the use of available knowledge and skills that allow the use of *facilities* to carry out business activities, chain process, and communication (Vorst van der et al. 2005). Where, facilities refer to any computers, peripherals, systems software, procedures, networks, technical information application packages and communication standards. In agriculture, ICT has found widespread applications in automation of production and marketing management processes including high speed information processing, storage, delivery, and retrieval leading to greater convenience and ease of accessibility. Globally, ICT are serving as cutting edge capital inputs that producers and consumers can find relevant and could provide sustainable solutions to almost all aspects of human endeavor. This paper examines opportunities and future prospects associated with the development of e-business and mobile-based ICT use in agribusiness and agriculture in general. Notably, developing countries and some emerging economies have experienced a major surge in the use of mobile technologies in agriculture and the agribusiness industry (Ayinde etal, 2004; Kantengeza[a] et al, 2011).

Over the past two decades, information technology (IT) systems were narrowly defined and limited to local area network (LAN) and wide area network (WAN) infrastructure where data carrying was considered the main function. But nowadays, IT is used with communication systems and is referred to more broadly as ICT. This modern ICT system integrates voice and data access, enabling instantaneous working environment between different geographically based coworkers, who can not only communicate with each other, but also access and share a wide variety of information. Moreover, this aptitude is not only limited to a fixed place, but can be performed from a wide variety of locations including wireless hot spot, Internet cafes, and even on the move (Lyons, 2005).

Information complexity in regard to business functions and processes is expected to increase as the agribusiness firm operations increase in scale. This information complexity can be the result of increased demand in information acquisition, storage and access regarding all business activities taking into account; customers, suppliers, products, employees, transactions, and internal processes. In this regard, ICT can facilitate information acquisition, processing and exchange in different ways. Some ICT facilitate information management using database and websites, some support supplier functions using websites and e-mails, some improve customer interaction using emails and blogs while others facilitate employee functions using list server, podcasts and intranets. Strategically, the intention and goal of the agribusiness sector will most likely determine the type of technology in use (Burke, 2010).

Given the recent outbreaks in food borne diseases (e.g., mad cow, melamine, etc.) resulting from food safety lapses and contamination, ICT is increasingly being relied upon as a means to provide potential solutions to food tracking and traceability. The increased concern of consumers to health, food safety and environmental issues has increased the stipulation of integrating ICT in food and agriculture quality control systems and related tracking and tracing systems of consumer goods in the supply chain (Vander Vorst et.al, 2002). High transaction costs associated with the transformation of subsistence agriculture to commercial agriculture are a major barrier to farmers. Farmers need to decide what, when and how much to produce and in which markets to sell. In this matter, ICT can play a major role in reducing cost of information search which is part of the transaction cost (Silva & Ratnadiwakara, 2010).

The aim of this chapter is to: 1) to examine trends in ICT use in agriculture, 2) identify key success factors for ICT utilization in agriculture, 3) investigate the implications of ICT-enabled supply chains for the agribusiness industry. The rest of the paper is organized as follows; Section two describes ICT uses in agriculture. Section 3 describes the key success factors in ICT uses in agriculture and the agribusiness industry. Section 4 discusses the implications of ICT deployment in agriculture and future directions for the agribusiness industry. Section 5 provides the conclusions.

2. ICT USE IN AGRICULTURE

One of the early uses of ICT has been in the area of e-business. E-business emerged in the late 1990s with increased use of electronic medium to exchange information among customers, business partners and the vendor (Ayinde et al. 2004). E-business emerged as a new philosophy and modern business practice concerned with the marketing of goods, services, ideas and information through the Internet and other electronic means (Lefebvre et.al, 2001).The term e-business refers to different business perspectives. For example, business-to-business (e.g. electronic marketplaces), business-to-authority (e.g. tax collection system, education), business to consumer (home shopping services), business-to-employee (e.g. tele-working system) approaches (Vander Vorst et.al, 2002). E-marketing has a broader scope than Internet marketing since Internet marketing only refers to marketing via the Internet, worldwide web, and e-mails while e-marketing in addition to that encompasses tools like; intranets, extranets and mobile phones. In converse, e-commerce and e-business have a broader scope than e-marketing (El-Gohary, 2010).

In agriculture, electronic commerce facilitates the exchange of market information by developing stronger buyer-supplier relationships and this can be witnessed by the current exponential growth of electronic commerce that most specialists have been forecasting in the last few years. From what has been seen, it is clear that the adoption of electronic commerce has continued, and is likely to indeed speed up, in the years to come (Lefebvre et.al 2001). As ICT diffusion grew within different sectors across many countries, it began to attract both researchers and policy analysts on its applicability in agricultural development and poverty alleviation (Kenny, 2002; Lio & Liu, 2006). These days, information system are used within agribusiness firms for different tasks such as; increasing productivity, reducing operational cost, increasing profits, enhancing management's ability to solve problems and decision making process and in maintaining a competitive advantage. Agribusiness firms can use a complete range of ICT in their effort to assure organizational objectives (Namani, 2009). The growth level of an agribusiness firm is influenced by external, organizational and technological factors and the firm's growth level in turn determines the type and extent of ICT implementation (Burke, 2010). Increased usage of the Internet by farmers facilitates the integration of e-business and e-commerce in agribusiness firms and affects their commercial activities and business strategies (Henderson et al. 2004). E-marketing could be considered as both software and hardware innovation in agriculture. As a software innovation, it provides farmers with the means to deliver information, communicate with customers about farm products, and enhance customer relationships. While as a hardware, it provides the means for advertising and sales to expand the scope of the farm operation (Baer and Brown, 2007). Although e-commerce started as a small part of the U.S economy, its exponential growth is enabling agribusiness firms to implement innovations that take advantage of this new economic movement (Henderson, 2004). E-marketing reduces the transaction costs related to physical trading environments such as ordering routing and other execution costs (Tompson and Kunda, 2000). For instance, the ability to trace and track the origin of agricultural inputs and final products throughout the supply chain has become critical in food quality and safety in the supply chains (Galliano & Orozco, 2008).

ICT also has major connotations in agriculture especially in the food industries as well as in the food supply chains. Based on Frank and Bunte (2009), implications of ICT in food supply chain are generalized into three categories, market structures, transaction costs and market transparency. With regards to market structures, ICT usage contributes to the performance level of a firm and the performance level in turn influences market structure since the level of performance determines whether firms enter or exit, grow or decline. ICT have direct impact on transaction costs as the new technologies open up new ways of conducting business transactions. Although computer or Internet-based e-commerce has been used in food retailing in particular, ICT has the potential to widen the scope of potential trade partners in the international market. ICT also enhances market transparency and consumer trust. ICT use in business enables markets to be transparent making information easily accessible, understandable, reliable and comparable. A transparent market enables consumers to purchase better quality products at fair prices.

2.1 Tangible Benefits Derived from ICT Use in Agriculture

The integration of ICT in agriculture production and marketing activities has resulted in tangible benefits for farmers and agribusiness firms. These benefits are described below under three main headings, market transparency, e-commerce, and applications in food traceability.

2.1.1 Role of ICT in Promoting Market Transparency

One of the key areas in which the adoption and diffusion of ICT has made an important contribution is in improving market transparency in agriculture value chains. Prior to the widespread use of ICT, agricultural markets were characterized by thinly traded volumes, information paucity, and general lack of trust between transacting partners. With the advent of ICT in agriculture, relatively larger volumes are now being traded in the marketplace. Customers and producers can now interact in real time and are able to send more accurate two way signals about product demand and supply and thus boosting information sharing regarding available product volumes and quality (Katengeza[b] et al. 2011). The use of ICT in agribusiness allows markets to be transparent in a way that makes information accessible, understandable, reliable and comparable across regions and countries.

Vander Vorst et.al (2002) classified the food web-business based on network differentiation and market segmentation, integrated quality, and network optimization. Network differentiation and market segmentation refers to the differentiation and segmentation of customers based on their preference. The study argues that such segmentation enables the market to meet the specific demands of customers (e.g., lowest cost, assortment, etc.) hence improve market transparency and trust. Integrated quality aims to meet the increasing concern of customers and governments towards food safety and to environmentally friendly produced products while network optimization refers to the effort towards cost minimization via a streamlined chain with ease of access and rational information supply.

2.1.2 E-Commerce Use by Farmers and Agribusiness Firms

With increasing broadband penetration, small farms are turning to e-commerce use to support their business operations (Baer and Brown, 2007). Overtime, the Internet has become a key option for agri-food businesses that want to sell "typical" local food products more successfully in an international arena. There is increased interest by consumers and retail distribution chains for such goods. A good example for such typical product web-marketing is the web-based wine marketing (Begalli et.al, 2009). E-commerce also at the same time creates the opportunity for suppliers to target specific customers based on their consumption preference (Bunte, 2009).

A general survey done within US agribusiness in 2000 showed that 87 percent of the respondents use personal computers. The most common applications software used were text processing and spreadsheets followed by payroll, database and inventory control applications. Among agribusinesses that have personal computers, 36 percent of them reported that firms were writing their own software applications among which financial or accounting programs were the most often developed application by the firms. In regard to network sophistication, a total of 38 percent of the respondents reported networking with computers among which 28 percent use Local Area Network (LAN) while 4.9 percent implemented Wide Area Networks (WAN). From the firms using computer network, 50 percent of them indicated Novell as their operating system. The results also indicated a positive and statistically significant relationship between software sophistication information and competitive advantage (Kagan, 2000).

According to Henderson et.al (2004), agricultural input firms in the US showed that 78 percent of the firms developed a foundation for further Internet activity by constructing a company web page. Of those without a website, over half were planning to develop a site by 2001. Though these agricultural input firms performed a wide variety of e-business transactions via their websites, the most common website features were the availability of company background information and technical product information which are the first steps towards e-commerce. In regards to e-business, among those who had a website, less than 20 percent of the agricultural input firms reported providing online ordering features of which less than 10 percent of the firms had online ordering and payment features available for customers. Moreover, the most common e-business transactions have been seen among agricultural input firms and their suppliers (B2B e-business transaction). Almost 20 percent of the respondents reported placing online orders with more than 5 percent of their suppliers while online orders from farm customers were only around 5 percent. Results from a 2003 study based on a random sample of Ohio farmers, showed that computer adoption stood at 44 percent. Among the farmers who adopted computers, 80 percent of them said they use Internet for communication, transactions processing, or for information retrieval though, *financial accounting* remains the most often used task among them. Moreover, computer adoption was found to increase with the farm size (sales) and education level of operator while it diminished with operator age (Batte, 2005).

A survey on ICT adoption by farmers on Big Island Hawaii who are known as producers of both flower and foliage growers, and fruit and vegetables showed that use of a computer was a standard way of doing their business. The research indicated 88 percent of the farmers use computer while only 30 percent of them have their own websites. Among the reasons the farmers mentioned why they didn't have their own website, 33 percent said their product is not suited for website while 37 percent said lack of technical knowledge. On the contrary, farmers who owned a website reported that website based sale accounts for more than 10 percent of the total firm sale indicating owning a website brings considerable revenue benefit (Burke, 2010).

Cloete and Doens (2008) stated the marginal impact of the Internet on developing countries' economies is considerably great. Despite the limited history of trade by such nations, the authors found 55.4 percent of agribusiness owners in South Africa make use of e-commerce. The study pointed out the fact that the adoption rate is lower than other sectors in the country but at the same time it showed that 80 percent of non-adopters had a plan to make use of e-commerce in the future. Majority of the agricultural products placed on e-market from the study are wine, grain and derivatives. A research undertaken in South

West Nigeria Agriculture based Food Industries (ABFI) also showed an increase in profits among the industries after ICT adoption. The improved market performance reported was largely due to a reduction in transaction cost for buyers and sellers and increased liquidity and efficiency. So, their study concluded that ICT usage was an effective way of marketing by the ABFIs in South West Nigeria (Ayinde et.al, 2004).

Taking into consideration poor Internet coverage in rural areas and the less enthusiastic trend of farmers in using computer and Electronic Data Inter-change (EDI), Lu et.al (2005) found out that mobile-commerce has the potential to solve such problems and improve efficiency in supply chains. In order for primary producers to use mobile-commerce, mobile companies have developed Short Message Service (SMS) targeting rural areas with timely information such as the provision of weather forecasts, harvest time, and pest management practices (Jonsson and Gunnarsson, 2005). Meanwhile, the researchers also found out that farmers are willing to pay for mobile-based information but they emphasized quality delivery for the paid service.

2.2.3 ICT Use in Food Traceability

As mentioned earlier, the ability to *trace the origin of inputs and food products* throughout the supply chain has become a key mechanism to ascertain food quality and safety in agri-food supply chain. Food marketing information during the past required less strategic communication and trust than information in the present. Food traceability has added a new dimension to the meaning of market transparency by emphasizing the need to prevent outbreaks of diseases, food safety lapses and preservation of the freshness of food as it flows through downstream industries to final consumers. Frank and Bunte (2009) identified- three types of transparency in the food supply chain networks (FSCN) namely *historical transparency, operational transparency* and *strategy transparency*. *Historical transparency* is a transparency through which information about the origin of the product and process history is captured in the food supply chain network. The main objective is to upgrade recall management and to prevent calamities. Radio Frequency Identification (RFID) technology is one of the tools used to make a detailed history transparency affordable in many agri-food businesses.

Traceability is a special case of history transparency. *Operational transparency* is about knowing what is happening in regards to logistics and other operational parameters across the food supply chain networks. It involves keeping partners informed and improves the efficiency by reducing waiting times and stocks. Moreover, it improves effectiveness and responsiveness of the food supply chain. *Strategy transparency* is about forecasting what may happen in the FSCN by investigating opportunities and threats so as to come up with adaptive response.

As explained above these transparency strategies allow tracing capability for any product at any stage within the supply chain to identify the initial source (i.e., backward tracing) and ultimately its final destination (i.e., forward tracing). The tracking capability helps to locate any product any given time. Backward tracing of deficient products allows the identification of the source of a problem from the point of origin. It is a precondition for the subsequent identification of products at later stages of the chain, which might have been affected by the contamination of the initial source (forward tracing) for call back or destruction (Schiefer, 2006). Without effective ICT deployment, developing a traceability system at any point is unattainable. ICT assures the recording and transmission of production data through the supply chain. Some of the technologies used for the identification and registration are *bar-coded tags,*

Radio Frequency Identification Device (RFID), ear tags and *DNA tracing*. In addition, *Enterprise Resource Planning System (ERP)* and *Warehouse Management System (WMS)* help in management of traceability in food and agribusiness industry while *Electronic Data Interchange (EDI)* is used in communicating traceability of data in food and agricultural marketing.

A number of previous studies have provided evidence of ICT integration in food traceability (Kakkainen, 2003; Gregor et al. (1997). An example of RFID application cited in the literature is a food manufacturing company called United Biscuits, a food processing firm that uses RFID technology to control raw materials movement and in weighing, mixing and baking process of the biscuits. The firm has reported improved efficiency in manufacturing process, improved information accuracy, better tracking of the food products and a decline in incidence of errors (Galliano & Orozco, 2008; Lu et.al, 2005; New, 2010). Similarly, an Internet-based Coffee Information System (CINFOS) is also one of the successful initiatives that prove how ICT can be applied to the case of coffee marketing and traceability. This system has been used to trace individual product batches indicating where and how they were produced, processed and distributed to the end consumer so as to facilitate identity preservation, which is the key to obtaining added value from differentiation (Niederhauser et al. 2008). As reported by OMAF (2004), *Geospatial Information System (GIS)* has also been used by dairy farms for shipping milk. The primary purpose of this system is to use GIS technology to efficiently route milk trucks, optimize resource allocation, promote effective planning and control strategies using factual and accurate information. As reported by IDRC (2008), GIS also has been applied in a number of initiatives in Sub-Saharan Africa. One example is the International Livestock Research Institute (ILRI) initiative that used GIS technology to map fences in its livestock-wildlife ecosystem. Table 1 provides a summary of some of the key examples of e-commerce applications in agriculture and agribusiness industry cited in the literature over the past decade.

2.1.4 Mobile Commerce Adoption and Use in Agriculture

Over the past decade, there has been a major increase in the adoption and use of mobile-based ICT in agriculture (Barnes & Scornavacca, 2004; Rao & Minakakis, 2003). Jensen (2010) also noted that the number of mobile subscriptions in developing countries increased significantly between the year 2003 and 2008. In Latin America, subscription increased by a factor of 3.6, from 127 to 460 million, and Asia (excluding Japan) it increased by a

Table 1. Summary of e-commerce applications in agriculture and agribusiness industry (1990-2000)

Agribusiness Sector	*Scope of ICT application*	*Author and Year*
Input	Business to business (B2B) e-marketing, E-business processes for input firms, and list server, intranets.	Allahyari & Chizari, 2010; Burke, 2010; Charles et al, 2003; Henderson et.al, 2004; Lal, 2008; Mueller, 2000; Vander Vorst et.al, 2002.
Production	E-procurement including communication, information retrieval, Database and inventory control, e-banking and financial accounting	Batte, 2005; Cloete & Doens, 2008; Kagan, 2000; Obayelu & Ibadan, 2006; Silva & Ratnadiwakara, 2010.
Marketing and Retailing	E-traceability- Radio Frequency Identification Device (RFID) and GIS technology, E-business, E-commerce, Customer Relationship Management(CRM)	Baer & Brown, 2007; El-Gohary, 2010; Galliano & Orozco, 2008; Fang & Lie 2006; Gregor et al. 1997; Henderson et.al, 2004; Lefebvre et.al 2001; Lu et.al, 2005; New, 2010; OMAF, 2004.

Source: Literature Survey, 2011

factor of 3.7, from 482 million to 1.8 billion. The study also identified a significant impact on output markets in forestry, fishery, and agricultural sectors in developing countries. Moreover, the empirical analysis justified a gain in consumer and /or producer welfare, particularly when there were no previous sources of information. A similar study done in Sri Lanka also emphasized how simple technologies such as mobile phones, Interactive Voice Recognition (IVR) systems, and text messaging have often been proven more effective than web applications (Silva and Ratnadiwakara, 2010). Mobile financial applications is one attribute of ICT-based services whereby agribusiness entrepreneurs derive tangible benefits through mobile banking and brokerage services, mobile money transfer, and mobile payments. These applications have transformed mobile device into a business tool (Varshney[2], 2003). Another example of mobile application in agriculture can be environmental monitoring. This technology enables farmers to easily monitor agricultural area in real time to avoid adverse effect of environmental changes. Moreover, this application helps in irrigation management, fertilizer determination, and planning harvest time. Precision agriculture is also another mobile data acquisition system to collect data for crop management including soil water availability, soil compaction, soil fertility, local climate data, and grain yield (Chigona et al, 2009). Table 2 summarizes various studies on the growing mobile ICT use in agriculture categorized by value creation source or stage along the agricultural value chain.

3. KEY SUCCESS FACTORS FOR ICT APPLICATION IN AGRIBUSINESS

Diffusion of a radical technology usually takes a lot of time; because the adoption of technology is not only one entity's decision but are rather a social phenomenon involving the interest of many people in an interdependent manner. Before people

Table 2. Summary of mobile commerce applications in agribusiness (2000-2011)

Value Creation Source	Mobile Application	Author and Year
Marketing	Short Message Service (SMS), Global Positioning System(GPS), commodity price applications e.g. Farmer's Partner, KIS futures	Barnes & Donner, 2006; Bauer et al. 2005; Rao & Minakakis, 2003; Scornavacca, 2004.
Distribution	E-traceability, GPS and RFID tags	Garcia et al. 2009; Huang et al, 2010; Karkkainen, 2003; Kumar & Stokkeland, 2003; Varshney[2], 2003.
Advertizing	User and location sensitive mobile advertising	Azni et al. 2008; Rao & Minakakis, 2003; Unni & Harmon, 2007;Varshney[1], 2003.
Production	Agronomy applications (i.e., weather monitoring, precision agriculture, precision irrigation). Example- Agrarian Mobile Information Center, Dekalb Plant Population Calculator, and wireless sensors	Anckar & Incau, 2002; Chigona et al. 2009; Wang, 2006.
Supply chain collaboration	Supply chain partnership (mobile data solutions), activity tracking and reporting, delivery management and verification, mobile logistic management, mobile inventory management	Africa Report,2011; Al, 2010; Dooli & Haj; Li, 2010; Ming et. al, 2008; Ogawara et. al, 2003.
Financial service delivery	Mobile banking, brokerage services, mobile money transfer, and mobile payment	Barnes & Corbitt, 2003; Hinman & Matovu, 2010; Varshney[1], 2003.
Market intelligence	Mobile agro-portal, wireless sensor	Garcia et. al, 2009; Katengeza et. al, 2011; Maumbe, 2010;.

Source: Literature Survey, 2011

decide to adopt a new technology, they need time to be convinced that the new technology is an improvement over the existing technology. And even after, the diffusion of technologies is typically slower than the diffusion of information. It might be relatively easy to buy and install an ICT application but using it effectively takes time and requires well-educated employees who can act independently as well as interdependently (Frank and Bunte, 2009). McGowan and Armstrong (2007) also support the dependence between ICT application and knowledge. As they mentioned, basic Internet applications require little financial or marketing involvement whereas understanding of website applications especially for business requires an advanced knowledge. Moreover, high level of understanding of a website efficacy for business results in consideration of strategic thinking which will derive effective website adoption and utilization.

A study on direct-market farms in the Northern United States conducted by Baer and Brown (2007) also found out that *the size of a farm business* has an effect on website adoption, verifying that the cost of a website influence larger farms to use it more often than smaller farms. Their research also indicated that having *access to high-speed Internet* increases the probability of website adoption. Therefore, they concluded since websites are likely more feasible for larger operations, any program encouraging small farms to use a website as part of their marketing system may need to consider subsidizing the cost. Supporting the above research outcome, Henderson et.al (2005) also emphasized how the size of the firm provides the largest determination of who does or does not implement an Internet strategy. In the same token, Del Aguila–Obra and Padilla-Melendez (2006) also conducted a bivariate analysis and again their results support the positive relationship between the *size of the organization* and the likelihood of having its own host. Moreover, they also found correlation between the presence of intranet and the geographical dispersion of the organization; and between firm size and the presence of mobile teleworkers. Namani (2009) also identified interdependency between *degrees of value-added product*, *firm size* and *ICT adoption* as, a firm that markets greater value-added product will get higher financial return from IT than a firm which is not positioned with a relatively large value-add mix of products.

According to McGowan and Armstrong (2007) study on the Northern Ireland agribusiness, lack of *marketing ability* and the *negative attitude* in relation to Internet marketing were found to be significant constraints on agri-food SME website adoption and utilization. This lack of marketing ability and negative attitudes especially hamper awareness of the efficacy of the Internet as a business tool for the individual business. Funk (2007) on his study on the Japanese market also found out that a lack of *basic infrastructure from roads and electricity* to *telephones* and *Internet connectivity* affects ICT adoption. On top of this, he also mentioned how a country's *legal and regulatory framework* hinders ICT development. Tompson and Kunda (2000) also pointed out how *geography* and *physical location* matters in agriculture e-marketing. They argue that since most agricultural products are bulky relative to value, efficiencies in logistic and handling may limit e-marketing of both agriculture input and output markets. On the contrary, they stated that *branded products* like; "Roundup Ready Soybeans" which are known to be genetically modified with a given biotechnology to resist roundup herbicide or Starbucks traded coffee beans can easily be e-marketed.

In regards to traceability, Angeles (2005) mentioned the *cost factor* as a major issue with the RFID tagging. The study indicated that the most inexpensive tag costs about 50 cents in very large quantities and as a result, keeping the costs down definitely will encourage widespread future adoption. A study conducted by Galliano and Orozco (2008) on French agribusiness showed the probability to adopt an electronic based traceability system is strongly and linearly correlated to the increase in the firm's size. The larger the firm is, the greater the implementation of electronic-based traceability system. Their result justifies the reason why a wide number of SME rely on paper-based traceability practices.

On the other side Burke (2010) also tried to relate ICT adoption with the *growth stage of a firm*. The stages model suggests that firms in the *existence and survival stage* would be less likely to employ extensive or more sophisticated ICTs while as firm grows, however, it can be expected to explore the use of more complex ICTs to better manage customer and supplier relationships and reduce operational inefficiencies. As a result, small SMEs are identified with the existence and survival growth stage while larger SMEs are more likely be classified in the *success, take off or mature growth stages*. Therefore, growth in small firms makes adoption of new and more advanced ICT possible.

Some of the inhibiting factors to the expansion of ICT usage in South West Nigeria were *lack of basic infrastructure, favorable business environment*, especially related to unstable economy of the country and *high cost* of ICT devices (Ayinde et.al, 2004). Results from a study conducted in Singapore by Liao and Cheung (2001) showed that *consumer's attitude* towards the content of products, transactions security, retail price, service quality of vendor, IT education and Internet usage affects the initial willingness to adopt e-marketing. As mentioned by Namani (2009) in a study on Kosovo, level of IT adoption is dependent on *lack of access to credit, inadequate information, expertise* and *high cost of access to information technology*. The key success factors based on the literature survey are categorized by source of the influential force (i.e., internal and external) and summarized in Table 3.

4. IMPLICATION OF ICT USE IN AGRICULTURE AND AGRIBUSINESS INDUSTRY

Internet has been used as a commercial means for business transaction worldwide in computer-mediated environment (Armesh et.al, 2010). E-commerce and m-commerce as information systems are expected to continue growing while individual firm websites will remain as an important communication channel. The growth in new generation smart phones has led to increased functionality and potential benefits for the agriculture community. Agribusiness firms around the world are positioning themselves on the global

Table 3. Summary of success factors in ICT use in agriculture and agribusiness industry

ICT Source Success Factors		Author(s) and Year
Internal Factors	Size of firm (diseconomies of scale)	Aguila-Obra & Padilla-Melendez, 2006; Baer & Brown, 2007; Henderson et.al, 2005; Namani, 2009.
	ICT Skill and knowledge,	Armesh, et al, 2010; Frank & Bunte, 2009; McGowan & Armstrong, 2007.
External Factors	Infrastructure and Technology	Ayinde et.al, 2004; Funk, 2007; Hamilton & Asundi, 2008; McGowan & Armstrong, 2007; Namani, 2009.
	Access to credit	Allahyari & Chizari, 2010; Mutula & Brakel, 2006; Namani, 2009.
	Trust and negative attitude towards Internet marketing	Belkhamza & Wafa, 2009; Liao & Cheung, 2001; McGowan & Armstrong, 2007; Zeinab, 2006.

Source: Literature Survey, 2011

marketplace using these modern technologies. This section discusses briefly the implication of ICT use in agriculture and agribusiness by examining the following key factors; trust, security and privacy, trust, ICT literacy, traceability, and risk management in the knowledge economy.

Given the rise in cybercrime, *security and privacy issues* (i.e., personal and agribusiness) have emerged among the leading barriers in the development of both e-commerce and mobile commerce applications (Belkhamza & Wafa, 2009). The problems of security and individual privacy issues started to come up when new information technologies advanced in the collection, storage, use and sharing of personal information. One example of this information abuse could be; since the Internet connection can cross geographic boundaries, an abuser be it an individual or agribusiness firm, may move outside of a specific country and continue to misuse the personal information that consumers provided previously (Zeinab, 2006). On top of that uncertainty about the identity of online trading parties or product quality is another big issue in online marketing. This is particularly problematic given that online trading parties can easily remain anonymous or change their identity (Belkhamza & Wafa, 2009; Tsai & Yeh, 2010).

A study done in Singapore showed that among the factors which affect initial willingness of customers to e-shop via the Internet, *trust* issues like; consumer's attitude towards the content of products, transactions security and service quality were mentioned (Liao & Cheung, 2001). Therefore, in order for buyers to shop in the online marketplace, they have to rely on the website without having the ability to physically inspect the product (Belkhamza & Wafa 2009). This consumer trust issue has led to the growing demand for the implementation of backward and forward traceability systems along the agriculture value chains.

A study on Malaysia found out that young people have been observed to be major users of Internet and e-commerce services. In retailing activities, individuals aged 30 and below account for about 76.3 percent of all online purchases. Moreover, *educational level* has also been identified as key factor with a major impact online shopping behavior. As noted from the results, 78 percent of the customers had at least a diploma while the rest 22 percent accounted for other educational levels (Armesh, et.al, 2010). Glory & Akridge (2000) also affirmed age and education as important factors in the adoption of ICT. Moreover, their study found a strong relationship between profit maximizing goal and computer adoption which led them to conclude that farm managers seem to perceive the PC as a profit maximizing input. On the contrary, Internet adoption was not strongly related to profit maximization, or production maximization, or an increase in free time as farm goals which indicates farmers' uncertainty on how the Internet might be used to achieve their goals.

Moreover, a cross cultural study done between United States and Korea by Choi and Lee (2003) showed that the risk associated with online shopping perceived by consumers appear to be significantly different. Overall, Korean respondents tend to have higher level of perceived risk toward online shopping than Americans do. Specifically, they are more concerned about product delivery, transaction security, and customer service issues when dealing with products online than Americans. On the contrary, the American online purchases tend to be influenced by all the risk factors except by customer service. To play an effective role in this knowledge economy and gain better advantages out of the digital era, firms need to understand how far perceived risk may affect executive managers' decisions to engage in e-commerce and mobile commerce applications (Belkhamza & Wafa, 2009). The Federal Trade Commission (FTC) in the US recommends participation in online privacy seal programs as a certification for secure system. This program requires licensees to implement certain fair information practices and submit to compliance testing in order to display a privacy seal on their web site (Zeinab, 2006). Finally, the future success of ICT applications in agribusiness value chains will depend on the extent and degree to which key challenges related to security, privacy, literacy, trust, and food traceability concerns among others are resolved. The steady growth of e-commerce and the emergence of mobile commerce especially over the past decade indicate rising interest among global consumers to use modern computers and mobile-based ICT to purchase new products and services on the virtual marketplace in addition to their traditional roles as communication tools.

5. CONCLUSION

As seen from the literature review, Information Communication Technologies (ICT) use in agriculture and agribusiness firms is diverse and growing globally. The various areas of modern ICT applications are designed to reduce operational cost, increase profits, enhance the ability to use information management to solve problems and decision making process, improving customer interaction, facilitate employee functions, promote automation, and develop high speed information delivery and retrieval. In addition, ICT has enhanced ease of accessibility to various information systems (i.e., traceability, environmental monitoring, location, etc.), helped reduce consumer concerns for food safety and environmental contamination issues, and continues to play strategic role in improving and maintaining agribusiness competitiveness and sustainability. In line with the mentioned different roles of ICT, agribusiness firms can use a wide range of new generation technologies to fulfill their organizational goals and objectives.

Among the leading areas of ICT in agriculture and agribusiness, we noted that input procurement, production, marketing, food traceability, and financial service delivery are some of the growing areas of application. Producers are increasingly seeking ways to add or create value to their businesses by integrating ICT in the value chain. Similarly, consumers are becoming more knowledgeable about their preferences and how they could use ICT to articulate their interests. Given cell phones, radios, digital cameras, geographical information systems (GIS), computing and tracking mechanisms among others, ICT will play an increasingly important role in improving the performance of agricultural value chains in future.

Despite the increasingly strategic role being played by ICT in agriculture and agribusiness industry, it is also seen that there are still some major factors that hinder its adoption and efficient utilization such as firm size, knowledge level, Internet access, infrastructure, geographical dispersion of the firm, degree of value-added product, legal and regulatory framework, product bulkiness relative to value, transaction security, marketing ability, negative attitudes of customers, relatively high access cost, as well as privacy and confidentiality issues. Despite all these constraints, and as seen from the key success factors arising from different countries ICT adoption and diffusion led by Internet, website, and mobile technologies has shown an increasing trend over time. Therefore, the future success of integrating ICT in agriculture and agribusiness industry requires the realization and understanding that key constraints facing ICT are not necessarily common to all places. Significant differences have been seen depending on geographic locations, types of production or farming systems, and relative level or scope of government intervention. As a result, we recommend that context specific empirical studies be conducted in order to come up with relevant policy recommendations and practical solutions that make it possible for the successful deployment and use of ICT in agriculture and agribusiness industries globally.

REFERENCES

Allahyari, M., & Chizari, M. (2010). Technologies (ICT) in agriculture sector. *Journal of Agricultural Science and Technology*, *4*(4), 115–120.

Anckar, B., & D'Incau, D. (2002). Value creation in mobile commerce: findings from a consumer survey. *Journal of Information Technology Theory and Application*, *4*(1), 43–64.

Angele, R. (2005). RFID technologies: Supply chain applications and implementation issues. *Information Systems Management*, *22*(1), 51–65. doi:10.1201/1078/44912.22.1.20051201/85739.7

Armesh, H., Salarzehi, H., Yaghoobi, N., Heydari, A., & Nikbin, D. (2010). The effect of security and privacy information on trust and trust worthiness and loyalty in online marketing in Malaysia. *International Journal of Marketing Studies*, *2*(2), 223–234.

Ayinde, I., Agbonlahor, M., Mafimisebi, T., & Adebayo, O. (2004). *Assessment of the use of information communication technologies on the economic performance of agro based food industries in south west Nigeria*. Paper prepared for presentation at the Farm Management Association of Nigeria Conference, Abuja, Nigeria.

Azni, H., Ahmad, H., & Selviawati, T. (2008). *Bluetooth mobile advertising system using pull-based approach*. International Symposium on Information Technology Information Technology, Kuala Lumpur, Malaysia Aug 26-28.

Baer, A., & Brown, C. (2007). Adoption of e-marketing by direct-market farms in the Northern United States. *Journal of Food Distribution Research*, *38*(2), 1–11.

Barnes, S., & Corbitt, B. (2003). Mobile banking: Concept and potential. *International Journal of Mobile Communications*, *1*(3), 273–288. doi:10.1504/IJMC.2003.003494

Barnes, S., & Scornavacca, E. (2004). Mobile marketing: The role of permission and acceptance. *International Journal of Mobile Communications*, *2*(2), 128–139. doi:10.1504/IJMC.2004.004663

Batte, M. (2005). Changing computer use in agriculture evidence from Ohio. *Computers and Electronics in Agriculture*, *47*(1), 1–13. doi:10.1016/j.compag.2004.08.002

Bauer, H., Reichardt, T., Barnes, S., & Neumann, M. (2005). Driving consumer acceptance of mobile marketing: A theoretical framework and empirical study. *Journal of Electronic Commerce Research*, *6*(3), 181–191.

Begalli, D., Condurri, S., & Gaeta, D. (2009). Wine and web marketing strategies. The case study of Italian specialty wineries. *British Food Journal*, *111*(6), 598–619. doi:10.1108/00070700910966041

Belkhamza, Z., & Wafa, A. (2009). The effect of perceived risk on the intention to use e-commerce: The case of Algeria. *Journal of Internet Banking and Commerce*, *14*(1), 1–10.

Burke, K. (2010). The impact of Internet and ICT use among SME agribusiness growers and producers. *Journal of Small Business and Entrepreneurship*, *23*(2), 173–194.

Chigona, W., Beukes, D., Vally, J., & Tanner, M. (2009). Can mobile Internet help to alleviate social exclusion in developing countries? *The Electronic Journal of Information Systems in Developing Countries*, *36*(7), 1–16.

Choi, J., & Lee, K. (2003). Risk perception and e-shopping cross-cultural study. *Journal of Fashion Marketing and Management*, *7*(1), 49–64. doi:10.1108/13612020310464368

Chung, K., & Pauline, F. (2010). *Evaluating the impact of ICT on trade of fruits and vegetables within the APEC countries*. 54th Australian Agricultural and Resource Economics Society, February 10-12, Adelaide, Australia.

Cloete, E., & Doens, M. (2008). B2B e-marketplace adoption in South African agriculture. *Information Technology for Development*, *14*(3), 184–196. doi:10.1002/itdj.20105

Del Aguila–Obra, A., & Padilla-Melendez, A. (2006). Organizational factors affecting Internet technology adoption. *Internet Research*, *16*(1), 94–110. doi:10.1108/10662240610642569

Donner, J. (2006). The use of mobile phones by micro entrepreneurs in Kigali, Rwanda: Changes to social and business networks. *Information Technologies and International Development*, *3*(2), 3–19. doi:10.1162/itid.2007.3.2.3

Doolin, B., & Ali, E. (2008). Adoption of mobile technology in the supply chain: An exploratory cross-case analysis. International. *Journal of Business Research*, *4*(4), 1–16. doi:10.4018/jebr.2008100101

El-Gohary, H. (2010). E-marketing-A literature review from a small business perspective. *International Journal of Business and Social Science*, *1*(1), 214–244.

Fang, C., & Lie, T. (2006). Assessment of Internet marketing and competitive strategies for leisure farming industry in Taiwan. *Journal of American Academy of Business*, *8*(2), 296–300.

Frank, H., & Bunte, L. (2009). *Thought for food: The impact of ICT on agribusiness*. Wageningen University and Research, the Netherlands. Paper prepared for presentation at the 113th EAAE Seminar-A Resilient European Food Industry and Food Chain in a Changing World. Chaina, Crete and Greece.

Funk, J. (2007). The future of mobile shopping: The interaction between lead users and technological trajectories in the Japanese market. *Technological Forecasting and Social Change*, *74*(3), 341–356. doi:10.1016/j.techfore.2005.11.007

Galliano, D., & Orozco, L. (2008). *Intra and inter organizational determinants of electronic based traceability adoption: Evidence from the French agri-food industry*. 12th Congress of the European Association of Agriculture Economics, Ghent, Belgium.

Garcia, L., Lunadei, L., Barreiro, P., & Robla, I. (2009). A review of wireless sensor technologies and applications in agriculture and food industry: State of the art and current trends. *Sensors (Basel, Switzerland)*, *9*(6), 4728–4750. doi:10.3390/s90604728

Glory, B., & Akridge, J. (2000). Computer and Internet adoption on large U.S farms. *International Food and Agribusiness Management Review*, *3*, 323–338. doi:10.1016/S1096-7508(01)00051-9

Gregor, S., Newman, S., & Larner, T. (1997). *The Australian meat industry, the EM hypothesis and Internet marketing*. 10th International Bled Electronic Commerce Conference, Bled, Slovenia 9-11 June.

Henderson, J., Dooley, F., & Akridge, J. (2004). Internet and e-commerce adoption by agricultural input firms. *Review of Agricultural Economics*, *26*(4), 505–520. doi:10.1111/j.1467-9353.2004.00196.x

Henderson, J., Dooley, F., Aridge, J., & Carerre, A. (2005). Adoption of Internet strategies by agribusiness firms. *International Food and Agribusiness Management Review*, *8*(4), 42–61.

Hinman, R., & Matovu, J. (2010). Opportunities and challenges for mobile-based financial services in rural Uganda. In *Proceedings of the 28th of The International Conference Extended Abstracts on Human Factors in Computing Systems*. New York, NY: ACM.

Hossain, M., & Quaddus, M. (2010). An adoption diffusion model of RFID-based livestock management system in Australia. *Human Benefit through the Diffusion of Information Systems Design Science Research, 318*(1), 179-191.

Huang, Y., Kuo, W., & Kao, M. (2010). A RFID field operation system design for agricultural traceability. In *Proceedings of the 2010 First International Conference on Pervasive Computing, Signal Processing and Application.* Washington, DC: IEEE Computer Society.

Jensen, R. (2010). Information, efficiency, and welfare in agricultural market. *Journal of International Association of Agricultural Economist, 41*(1), 203–216.

Jonsson, S., & Gunnarsson, C. (2005). Internet technology to achieve supply chain performance. *Business Process Management Journal, 11*(4), 403–417. doi:10.1108/14637150510609426

Kagan, A. (2000). Information systems implementation within US agribusiness: An applications approach. *Computers and Electronics in Agriculture, 28*(3), 207–228. doi:10.1016/S0168-1699(00)00144-7

Karkkainen, M. (2003). Increasing efficiency in the supply chain for short life goods using RFID tagging. *International Journal of Retail and Distribution Management, 31*(10), 529–536. doi:10.1108/09590550310497058

Katengeza, S. P., Kiiza, B., & Okello, J. (2011). The role of ICT-based market information services in spatial food market integration: The case of Malawi agricultural commodity exchange. *International Journal of ICT Research and Development in Africa, 2*(1), 1–14. doi:10.4018/jictrda.2011010101

Katengeza, S. T., Okello, J. J., & Jambo, N. (2011). Use of mobile phone technology in agricultural marketing. *International Journal of ICT Research and Development in Africa, 2*(2), 14–25. doi:10.4018/jictrda.2011070102

Kenny, C. (2002). Information and communication technologies for direct poverty alleviation, costs and benefits. *Development Policy Review, 20*(2), 141–157. doi:10.1111/1467-7679.00162

Kumar, S., & Stokkel, J. (2003). Evolution of GPS technology and its subsequent use in commercial markets. *International Journal of Mobile Communications, 1*(1), 180–193. doi:10.1504/IJMC.2003.002465

Lal, D., Renee, P., & David, N. (2008). *How can ICTs be used and appropriated to address agricultural information needs of Bangladeshi farmers?* GlobDev, Paper 21.

Lefebvre, A., Cassivi, L., & Lefebvre, E. (2001). E-commerce transition mode for supply chain management. *Chain and Network Science, 1*(1), 23–32. doi:10.3920/JCNS2001.x003

Li, L., Guo, W., & Wang, Y. (2010). *Mobile supply chain management: Theory and method.* The International Conference on E-Product, E-Service and E-Entertainment (ICEEE) Nov. 7-9 Henan, China.

Liao, Z., & Cheung, M. (2001). Internet-based e-shopping and consumer attitudes: An empirical study. *Information & Management, 38*, 299–306. doi:10.1016/S0378-7206(00)00072-0

Lio, M., & Liu, M. (2006). ICT and agricultural productivity: Evidence from cross-country data. *Agricultural Economics, 34*(3), 221–228. doi:10.1111/j.1574-0864.2006.00120.x

Lu, N., Swatman, C., & Perreau, S. (2005). Sustainable agriculture mobile commerce– Extending the rural grocery supply chain through m-commerce. *TENCON (2005). IEEE Region, 10*, 1–6.

Lyons, M. (2005). Future ICT system-Understanding the business drivers. *BT Technology Journal, 23*(3), 11–23. doi:10.1007/s10550-005-0026-1

Mahatanankoon, P., Joseph, H., & Wen, B. (2005). Mobile commerce: A new frontier in e-commerce and wireless communication consumer-based m-commerce: Exploring consumer perception of mobile applications. *Computer Standards & Interfaces*, *27*(4), 347–357. doi:10.1016/j.csi.2004.10.003

Maumbe, B. M. (2010). Mobile agriculture in South Africa: Implementation, framework, value-added services and policy implications. *International Journal of ICT Research and Development in Africa*, *1*(2), 35–59. doi:10.4018/jictrda.2010040103

McGowan, P., & Armstrong, G. (2007). Determinants of Internet adoption by SME agri-food companies. *Journal of Small Business and Enterprise Development*, *14*(4), 620–640. doi:10.1108/14626000710832730

Ming, Z., Xin, L., & Zhijun, Z. (2008). *Research on mobile supply chain management based ubiquitous network*. 4th International Conference on Mobile Commerce and Service, May 25-28, Kunming, Chaina.

Moss, C., Schmitz, T., Kagan, A., & Schmitz, A. (2003). Institutional economics and the emergence of e-commerce in agribusiness. *Journal of Agribusiness*, *21*(1), 83–101.

Mueller, E. (2000). Emergent e-commerce in agriculture. *Agriculture Issues Center Brief, 14*.

Mueller, R. (2001). E-commerce and entrepreneurship in agricultural markets. *American Journal of Agricultural Economics*, *83*(5), 1243–1249. doi:10.1111/0002-9092.00274

Mutula, S., & Brakel, P. (2006). E-readiness of SMEs in the ICT sector in Botswana with respect to information access. *The Electronic Library*, *24*(3), 402–417. doi:10.1108/02640470610671240

Namani, M. (2009). *The role of information technology in small and medium sized enterprises in Kosova*. Fulbright Academy Conference, Skobje, Republic of Macedonia

New, S. (2010). Get the best from your people: The transparent supply chain. *Harvard Business Review*, *88*(10), 76–80.

Niederhauser, N., Oberthur, T., Kattning, S., & Cock, J. (2008). Information and its management for differentiation of agricultural products: The example of specialty coffee. *Computers and Electronics in Agriculture*, *61*(2), 241–253. doi:10.1016/j.compag.2007.12.001

Obayelu, A., & Ibadan, I. (2006). Analysis of information communication technology (ICT) for gender empowerment and sustainable poverty alleviation in Nigeria. *International Journal of Education and Development using ICT, 2*(3), 45-69.

Ogawara, S., Chen, J., & Zhang, Q. (2003). Internet grocery business in Japan: Current business models and future trends. *Industrial Management (Des Plaines)*, *103*(9), 727–735.

Rao, B., & Minakakis, L. (2003). Evolution of mobile location based service. *Communications of the ACM*, *46*(12), 61–65. doi:10.1145/953460.953490

Schaefer, G. (2008). Tracking and tracing-A challenge for system organization and IT. *Journal of Information Technology in Agriculture*, *3*(1), 19–25.

Silva, H., & Ratnadiwakara, D. (2010). ICT policy for agriculture based on a transaction cost approach: Some from Sri Lanka. *International Journal of ICT Research and Development in Africa*, *1*(1), 51–64. doi:10.4018/jictrda.2010010104

Sultan, F., & Rohm, A. (2004). The evolving role of the Internet in marketing strategy: An exploratory study. *Journal of Interactive Marketing*, *18*(12), 6–19. doi:10.1002/dir.20003

Tenhunen, S. (2008). Mobile technology in the village: ICT, culture, and social logistic in India. *Journal of the Anthropological Institute*, *14*(3), 514–534.

Thompson, S., & Sonka, S. (1979). Potential effect of information technologies on the economic performance of agricultural and food markets. *American Journal of Agricultural Economics*, *79*(2), 657–662. doi:10.2307/1244167

Tollens, E. (2006). *Market information systems in sub-Saharan Africa: Challenges and opportunities*. Poster paper prepared for presentation at the International Association of Agricultural Economists Conference, Gold Coast, Australia.

Tompson, S., & Kunda, E. (2000). *E-commerce and agricultural markets*. OFOR Paper Number 00-03.

Tsai, Y., & Yeh, J. (2010). Perceived risk of information security and privacy in online shopping: A study of environmentally sustainable products. *African Journal of Business Management*, *4*(18), 4057–4066.

Unni, R., & Harmon, R. (2007). Perceived effectiveness of push vs. pull mobile location-based advertising. *Journal of Interactive Advertising*, *7*(2), 28–40.

Van der Vorest, J., Van Dongen, S., Nouguier, S., & Hilhorst, R. (2002). E-business initiatives in food supply chains: Definition and typology of electronics business models. *The Journal of Logistics: Research and Applications*, *5*(2), 119–138. doi:10.1080/13675560210148641

Vander Vorst, J., Dongen, S., Nouguier, S., & Hilhorst, R. (2002). E-business initiatives in food supply chains: Definition and typology of electronic business models. *International Journal of Logistics: Research and Applications*, *5*(2), 119–138.

Varshney, U. (2003). Location management for mobile commerce applications in wireless Internet environment. *Transactions on Internet Technology*, *3*(3), 236–255. doi:10.1145/857166.857169

Varshney, U. (2003). Mobile and wireless information systems: Applications, networks, and research problems. *Communications of the Association for Information Systems*, *12*(11), 155–165.

Vorst van der. J., Beulens, M., & Van Beek, P. (2005). Innovation in logistics and ICT in food supply chain networks (pp. 245-292). Wagengen, The Netherlands: Wagengen Academic Publisher.

Wang, N., Zhang, N., & Wang, M. (2006). Wireless sensors in agriculture and food industry- Recent development and future perspective. *Computers and Electronics in Agriculture*, *50*, 1–14. doi:10.1016/j.compag.2005.09.003

Zeinab, K. (2006). Trust, privacy and security in electronic business: The case of GCC countries. *Information Management & Computer Security*, *14*(3), 270–283. doi:10.1108/09685220610670413

Zeinab, K. (2008). *Final report: ICT and small scale agriculture in Africa: A scoping study*. The International Development Research Centre.

Zeinab, K. (2011, February 25). Mobile applications help farmers profit. *Africa Report*.

Compilation of References

Aboyade, B. O. (1984). *The provision of information for rural development* (pp. 126–137). Ibadan Foundation Publication.

Action for Enterprise. (2010). *Using ICT to provide agriculture market price in Africa*. Briefing paper supported by USAID's Fostering Agriculture Competitiveness Employing Information Communication Technologies (FACET) project. Washington, DC: United States Agency for International Development (USAID).

Adesina, F. A. (2005) *Geoinformation and natural resources exploitation in Africa*. Economic Commission for Africa, Fourth Meeting of the Committee on Development Information (CODI IV) Addis Ababa, Ethiopia 23 – 28 April. [Online]. Retrieved from www.uneca.org/.../Geoinformation%20and%20natural20resources%20 exploitation.doc

Adhau, B. P. (2010). Contribution of agriculture in the development of India. *International Research Journal, 1*(17).

Agrawal, G. K. (2008). Village knowledge centers / common service centers, information technology in developing countries. *Newsletter, 18*(1). Retrieved on February 9, 2011, from http://www.iimahd.ernet.in/egov/ifip/feb2008/gk-agrawal.htm

Ahmed, R. (Ed.). (2009). *Globalisation and economic development –Part – 1*. New Delhi, India: Mittal Publications.

Aker, J. C. (2008). *Does digital divide or provide? The impact of cell phones on grain markets in Niger*. BREAD working paper number, 177.

Aker, J. C. (2008). *Does digital provide or divide? The impact of cell phones in grain markets in Niger.* Center for Global Development, Working Paper 154.

Aker, J., & Mbiti, I. (2010). *Mobile phones and economic development in Africa*. CGD working paper 211.

Aker, J. (2008). *Can you hear me now? How cell phones are transforming markets in Sub-Saharan Africa*. USA: Center for Global Development.

Aker, J. C. (2008). *Does digital divide or provide? The impact of cell phones on grain markets in Niger. Job Market Paper*. Berkeley: University of California.

Alam, S., & Noor, M. (2009). ICT adoption in small and medium enterprises: an empirical evidence of service sectors in Malaysia. *International Journal of Business and Management, 4*(2), 112–125.

Alanana, O. O. (2003). Youth unemployment in Nigeria: Some implication for the third millennium. *Global Journal of Socila Sciences, 2*(1), 21–26.

Alberta Forest Extension. (2009). *An introduction to the Alberta Forest Extension Network*. Retrieved from www.afex.ca/images/stories/docs/afex_powerpoint_for_website_june_2009.pdf

Alberta Forest Products Association. (2010). *Training.* Retrieved from http://www.albertaforestproducts.ca/resources/training.aspx

Allahyari, M., & Chizari, M. (2010). Technologies (ICT) in agriculture sector. *Journal of Agricultural Science and Technology, 4*(4), 115–120.

Amarach Consulting. (2007). *A research de-brief for Comreg.* Prepared by Amárach Consulting. Retrieved from http://www.comreg.ie/_fileupload/publications/ComReg0708.pdf

Amit, R., & Zott, C. (2001). Value-creation in e-business. *Strategic Management Journal*, *22*, 493–520. doi:10.1002/smj.187

Anckar, B., & D'Incau, D. (2002). Value creation in mobile commerce: findings from a consumer survey. *Journal of Information Technology Theory and Application*, *4*(1), 43–64.

Angele, R. (2005). RFID technologies: Supply chain applications and implementation issues. *Information Systems Management*, *22*(1), 51–65. doi:10.1201/1078/44912.22.1.20051201/85739.7

An, K., Xie, G., Leng, Y., & Xiao, Y. (2003). Design of farmland GIS for precision agriculture. *Chinese Geographical Science*, *13*, 20–24. doi:10.1007/s11769-003-0079-3

Annamalai, K., & Rao, S. (2003). *What works: ITC's E – CHOUPAL and profitable rural transformation web – based information and procurement tool for Indian farmers*. Retrieved February 6, 2011, from http://www.bdsknowledge.org/dyn/bds/docs/292/ITC e-choupalIndia.pdf

Antonopoulou, E., Karetsos, S. T., Maliappis, M., & Sideridis, A. B. (2010). Web and mobile technologies in a prototype DSS for major field crops. *Computers and Electronics in Agriculture*, *70*(2), 292–301. doi:10.1016/j.compag.2009.07.024doi:10.1016/j.compag.2009.07.024

Anupindi, R. (2004). *Going direct to farmer: ITC's e – Choupal initiative*. Retrieved February 6, 2011, from http://www.mba.tuck.dartmouth.edu/digital/Programs/CorporateEvents/SupplyChainThoughtLeaders/session6.pdf

Anyanwe, A. C., & Maxwell, U. C. (2007). *A textbook of agricultural science for school and college* (p. 237). Benin-City, Benin: Justice Jeco Press and Publishers Ltd.

Apud, E. (1989). Human biological methods for ergonomics research in forestry. In Apud, E., Bostrand, L., Mobbs, I. D., & Strehlke, B. (Eds.), *Guide-lines on ergonomic study in forestry*. Geneva, Switzerland: International Labour Organization.

Arabi, U. (2009). State of Indian agriculture sustainability under changing technology: Emerging policy agenda. In Ahmed, R. (Ed.), *Globalisation and economic development –Part – 1* (pp. 49–71). New Delhi, India: Mittal Publications.

Aris, N. M. (2006). *Building blocks for economic growth*. Paper presented at the National statistics Conference.

Armesh, H., Salarzehi, H., Yaghoobi, N., Heydari, A., & Nikbin, D. (2010). The effect of security and privacy information on trust and trust worthiness and loyalty in online marketing in Malaysia. *International Journal of Marketing Studies*, *2*(2), 223–234.

Aronoff, S. (1989). *Geographic information system: A management perspective* (pp. 24–37). Ottawa, Canada: WDL.

Ascough, J. C., Dunn, G. H., McMaster, G. S., Ahuja, L. R., & Andales, A. A. (2005). Producers, decision support systems and GPFARM: Lessons learned from a decade of development in Zerger. *Proceedings of International Congress on Modelling and Simulation 2005*, (pp. 170–176).

Assadi, M. (1998). Farmer's suicide –Signs of distress in rural economy. *Economic and Political Weekly*, *23*(14), 747–748.

Assisi, C., & Gupta, I. (2003). ITC's rural symphony. *Business World*, *22*(34), 30–37.

Avouris, N., Solomos, K., & Tselios, N. (2001). Internet as a tool for open and remote teaching: Alternative mechanisms of quality control. In *Proceedings of the First Panhellenic Conference on Open and Remote Teaching*. Retrieved from http://www.eap.gr/news/EXAGGELIA_SYNEDRIOU/synedrio/html/sect5/22.htm

Ayanz, J. S. M., Barbosa, P. M., Schmuck, G., & Liberta, G. (2003). *The European forest fire information system (EFFIS).* European Commission. JIR Institute for Environment and sustainability. Retrieved from http://natural-hazards.jrc.it/documents/fires/2003- 591publications/EFFIS-earsel-paper.pdf

Ayinde, I., Agbonlahor, M., Mafimisebi, T., & Adebayo, O. (2004). *Assessment of the use of information communication technologies on the economic performance of agro based food industries in south west Nigeria*. Paper prepared for presentation at the Farm Management Association of Nigeria Conference, Abuja, Nigeria.

Azni, H., Ahmad, H., & Selviawati, T. (2008). *Bluetooth mobile advertising system using pull-based approach.* International Symposium on Information Technology Information Technology, Kuala Lumpur, Malaysia Aug 26-28.

Badgaiyan, B. D. (2002). Avoiding crisis in rural areas through information technology. *Information Technology in Developing Countries, 12*(2). Retrieved February 9, 2011, from http://www.iimahd.ernet.in/egov/ifip/aug2002/article4.htm

Badgaiyan, B. D. (2002). On sustainability of a developmental website: Is it viable?*Information Technology in Developing Countries, 12*(3). Retrieved February 9, 2011, from http://www.iimahd.ernet.in/egov/ifip/dec2002/article3.htm

Baer, A., & Brown, C. (2007). Adoption of e-marketing by direct-market farms in the Northern United States. *Journal of Food Distribution Research, 38*(2), 1–11.

Bagetoft, P., & Olesen, H. B. (2004). *The design of production contracts*. Copenhagen, Denmark: Copenhagen Business School Press.

Bahl, M. (2008). *ICT in agricultural marketing*. India Science and Technology Report: 2008. New Delhi, India: NISTADS. Retrieved February 6, 2011, from http://www.nistads.res.in/indiasnt2008/India-S&T-2008-Full.pdf

Bakari, J. K., Tarimo, C. N., & Mutagahywa, B. (2006). Issues and challenges to be addressed in e-government from an information security point of view. In *Proceedings of the IST Africa Conference*, 3-5th May, CSIR International Conventions Center, Johannesburg, South Africa.

Balakrishnan, P., Golait, R., & Kumar, P. (2008). *Agriculture growth in India since 1991*. Study number 27, DEAP, Reserve Bank of India, Mumbai. Retrieved February 7, 2011, from http://rbidocs.rbi.org.in/rdocs/content/pdfs/85240.pdf

Balanskat, A., Blamire, R., & Kefala, S. (2006). *The ICT Impact Report: a review of studies of ICT impact on schools in Europe*. Brussels, Belgium: European Communities.

Ballantyne, P., Maru, A., & Porcari, E. (2010). *Crop Science Society of America, 50*(10).

Barnes, S., & Corbitt, B. (2003). Mobile banking: Concept and potential. *International Journal of Mobile Communications, 1*(3), 273–288. doi:10.1504/IJMC.2003.003494

Barnes, S., & Scornavacca, E. (2004). Mobile marketing: The role of permission and acceptance. *International Journal of Mobile Communications, 2*(2), 128–139. doi:10.1504/IJMC.2004.004663

Barrett, C. (2008). Smallholder market participation: Concepts and evidence from eastern and southern Africa. *Food Policy, 34*, 299–317. doi:10.1016/j.foodpol.2007.10.005

Barry, C. A. (1997). The research activity timeline: A quantitative tool for information research. *Library & Information Science Research, 19*(2), 153–179. doi:10.1016/S0740-8188(97)90041-4

Batte, M. (2005). Changing computer use in agriculture evidence from Ohio. *Computers and Electronics in Agriculture, 47*(1), 1–13. doi:10.1016/j.compag.2004.08.002

Bauer, H., Reichardt, T., Barnes, S., & Neumann, M. (2005). Driving consumer acceptance of mobile marketing: A theoretical framework and empirical study. *Journal of Electronic Commerce Research, 6*(3), 181–191.

Bayo-Moriones, A., & Lera-Lopez, F. (2007). A firm-level analysis of determinants of ICT adoption in Spain. *Technovation: The International Journal of Technological Innovation. Entrepreneurship and Technology Management, 27*, 352–366.

Beavers, A. M., Ruzycki, T. S., & Kunze, M. & CEMML. (2007). *Wildland fire risk assessment of western military installations.* Colorado State University, Department of Defense, Legacy Resource Management Program. (Project Number 04-230).

Begalli, D., Condurri, S., & Gaeta, D. (2009). Wine and web marketing strategies. The case study of Italian specialty wineries. *British Food Journal, 111*(6), 598–619. doi:10.1108/00070700910966041

Belkhamza, Z., & Wafa, A. (2009). The effect of perceived risk on the intention to use e-commerce: The case of Algeria. *Journal of Internet Banking and Commerce, 14*(1), 1–10.

Bennet, A. C., Price, A. J., Sturgill, M. C., Buol, G. S., & Wilkerson, G. G. (2003). HADSS, pocket HERB, and WebHADSS: Decision aids for field crops. *Weed Technology*, *17*(2), 412–420. doi:10.1614/0890-037X(2003)017[0412:HPHAWD]2.0.CO;2doi:10.1614/0890-037X(2003)017[0412:HPHAWD]2.0.CO;2

Berisha-Namani, M. (2009). *The role of information technology in small and medium sized enterprises in Kosova*. Paper presented at the Fulbright Academy 2009 Conference Small Places can Change the World.

Besley, T. (1998). How do financial failures justify interventions in rural credit markets? In Eicher, C. K., & Staatz, J. M. (Eds.), *International agricultural development* (3rd ed., pp. 370–389).

Besser, H. (2001). *The next digital divides*. Retrieved from http://www.tcla.gseis.ucla.edu/divide/politics/besser.html

Bhatnagar, S., Dewan, A., Torres, M. M., & Kanungo, P. (2004). *E–Choupal: ITC's rural networking project*. Retrieved February 6, 2011, from http://www.siteresources.worldbank.org/INTEMPOWERMENT/Resources/14647_E-Choupal_web.pdf

Biggs, N. L. (1979). The roots of combinatorics. *Historia Mathematica*, *6*, 109–136. doi:10.1016/0315-0860(79)90074-0

Blanchard, O., & Gali, J. (2007). Real wages rigidity and the new Keynesian model. *Journal of Money, Credit and Banking*, *39*(1), 35–65. doi:10.1111/j.1538-4616.2007.00015.x

Blumenstock, R. (1987). Reaching the woodlot owners with the use of mass media. In *Proceedings of the Seminar on Extension activities for owners of small woodlands* (pp. 98-109). Fredericton, Canada: Canadian Forestry Service.

Boehlje, M., Akridge, J., & Downey, D. (1995). Restructuring agribusiness for 21st century. *Agribusiness*, *11*(6), 493–500. doi:10.1002/1520-6297(199511/12)11:6<493::AID-AGR2720110602>3.0.CO;2-G

Bolarin, O., & Ayanlade, A. (2010). Utilization of satellite imaging and information communication technology in agricultural information dissemination in Nigeria. *Journal of Agricultural & Food Information*, *11*, 351–357. doi:10.1080/10496505.2010.520987

Borwein, P. (1985). On the complexity of calculating factorials. *Journal of Algorithms*, *6*, 376–380. doi:10.1016/0196-6774(85)90006-9

Bowonder, B., & Yadav, Y. (2003). *Developing ICT platform for enhancing agricultural productivity: Case study of EID Parry*. Retrieved February 06, 2011, from http://www.anningcommission.nic.in/reports/sreports/ser/stdy_ict/6_eid.pdf

Bowonder, B., Gupta, V., & Singh, A. (2004). *Developing a rural market e-hub: The case study of e-Choupal experience of ITC*. Retrieved February 6, 2011, from http://www.planningcommission.nic.in/reports/sreports/ser/stdy_ict/4_e-choupal.pdf

Boxwall, J. (2005). The nature of geospatial information and security. *Government Information Quarterly*, *22*, 644–662. doi:10.1016/j.giq.2005.09.001

Boyd, G. D., Daniels, B. J., Fallon, R., & Hyde, W. F. (1988). Measuring the effectiveness of public forestry assistance programs. *Forest Ecology and Management*, *23*(4), 297–309. doi:10.1016/0378-1127(88)90060-6

Braun, M., & Pantel, A. (2004). GIS for sustainable agriculture. *ArcNews Magazine*, pp. 26-39. Retrieved from http://www.esri.com/library/bestpractices/sustainable-agriculture.pdf

Brillinger, D. R., Preisler, H. K., & Benoit, J. W. (1986). *Risk assessment: A forest fire example*. University of California, Department of Statistics. Retrieved from http://stat-www.berkeley.edu/users/brill/Papers/speed23.pdf

Broadbent, K. P. (1987). Mobilizing local capabilities for more effective exchange of information in developing countries: Agriculture as an example. *Quarterly Bulletin of IAALD*, *32*(2), 84–93.

Brodal, G., & Rafoss, T. (2006). Forecasting pests and diseases in Norway. *Proceedings 2006 Bioforsk Plantehelse - Ås Plant Health and Plant Protection.*

Brodie, M. L. (1993). The promise of distributed computing and the challenges of legacy information systems. In D. K. Hsiao, E. J. Neuhold, & R. Sacks-Davis(Eds.), *Interoperable database systems (DS-5)*, (pp. 1-31). North-Holland, The Netherlands: Elsevier.

Bryceson, K. P. (2006). *E-issues in agribusiness: The what, why and how*. USA: University of Queensland, CAB International. doi:10.1079/9781845930714.0000

Burke, K. (2010). The impact of Internet and ICT use among SME agribusiness growers and producers. *Journal of Small Business and Entrepreneurship*, *23*(2), 173–194.

Butcher, N. (2003). *Technological infrastructure and use of ICT in education in Africa: An overview*. Paris, France: Association for the Development of Africa.

Caidi, N., & Ross, A. (2005). Information rights and national security. *Government Information Quarterly*, *22*, 663–684. doi:10.1016/j.giq.2005.11.003

Calvo, T., Mayor, G., & Mesira, R. (2002). *Aggregation operators: New trends and applications*. Heidelberg, Germany: Physica-Verlag Studies in Fuzziness and Soft Computing. doi:10.1007/978-3-7908-1787-4

Cao, W., Tang, L., Zhu, Y., Pan, J., Li, W., & Chen, B. (2006). Development of growth model-based decision support system for crop management. *Frontiers of Agriculture in China*, *1*(3).

Castro, C. (Ed.). (1998). *Education in the information age*. Washington, DC: Inter-American Development Bank.

Cetin, B., Akmar, A., & Ozsaym, D. (2004). The uses of information and communication technologies as critical success factor for marketing Turkish agri-food companies. *Food Reviews International*, *20*(3), 221–228. doi:10.1081/FRI-200029420

Chakracarty, R. (2000). *IT at milk collection centers in cooperative dairies: The National Dairy Development Board Experience*.

Chambers Ireland. (2007). *E-business survey 2007*, Vol. 1. Co Partnered Chambers Ireland and Comreg Dublin. Retrieved from http://www.chambers.ie/ktmlpro/files/uploads/eBusinessSurvey%2007.pdf

Chand, R. (Ed.). (2004). *India's national agricultural policy: A critique. India's agricultural challenges: Reflections on policy, technology and other issues*. New Delhi, India: Centre for Trade and Development. Retrieved February 9, 2011, from www.iegindia.org/dis_rc_85.pdf

Chang, I., Hwang, H., Yen, D. C., & Huang, H. (2006). An empirical study of the factors affecting internet security for the financial industry in Taiwan. *Telematics and Informatics*, *23*, 343–364. doi:10.1016/j.tele.2005.11.001

Chang, K. L., Lee, S., & Kang, I. (2005). KMPI: Measuring knowledge management performance. *Information & Management*, *42*(8), 1149. doi:10.1016/j.im.2005.10.003

Chang, T. C., Hasegawa, K., & Ibbs, C. W. (1991). The effects of membership function on fuzzy reasoning. *Fuzzy Sets and Systems*, *44*(2), 169–186. doi:10.1016/0165-0114(91)90001-7

Chauhan, N. (2005). Information technology for agricultural development in India. In Dipak, D., & Jirli, B. (Eds.), *Communication support for sustainable development*. Jangamawadi Math, India: GangaKaveri Publishing House.

Chawala, R., & Bhatnagar, S. (2004). *Online delivery of land titles to rural farmers in Karnataka, India*. Presented at Scaling up Poverty Reduction: A Global Learning Process and Conference, May 25 – 27, 2004, Shanghai. Retrieved February 6, 2011, from http://www.siteresources.worldbank.org/

Cheney, P. (1991). Models used for fire danger rating in Australia. In N. P. Cheney & A. M. Gill (Eds.), *Proceedings Conference on Bushfire Modeling and Fire Danger Rating Systems*, CSIRO Division of Forestry, Canberra, (pp. 119-126).

Chhiber, N. (2004). Enhancing developmental opportunities by promoting ICT use: Vision for rural India. *The Journal of E-Government Policy and Regulation*, *27*(3).

Chigona, W., Beukes, D., Vally, J., & Tanner, M. (2009). Can mobile Internet help to alleviate social exclusion in developing countries? *The Electronic Journal of Information Systems in Developing Countries*, *36*(7), 1–16.

Choi, J., & Lee, K. (2003). Risk perception and e-shopping cross-cultural study. *Journal of Fashion Marketing and Management*, *7*(1), 49–64. doi:10.1108/13612020310464368

Chung, K., & Pauline, F. (2010). *Evaluating the impact of ICT on trade of fruits and vegetables within the APEC countries*. 54th Australian Agricultural and Resource Economics Society, February 10-12, Adelaide, Australia.

CIDA. (2003). *Knowledge sharing: Methods, meetings and tools.* Canadian International Development Agency (CIDA) Knowledge Management Secretariat, November 2003. Retrieved from km.fao.org/uploads/media/CIDA_guide_Knowledge.pdf

Cloete, E., & Doens, M. (2008). B2B e-marketplace adoption in South African agriculture. *Information Technology for Development, 14*(3), 184–196. doi:10.1002/itdj.20105

Comptroller & Auditor General. (2008). *Special report: eGovernment (2007).* Government of Ireland. Retrieved from http://www.audgen.gov.ie/documents/vfmreports/58_eGovernment.pdf

Comreg. (2010). *Irish communications market: Quarterly key data report; Data as of second quarter 2010.* Document no 10/73R. Commission for Communication Regulation, Dublin.

Comreg. (2010). *Irish communications market: Quarterly key data report; Data as of first quarter 2010.* Document no 10/43. Commission for Communication Regulation, Dublin.

Conforte, D. (2011). The nature of agribusiness management research. *International Food and Agribusiness Management Review, 14*(2), 143–147.

Cox, E. (2005). *Fuzzy modeling and genetic algorithms for data mining and exploration.* Elsevier Science, USA.

Craig, W. J. (1995). Why we can't share data: Institutional inertia. In Onsrud, H. J., & Rushton, G. (Eds.), *Sharing geographic information.* New Brunswick, NJ: Center for Urban Policy Research, Rutgers University.

Croswell, P. L. (1989). Facing reality in GIS implementation: Lessons learned and obstacles to be overcome. In M. J. Salling & W. F. Gayk (Eds.), *Proceedings of the 1989 Annual Conference of the Urban and Regional Information Systems Association (URISA)*, Vol. IV (pp. 15-35). Washington, DC: URISA.

CSO. (2010, June). *CSO output, input and income in agriculture, 2009: Final estimate.* Dublin, Ireland: Central Statistics Office.

CTA. (1998). Assessing the impact of information and communication management on institutional performance. In K. Batjes-Sinclair (Ed.), *Proceedings of a CTA Workshop*, January 27-29, 1998, Wageningen, The Netherlands.

Dale, P. (1999). *Is technology a blessing or a curse in land administration?* Presented at the UN-FIG Conference on Land Tenure and Cadastral Infrastructure for Sustainable Development, Melbourne, Australia, October 1999.

Dale, P. (2000). *The importance of land administration in the development of land markets - A global perspective. Land markets and land consolidation in Central Europe.* TU Delft – Report of UDMS 2000.

Dangi, N., & Singh, H. (2010). e-Choupal: Hope or hype? *American Journal of Economics and Business Administration, 2*(2), 179–184. doi:10.3844/ajebasp.2010.179.184

Dar, W. (2003). *Technology for poverty reduction: Working towards a South Asia strategy.* Intermediate Technology Development Group-South Asia (ITDG-South Asia), British Council, New Delhi, 10-11 October, 2003, South Asia Conference on Technologies for Poverty Reduction, New Delhi.

De Silva, H., & Ratnadiwakara, D. (2010). *Using ICT to reduce transaction costs in agriculture through better communication: A case-study from Sri Lanka.* LIRNEasia, Colombo.

De Silva, H. (2010). ICT policy for Sri Lanka. *International Journal of ICT and Research Development, 1*(39).

De Silva, H. (2010). *The Govi Gnana Servic: An ICT info structure to fight agricultural poverty in Sri Lanka. Report of E Development Labs.* Public Interest Program Unit.

Debrah, S. K. (2007). *Final report of the Market Information Systems for Trade Organizations for West Africa (MISTOWA) project.* Unpublished IFDC report submitted to USAID under the USAID Cooperative Agreement No. 641-A-00-04-00230-00, November 15, 2007.

Debrah, S. K. (2007). *Compiled testimonies of beneficiaries of the MISTOWA project interventions.* Unpublished IFDC report submitted to USAID under the USAID Cooperative Agreement No. 641-A-00-04-00230-00, October 25, 2007.

Deeming, J., Burgan, R., & Cohen, J. (1978). *The national fire danger rating system.* U.S. Department of Agriculture.

Del Aguila–Obra, A., & Padilla-Melendez, A. (2006). Organizational factors affecting Internet technology adoption. *Internet Research, 16*(1), 94–110. doi:10.1108/10662240610642569

Delhaxhe, A. (2004). *Key data on information and communication technology in schools in Europe*. Brussels, Belgium: EURYDICE.

Department of Agriculture and Food. (1999). *Ensuring the future- A strategy for rural development: A white paper on rural development in Ireland*. Dublin, Ireland: Author.

Department of Agriculture. Fisheries and Food. (2010). *Annual review and outlook for agriculture, fisheries and food 2009/2010.* Dublin, Ireland: Author.

Department of Agriculture. Fisheries and Food. (2010). *Factsheet on Irish agriculture – August 2010*.

De, R. (2005). E-government systems in developing countries: Stakeholders and conflict. In Wimmer, M. A. (Eds.), *E-Gov - 2005, LNCS 3591* (pp. 26–37). doi:10.1007/11545156_3

De, R. (2006). The impact of e-government initiatives: Issues of poverty and vulnerability reduction. *Regional Development Dialogue*, *27*(2), 88–100.

Descombes, P. (1998). Flexiform – Self training tools suited to the profession. In *Forestry training for target groups that are hard to reach* (pp. 327-344). La Bastide-des-Jourdans, France: Centre Forestier de la Région Provence Alpes Côte d'Azur.

Dey, S. (2010, June 12). M-Choupal. *Business Outlook*. Retrieved February 16, 2011, from http://business.outlookindia.com/article.aspx?265712

Dimitrakopoulos, A. (1994). *Analysis of forest fire causes in Greece*. Edition of MAI of Chania Greece

Division of Agricultural Economics, IARI. (2006). *Agricultural Economics Research Review, 19*.

Donner, J. (2006). The use of mobile phones by micro entrepreneurs in Kigali, Rwanda: Changes to social and business networks. *Information Technologies and International Development*, *3*(2), 3–19. doi:10.1162/itid.2007.3.2.3

Doolin, B., & Ali, E. (2008). Adoption of mobile technology in the supply chain: An exploratory cross-case analysis. International. *Journal of Business Research*, *4*(4), 1–16. doi:10.4018/jebr.2008100101

Dowuona, S. N. N. (2011, February 4). Mobile telecom market share: Three gain, two lose. *MyJoyonline*. Retrieved from http://news.myjoyonline.com/business/201102/60519.asp

Dwivedi, S. (2008). *Challenges of e-government implementation in India: Emerging technologies in e-government* (pp. 210–215). E-Governance Publication.

Dyson, R. G. (2004). Strategic development and SWOT analysis at the University Warwick. *European Journal of Operation Research. Journal of International Management*, *152*(3), 631–640.

Economics and Planning Division, Dublin, Ireland: Author European Commission. (2007). *Europa: Activities of the European Union, summary legislation.* Retrieved from: http://europa.eu/legislation_summaries/information_society/index_en.htm

Edens, K. M. (2000). How do adolescents process advertisements? The influence of ad characteristics, processing objectives, and gender. *Contemporary Educational Psychology*, *25*, 450–463. doi:10.1006/ceps.1999.1031

Eechambadi, N. V. (1994). Does advertising work? *The McKinsey Quarterly*, 3.

Efthymiou, P. N. (1989). Ergonomic problems and accidents in the mountain logging of Greece. In *Seminar on the mechanization of Harvesting Operations in mountainous terrain.* Antalya/Turkey 20-24 Nov. 1989.

Efthymiou, P. N. (2001). Forest protection and forest work. In Konstantinidis, P. N. (Ed.), *Development and protection of forests – Forest work*. Thessaloniki, Greece: Ministry of Agriculture, Forest Research Institute.

Efthymiou, P. N. (2003). Efficiency problems in harvesting small dimensioned wood. In Arzberger, U., & Grimoldi, M. (Eds.), *New trends in wood harvesting with cable systems for sustainable forest management in the mountains* (pp. 173–178). Rome, Italy: Food and Agriculture Organization, United Nations Economic Commission for Europe, International Labour Organization, International Union of Forest Research Organizations.

Egan, A. (2005). Training preferences and attitudes among loggers in northern New England. *Forest Products Journal*, *55*(3), 19–26.

Egan, A., & Taggart, D. (2009). Public perception of the logging profession in Maine and implications for logger recruitment. *Northern Journal of Applied Forestry*, *26*(3), 93–98.

El-Gohary, H. (2010). E-marketing-A literature review from a small business perspective. *International Journal of Business and Social Science*, *1*(1), 214–244.

Elliott, N. C., Royer, T. A., Giles, K. L., Kindler, S. D., Porter, D. R., Elliott, D. T., & Waits, D. A. (2004). *A web-based decision support system for managing greenbugs in wheat*. Crop Management. doi:10.1094/CM-2004-1006-01-MGdoi:10.1094/CM-2004-1006-01-MG

Ellis, E. A., Nair, P. K. R., & Jeswani, D. S. (2005). Development of a web-based application for agroforestry planning and tree selection. *Computers and Electronics in Agriculture*, *49*(1), 129–141. doi:10.1016/j.compag.2005.02.008doi:10.1016/j.compag.2005.02.008

Esoko. (2010, October 14). Esoko: The market on your mobile. *Esokonetworks*. Retrieved from http://www.youtube.com/user/esokonetworks

European Centre for the Development of Vocational Training. (2002). *Common vocational certificates project – Methodology for construction – Basic methodology*. Retrieved from www.trainingvillage.gr

European Centre for the Development of Vocational Training. (2002). *Elearning Europe 2002: Survey*. Retrieved from www.trainingvillage.gr

European Commission. (2002). *Leonardo da Vinci series good practices: Distance learning. Luxemburg*. Office of the Official Publications of the European Communities.

European Training Thesaurus. (2010). Retrieved from http://libserver.cedefop.europa.eu/ett/en

Evju, O. (1989). Further training of professional forest workers. In *Proceedings of the Joint FAO/ECE/ILO Committee Seminar "The Training of Professional Forest Workers"* (pp. 129-142). Helsinki, Finland: The Finnish Government Printing Centre.

Ewing, J. (2007, September 13). Upwardly mobile in Africa. *Business Week*. GSMA Development Fund. 2008.

Fabiani, S., Schivardi, F., & Trento, S. (2005). ICT adoption in Italian manufacturing: Firm-level evidence. *Journal of Industrial and Corporate Change*, *14*(2), 225–249. doi:10.1093/icc/dth050

Fafchamps, M., & Gabre Madhin, E. (2006). Agricultural markets in Benin and Malawi. *African Journal of Agricultural and Resource Economics*, *1*, 67–94.

Fafchamps, M., & Hill, R. V. (2005). Selling at the farm gate or traveling to the market. *American Journal of Agricultural Economics*, *87*, 717–734. doi:10.1111/j.1467-8276.2005.00758.x

Fakiyesi, T. (2010). ICT raise Nigerian GDP. *ITRealms Link*.

Fang, C., & Lie, T. (2006). Assessment of Internet marketing and competitive strategies for leisure farming industry in Taiwan. *Journal of American Academy of Business*, *8*(2), 296–300.

FAO. (1993). *The potentials of microcomputers in support of agricultural extension, education and training* (pp. 1–173). Rome, Italy: Food and Agriculture Organisation of the United Nations.

FAO. (2003). *Statement at the ministerial conference of the WTO*. Fifth Session, Cancun, 10-14 September, Doc No: WT/MIN(03)/ST/61.

FAO. (2010). *Crop calendar – A crop production information tool for decision making*. Rome, Italy: FAO. Retrieved from http://www.fao.org/agriculture/seed/cropcalendar

Fasheun, K. P. (2001). Problems and prospects of livestock production in information age. *Delta State College of Agriculture Newsletter*, *12*(4), 23–27.

Fischer, M. M., Scholten, H. J., & Unwin, D. (1996). Geographic information systems, spatial data analysis and spatial modeling: An introduction. In Fischer, M. M., Scholten, H. J., & Unwin, D. (Eds.), *Spatial analytical perspectives on GIS*. Taylor and Francis.

Fisher, I. (1973). I discovered the Phillips curve: A statistical relationship between unemployment and price changes. *The Journal of Political Economy*, *81*(2), 496–502. doi:10.1086/260048

Flamm, K. (2003). Moore's law and the economics of semiconductor price trends. *International Journal of Technology. Policy and Management*, *3*(2), 127–141.

Food and Agriculture Organization. (2010). *Equitable partnerships, markets and forest extension*. Retrieved from http://www.fao.org/forestry/partnerships/en/

Food and Agriculture Organization. (2010). *Criteria and indicators for sustainable forest management*. Retrieved from http://www.fao.org/forestry/ci/en/

Frank, H., & Bunte, L. (2009). *Thought for food: The impact of ICT on agribusiness*. Wageningen University and Research, the Netherlands. Paper prepared for presentation at the 113th EAAE Seminar-A Resilient European Food Industry and Food Chain in a Changing World. Chaina, Crete and Greece.

Fuchs, C., & Horak, E. (2008). Africa and the digital divide. *Telematics and Informatics*, *25*, 99–116. doi:10.1016/j.tele.2006.06.004

Fukatsu, T., & Hirafuji, M. (2003). Development of field servers for a field monitoring system. *Agricultural Information Research*, *12*, 1–12.

Funk, J. (2007). The future of mobile shopping: The interaction between lead users and technological trajectories in the Japanese market. *Technological Forecasting and Social Change*, *74*(3), 341–356. doi:10.1016/j.techfore.2005.11.007

Galbi, D. A. (2001). E-government: Developing state communications in a free media environment. *Journal of Futures Studies. Strategic Thinking and Policy*, *3*(1), 9–13. doi:10.1108/14636680110803012

Galliano, D., & Orozco, L. (2008). *Intra and inter organizational determinants of electronic based traceability adoption: Evidence from the French agri-food industry*. 12th Congress of the European Association of Agriculture Economics, Ghent, Belgium.

Garcia, L., Lunadei, L., Barreiro, P., & Robla, I. (2009). A review of wireless sensor technologies and applications in agriculture and food industry: State of the art and current trends. *Sensors (Basel, Switzerland)*, *9*(6), 4728–4750. doi:10.3390/s90604728

Gates, B. (2008, May 8). Promoting technology adoption—and prosperity—Through public-private partnerships. *Wall Street Journal Asia*.

Gebremedhin, B., Hoekstra, D., & Azage Tegegne, A. (2006). *Commercialization of Ethiopian agriculture: Extension service from input supplier to knowledge broker and facilitator*. IPMS (Improving Productivity and Market Success) of Ethiopian Farmers Project Working Paper 1. Nairobi, Kenya: International Livestock Research Institute. Retrieved from mahider.ilri.org/handle/10568/564

Gelb, E., & Offer, A. (2006). *ICT in agriculture: Perspectives of technological innovation*. Hebrew University of Jerusalem, Centre for Agricultural Economic Research.

Gerding, V. (2001). Die Learntec 2001 in Karlsruhe – Vom vernetzten Denken zu multimedialer Didaktik. *Forsttechnische Informationen*, *8*, 84–87.

Gilbert, L., Sim, Y. W., & Wang, C. (2005). An e-learning systems engineering methodology. In *The 5th IEEE International Conference on Advanced Learning Technologies*, Kaohsiung, Taiwan (pp. 150-154).

Glory, B., & Akridge, J. (2000). Computer and Internet adoption on large U.S farms. *International Food and Agribusiness Management Review*, *3*, 323–338. doi:10.1016/S1096-7508(01)00051-9

Golding, P., Donaldson, O., Tennat, V., & Black, K. (2008). *An analysis of factors affecting the adoption of ICT by SMEs in rural and urban Jamaica* [Electronic Version]. Retrieved from http://is2.lse.ac.uk/asp/aspecis/20080109.pdf

Gomez, R., & Baron-Porras, L. F. (2011). Does public access computing really contribute to community development? Lessons from libraries, tele-centers and cybercafés in Colombia. *Electronic Journal of Information Systems in Developing Countries*, *49*(2), 1–11.

Goodchild, M. F. (1993). The state of GIS for environmental problem- solving. In Goodchild, M. F. (Ed.), *Environmental modelling with GIS* (pp. 8–15). New York, NY: Oxford University Press.

Government of Ireland. (1999). *Implementing the information society in Ireland: An action plan.* (PN6727) Government of Ireland. Retrieved from http://www.taoiseach.gov.ie/attached_files/Pdf%20files/InfoSociety-ActionPlanJan99.pdf on 9 December, 2010

Gregor, S., Newman, S., & Larner, T. (1997). *The Australian meat industry, the EM hypothesis and Internet marketing.* 10th International Bled Electronic Commerce Conference, Bled, Slovenia 9-11 June.

Gröger, V., & Lewark, S. (2002). *Der arbeitende Mensch im Wald.* Dortmund, Germany: Bundesanstalt für Arbeitsschutz und Arbeitsmedizin.

Grover, V., & Davenport, T. (2001). General perspective on knowledge management: Fostering a research agenda. *Journal of Management Information Systems, 18*(1), 5–21.

Gupta, R. K. (2003). *E–Choupal so far and ahead: A case study in a village in M.P.* Theme paper, fieldwork segment, PRM 2002 – 2C IRMA, Anand, Gujrat, India, February.

Haddad, W. D., & Draxler, A. (2002). *Technologies for education. Paris, France: UNESCO.* Washington, DC: Academy for Educational Development.

Harren, P. C. (2004). *Agricultural science and application for schools* (pp. 271–278). London, UK: Heinemann Education Book.

Hartwich, F., Monge Perez, M., Ampuero Ramos, L., & Soto, J. L. (2007). Knowledge management for agricultural innovation: Lessons from networking efforts in the Bolivian Agricultural Technology System. *Knowledge Management for Development Journal, 3*(2), 21-37. Retrieved from journal.km4dev.org/index.php/km4dj/article/view/106/174

Harzbo, S., Arnela, C., & Chun-Yan, H. (2008). ICT adoption model of Chinese SMEs. *International Journal of Business Research, 44*, 161–165.

Hauser, S., Ndi, J. N., & Hulugalle, N. R. (2000). Yields of maize/cassava intercrops grown with hedgerows of three multipurpose trees on an acid Ultisol of Cameroon. *Agroforestry Systems, 49*, 111–122. doi:10.1023/A:1006351025490

Hawkins, R. (2004). *Ten lessons for information and communications technology in developing world.* Washington, DC: World Bank.

Helvoigt, T. L., Adams, D. M., & Ayre, A. L. (2003). Employment transitions in Oregon's wood products sector during the 1990s. *Journal of Forestry, 4*(1), 42–46.

Henderson, J., Dooley, F., & Akridge, J. (2004). Internet and e-commerce adoption by agricultural input firms. *Review of Agricultural Economics, 26*(4), 505–520. doi:10.1111/j.1467-9353.2004.00196.x

Henderson, J., Dooley, F., Aridge, J., & Carerre, A. (2005). Adoption of Internet strategies by agribusiness firms. *International Food and Agribusiness Management Review, 8*(4), 42–61.

Hermans, L. M., & Thissen, A. H. (2009). Actor analysis methods and their use for public policy analysis. *European Journal of Operational Research, 196*(2), 808–818. doi:10.1016/j.ejor.2008.03.040

Herremans, A. (1995). *New training technologies.* Turin, Italy: International Training Center of the ILO.

Hetemäki, L., & Nilsson, S. (2005). Introduction. In Hetemäki, L., & Nilsson, S. (Eds.), *Information technology and the forest sector* (*Vol. 18*, pp. 1–7). Vienna, Austria: IUFRO.

Hinman, R., & Matovu, J. (2010). Opportunities and challenges for mobile-based financial services in rural Uganda. In *Proceedings of the 28th of The International Conference Extended Abstracts on Human Factors in Computing Systems.* New York, NY: ACM.

Hossain, M., & Quaddus, M. (2010). An adoption diffusion model of RFID-based livestock management system in Australia. *Human Benefit through the Diffusion of Information Systems Design Science Research, 318*(1), 179-191.

Huang, C. F. (2005). Risk assessment of natural disasters-Theory & practice. Beijing, Chian: Science Press.

Huang, Y., Kuo, W., & Kao, M. (2010). A RFID field operation system design for agricultural traceability. In *Proceedings of the 2010 First International Conference on Pervasive Computing, Signal Processing and Application.* Washington, DC: IEEE Computer Society.

Huang, C., & Ruan, D. (2008). Fuzzy risks and an updating algorithm with new observations. *Risk Analysis-. International Journal (Toronto, Ont.), 28*(3), 681–694.

Hueth, B. (1999). Incentive instruments in fruits and vegetables: Input control monitoring, measuring and price risk. *Review of Agricultural Economics*, *2*, 374–398.

Hussain, I., Si, S., & Ahmed, A. (2010). Knowledge management for small and medium enterprise in developing countries. *Journal of Knowledge Management Practice*, *11*(2).

IFPRI. (2002). *Reaching sustainable food security for all by 2020: Getting priorities and responsibilities right.* Washington, DC: International Food Research Policy Institutes (IFPRI).

Igbal, F. (2010). *Remote sensing and GIS application for drought severity assessment: Using remote sensing and GIS to estimate drought severity* (pp. 45-56). VDM Verlag Dr. *Müller Inc.*, *ISBN-13*, 978–3639303261.

IICD. (2006). *ICTs for agricultural livelihoods: Impact and lessons learned from IICD supported activities*. Retrieved February 9, 2011, from http://www.iicd.org/files/IICD-agri-impact-2006.pdf

Iliadis, L. (2005, May). A decision support system applying an integrated Fuzzy model for long - term forest fire risk estimation. *Journal Environmental Modelling and Software, 20*(5), 613-621. ISSN: 1364-8152

Iliadis, L., & Zigkrika, N. (2010). Performing fuzzy multi-feature scenarios for the determination of forest fire risk Προφορική παρουσίαση. *Proceedings of the 3rd International Conference on Information and Communication Technologies in Agriculture, Food, Forestry and Environment* (ITAFFE 2010) (pp. 170-177).

Iliadis, L., Skopianos, S., Tachos, S., & Spartalis, S. (2010). A fuzzy inference system using Gaussian Distribution Curves for forest fire risk estimation. *Proceedings of the 6th International Conference of Artificial Intelligence Applications and Innovations Springer AIAI 2010, IFIP AICT 339*, (pp. 376–386). IFIP (International Federation for Information Processing).

Iliadis, L., Spartalis, S., Maris, F., & Marinos, D. (2004). A decision support system unifying trapezoidal function membership values using t-norms. *Proceedings International Conference in Numerical Analysis and Applied Mathematics* (ICNAAM). Weinheim, Germany: J. Wiley-VCH Verlag GmbH Publishing Co.

Iliadis, L. (2007). *Intelligent information systems and applications in risk estimation.* Thessaloniki, Greece: Stamouli Publishing.

Iliadis, L., Papastavrou, A., & Lefakis, P. (2002). A computer system that classifies areas of Greece in forest fire risk zones, using fuzzy sets. *Forest Policy and Economics*, *4*, 43–54. doi:10.1016/S1389-9341(01)00079-X

Information Security News. (2006). *Hackers attack South Africa's stats site to protest monopoly of Telcom.* Retrieved from www.Insecure.org/

International Centre for Technical and Vocational Education and Training. (2010). *What is TVET?* Retrieved on March 5, 2010, from http://www.unevoc.unesco.org/2.0.html?tx_drwiki_pi1[keyword]=What%20is%20TVET International Fertilizer Development Center. IFDC. (2010). *Biannual report of the Ghana Agro-Dealer Development Project (GADD)*, (p. 21). Unpublished report submitted by IFDC to the Alliance for a Green Revolution in Africa (AGRA), September 2010

International Labour Organization. (2001). *World employment report 2001: Life at work in the information economy.* Geneva, Switzerland: International Labour Organization.

International Labour Organization. (2003). *Learning and training for work in the knowledge society.* Report 91 IV. Retrieved January 3, 2010, from http://www.ilo.org/public/libdoc/ilo/2002/102B09_22_engl.pdf

International Organization for Standardization. (2001). *IEC 9126-1, Software engineering – Product quality – Part 1: Quality model.*

International Telecommunication Union. (2010). *Measuring the information society 2010.* Retrieved March 8, 2010, from http://www.itu.int/dms_pub/itu-d/opb/ind/D-IND-ICTOI-2010-SUM-PDF-E.pdf

International Telecommunication Union. (2011). *The world in 2010 ICT facts and figures: The rise of 3G.*

International Union of Forest Research Organizations. (2002). *Terminology project – New definitions.* Vienna, Austria: IUFRO.

InternetWorldStats. (2000). *World internet usage and population statistics.* Retrieved from http://www.internetworldstats.com/stats.htm 2011

INVERDE. (2010). *Forestweb – E-learning about sustainable forestry*. Retrieved March 9, 2010, from http://www.inverde.be/en/forestweb_987

IPMS. (2005). *Project implementation plan (PIP) submitted to CIDA*. Improving Productivity and Market Success (IPMS) of Ethiopian Farmers' Project, March 2005. Retrieved from http://www.ipms-ethiopia.org/content/files/Documents/pip/Pip%20final.doc

IPMS. (2010). *12th progress report* (April 2010 – September 2010). Improving Productivity and Market Success (IPMS) of Ethiopian Farmers Project. Retrieved from http://www.ipms-ethiopia.org/Documents-Publications/Reporting/Progress-Reports.asp

IPMS. (2010). *Monitoring and evaluation report of year 5* (2009/2010). Improving Productivity and Market Success (IPMS) of Ethiopian Farmers Project. Retrieved from http://www.ipms-ethiopia.org/Documents-Publications/M-&-E-Reports.asp

Isiaka, B. T. (2006). *Potentials and effectiveness of video self-learning method in agricultural information dissemination in Southwest Nigeria*. PhD Thesis, University of Agriculture, Abeokuta.

ISITA. (2003). *Identification of the obstacles to ICT uptake on farms and the strategic way forward*. Annual Technical Meeting of the Irish Society for Information Technology in Agriculture (ISITA), 23 October, 2003, Newbridge, Ireland. Retrieved from http://www.iol.ie/~harkin/isita.htm

Jackson, P. (1990). *Introduction to expert systems* (2nd ed.). Addison Wesley.

Jaffee, S. M. (1995). The many faces of success: The development of Kenyan horticultural exports. In Jaffee, S., & Morton, J. (Eds.), *Marketing Africa's high-value foods* (pp. 319–374). Washington, DC: World Bank.

Jensen, A. L., Thysen, I., Boll, P. S., Hansen, J. G., Secher, B. J. M., & Juhl, O. (1997). *Pl@nteInfo – Using the internet for custom tailored crop information*. Retrieved from www.planteinfo.dk/information/publikationer/efita97/

Jensen, R. (2007). The digital provide: Information (technology), market performance and welfare in the south Indian fisheries sector. *The Quarterly Journal of Economics*, *122*(3), 879–924. doi:10.1162/qjec.122.3.879

Jensen, R. (2010). Information, efficiency, and welfare in agricultural market. *Journal of International Association of Agricultural Economist*, *41*(1), 203–216.

Jensen, R. T. (2010). Information, efficiency, and welfare in agricultural markets. *Agricultural Economics*, *41*, 203–216. doi:10.1111/j.1574-0862.2010.00501.x

Jha, P., & Negre, M. (2007). *Indian economy in the era of contemporary globalization: Some core elements of the balance sheet*. Centre for Economic Studies and Planning, Jawaharlal Nehru University, New Delhi, India. Retrieved February 7, 2011, from http://www.macroscan.org/anl/may07/pdf/IndianEconomy.pdf

Jhunjhunwala, A., Ramachandran, A., & Ramachander, S. (2006). Connecting rural India: Taking a step back for two forward. *Information Technology in Developing Countries*, *16*(1). Retrieved on February 9, 2011, from http://www.iimahd.ernet.in/egov/ifip /feb2006/article1.htm

Jidaw System Limited (Online). (2011). *Internet service providers in Nigeria*. Retrieved from http://www.jidaw.com/isp.html

Johansson, K., & Strehlke, B. (1996). *Improving working conditions and increasing profits in forestry*. Geneva, Switzerland: ILO.

Joint FAO/ECE/ILO Committee on Forest Technology. Management and Training. (1997). *People, forests and sustainability: Social elements of sustainable forest management in Europe*. Geneva, Switzerland: International Labour Organization.

Jones, V. P., Brunner, J. F., Grove, G. G., Petit, B., Tangren, G. V., & Jones, W. E. (2010). A web-based decision support system to enhance IPM programs in Washington tree fruit. *Pest Management Science*, *66*(6), 587–595. PubMed

Jonsson, S., & Gunnarsson, C. (2005). Internet technology to achieve supply chain performance. *Business Process Management Journal*, *11*(4), 403–417. doi:10.1108/14637150510609426

Kagan, A. (2000). Information systems implementation within US agribusiness: An applications approach. *Computers and Electronics in Agriculture*, *28*(3), 207–228. doi:10.1016/S0168-1699(00)00144-7

Kaloudis, S., Tocatlidou, A., Lorentzos, N., Sideridis, A., & Karteris, M. (2005). Assessing wildfire destruction danger: A decision support system incorporating uncertainty. *Ecological Modelling, 181*(10), 25–38. doi:10.1016/j.ecolmodel.2004.06.021

Karkkainen, M. (2003). Increasing efficiency in the supply chain for short life goods using RFID tagging. *International Journal of Retail and Distribution Management, 31*(10), 529–536. doi:10.1108/09590550310497058

Karoulis, A., & Pombortsis, A. (2003). The cognitive transfer and the tutor's role in a CBL environment. [Vilnius, Ithuania: Institute of Mathematics and Informatics.]. *Informatics in Education, 2*, 1–15.

Kasibante, F., Mbowa, S., & Mugisha, J. (2004). Determinants of use of information and communication technologies by agribusiness firms in Uganda. *Eastern Journal of Rural Development, 20*(1), 3–11.

Katengeza, S. P., Kiiza, B., & Okello, J. (2011). The role of ICT-based market information services in spatial food market integration: The case of Malawi agricultural commodity exchange. *International Journal of ICT Research and Development in Africa, 2*(1), 1–14. doi:10.4018/jictrda.2011010101

Katengeza, S. T., Okello, J. J., & Jambo, N. (2011). Use of mobile phone technology in agricultural marketing. *International Journal of ICT Research and Development in Africa, 2*(2), 14–25. doi:10.4018/jictrda.2011070102

Kebede, G. (2010). Knowledge management: An information science perspective. *International Journal of Information Management, 30*(5), 416–424. doi:10.1016/j.ijinfomgt.2010.02.004

Kecman, V. (2001). *Learning and soft computing*. London, UK: MIT Press.

Keegan, V. (2000, 14 December). Divide and rule out. *Guardian Unlimited Net News*.

Kenny, C. (2002). Information and communication technologies for direct poverty alleviation, costs and benefits. *Development Policy Review, 20*(2), 141–157. doi:10.1111/1467-7679.00162

Key, N., & Runsten, D. (1999). Contract farming, smallholders, and rural development in Latin America: The organization of agro-processing firms and scale of outgrower production. *World Development, 27*, 381–401. doi:10.1016/S0305-750X(98)00144-2

Koohang, A. A. (1989). A study of attitudes toward computers: Anxiety, confidence, liking, and perception of usefulness. *Journal of Research on Computing in Education, 20*, 137–151.

Korsten, H. (2005). Evaluating South African government web sites: Methods, findings and recommendations (part 1). *South African Journal of Information Management, 7*(2).

Kotler, P. (1997). *Marketing management: Analysis, planning, implementation and control* (9th ed.). Prentice Hall.

Kotrlik, J. W., Harrisson, B. C., & Redmann, D. H. (2000). A comparison of information technology training sources, value, knowledge and skills for Louisiana's secondary vocational teachers. *Journal of Vocational Education Research, 25*(4). Retrieved from http://scholar.lib.vt.edu/ejournals/JVER/v25n4/doi:10.5328/JVER25.4.396

Kufoniyi, O., & Agbaje, G. I. (2005). *National geospatial data infrastructure development in Nigeria: The journey so far* (p. 3). A paper presented at Pharaohsto Geoinformatics FIG Working Week 2005 and GSDI – 8 Cairo, Egypt.

Kumar, R. (2004). E-Choupal: A study on the financial sustainability of village Internet centers in rural Madhya Pradesh. *Information Technology and International Development, 2*(1), 45–73. doi:10.1162/1544752043971161

Kumar, S., & Stokkel, J. (2003). Evolution of GPS technology and its subsequent use in commercial markets. *International Journal of Mobile Communications, 1*(1), 180–193. doi:10.1504/IJMC.2003.002465

Kwesiga, J. C. (2000, April). *Women in agriculture: The vital contribution of the rural women farmers to African economies and faculty sustenance farming*, (pp. 4–7).

Kydd, J., & Dorward, A. (2004). Implications of market and coordination failures for rural development in least developed countries. *Journal of International Development, 16*, 951–970. doi:10.1002/jid.1157

Lal, D., Renee, P., & David, N. (2008). *How can ICTs be used and appropriated to address agricultural information needs of Bangladeshi farmers?* GlobDev, Paper 21.

Lavacic, R., & Rebmann, A. (Eds.). (1982). *Macroeconomics*. London, UK: Macmillan.

Lawal-Adebowale, O. A. (2009). *Information and communication technology usage in research-extension-farmer linkage system for agricultural development in Southwest Nigeria*. PhD Thesis, University of Agriculture, Abeokuta.

Lawal-Adebowale, O. A., & Adebayo, K. (2006). Information and communication technologies: Pathways to effecting feedback in mass extension method of message delivery. *Proceedings of Agricultural Extension Society of Nigeria*, (pp. 184–190).

Leary, J., & Berge, Z. (2006). Trends and challenges of elearning in national and international agricultural development. *International Journal of Education and Development using ICT, 2*(2), 51-59.

Lee, H., & Choi, B. (2003). Knowledge management enablers, processes, and organizational performance: An integrative view and empirical examination. *Journal of Management Information Systems, 20*(1), 179–228.

Lefebvre, A., Cassivi, L., & Lefebvre, E. (2001). E-commerce transition mode for supply chain management. *Chain and Network Science, 1*(1), 23–32. doi:10.3920/JCNS2001.x003

Leiken, S. (2002). *Extending the information revolution – Secondary and Post-secondary education.* Retrieved from www.digitaldividenetwork.org/content/stories/index.cfm?key=226

Lemma, T. T., Sehai, E., & Hoekstra, D. (2010). *Knowledge centers in the pilot learning woredas of improving productivity and market success project: Utilization, relevance and effectiveness*. Retrieved from www.ipms-ethiopia.org/IPMS

Leondes, C. (1998). *Fuzzy logic and expert systems applications*. USA: Academic Press.

Levedys, A. (2004). *Trends and current status of the contribution of the forestry sector to national economies.* FAO working paper on financing sustainable forest management: FSFM/ACC/07.

Lexer, M. J., Vacik, H., Hönninger, K., & Unegg, F. (2000). Implementing a decision support system for silvicultural decision making in low-elevation Norway spruce forests. In E. Klimo, H. Hager, & J. Kulhavy, (Eds.), *EFI Proceedings on Spruce Monocultures in Central Europe: Problems and Prospects,* Vol. 33 (pp. 119–134).

Li, L., Guo, W., & Wang, Y. (2010). *Mobile supply chain management: Theory and method.* The International Conference on E-Product, E-Service and E-Entertainment (ICEEE) Nov. 7-9 Henan, China.

Liao, Z., & Cheung, M. (2001). Internet-based e-shopping and consumer attitudes: An empirical study. *Information & Management, 38*, 299–306. doi:10.1016/S0378-7206(00)00072-0

Lieberg, S., & Morgan, J. (1997). *Information and communication technologies in the Danish school system: A state of art evaluation in 6 areas of government action.* Copenhagen, Denmark: Ministry of Education.

Lio, M., & Liu, M. (2006). ICT and agricultural productivity: Evidence from cross-country data. *Agricultural Economics, 34*(3), 221–228. doi:10.1111/j.1574-0864.2006.00120.x

Lionarakis, A. (2001). What kind of distance education are we talking about? In *Proceedings of the First Panhellenic Conference on Open and Remote Teaching*. Retrieved from http://www.eap.gr/news/EXAGGELIA_SYNEDRIOU/synedrio/html/sect2/34.htm

Lipsmeier, A. (1991). *Wie konzipiert man Fernunterricht in der Beruflichen Bildung?* Mannheim, Germany: Zentralstelle für Gewerbliche Berufsförderung.

Lu, N., Swatman, C., & Perreau, S. (2005). Sustainable agriculture mobile commerce– Extending the rural grocery supply chain through m-commerce. *TENCON (2005). IEEE Region, 10*, 1–6.

Lyons, M. (2005). Future ICT system-Understanding the business drivers. *BT Technology Journal, 23*(3), 11–23. doi:10.1007/s10550-005-0026-1

Mahatanankoon, P., Joseph, H., & Wen, B. (2005). Mobile commerce: A new frontier in e-commerce and wireless communication consumer-based m-commerce: Exploring consumer perception of mobile applications. *Computer Standards & Interfaces, 27*(4), 347–357. doi:10.1016/j.csi.2004.10.003

Malhan, I. V. (2007). Agricultural knowledge transfer in India: A study of prevailing communication channels. *Library Philosophy and Practice,* 2007, February.

Maris, F., Iliadis, L., Tachos, S., Loukas, A., Spartali, I., Vassileiou, A., & Pimenidis, E. (2010). Support vector machines-Kernel algorithms for the estimation of the water supply in Cyprus. *Proceedings of the 20th International Conference ICANN (International Conference of Artificial Neural Networks), Lecture Notes in Computer Science, Vol. 6353, Part* II, (pp. 21-29).

Markin, R. J. (1982). *Marketing strategy and management* (2nd ed.). New York, NY: John Wiley and Sons.

Martinetz, S. (2005). *Vertical coordination of marketing systems: Lessons learned from the poultry, egg and pork industries*. United States Department of Agriculture. ERS Agricultural Economic Report No. 708.

Maru, A. (2003). *Information and communication technology use in agriculture extension in India*. CTA. Retrieved February 6, 2011, from http://www.cta.int/en/content/download/2811/..../Case_Study_india.pdf

Maru, A. (2004). *ICT enabled information systems for agricultural development in India: Status, issues and future directions*. Presented at IAITA Second National Workshop on ICT in Agriculture and Rural Development, Dhirubhai Ambani Institute for Information and Communications Technology (DA-IICT), Gandhinagar.

Masiero, S. (2011). Financial versus social sustainability of tele-centers: Mutual exclusion or mutual reinforcement? *Electronic Journal of Information Systems in Developing Countries*, *45*(3), 1–23.

Matsumoto, K., & Machida, T. (2002). Studies on development of work journal supporting systems with speech recognition. *Agricultural Information Research*, *11*, 255–262.

Maumbe, B. M. (2010). Mobile agriculture in South Africa: Implementation, framework, value-added services and policy implications. *International Journal of ICT Research and Development in Africa*, *1*(2), 35–59. doi:10.4018/jictrda.2010040103

Maumbe, B. M. (Ed.). (2010). *E-agriculture and e-government for global policy development: Implications and future directions*. Hershey, PA: IGI Global.

Maumbe, B. M., & Okello, J. (2010). Uses of information and communication technology (ICT) in agriculture and rural development in sub-Saharan Africa: Experiences from South Africa and Kenya. [IJICTRDA]. *International Journal of ICT Research and Development in Africa*, *1*(1), 1–22. doi:10.4018/jictrda.2010010101

Maumbe, B. M., Owei, V., & Alexander, H. (2008). Questioning the pace and pathway for e-government development in Africa: Case of South Africa's Cape Gateway Project. *Government Information Quarterly*, *25*, 757–777. doi:10.1016/j.giq.2007.08.007

Maumbe, B. M., Taylor, W. J., Erwin, G., & Wesso, H. (2007). E-value creation in a government web portal in South Africa. In Tutnall, A. (Ed.), *Encyclopedia of portal technologies and applications*. Melbourne, Australia: Victoria University. doi:10.4018/978-1-59140-989-2.ch065

Maumbe, B. M., Taylor, W. J., Wesso, H., & Erwin, G. (2006). E-value creation in a government web portal in South Africa. In Khosrow-Pour, M. (Ed.), *Encyclopaedia of e-commerce, e-government, and mobile commerce*. Hershey, PA: IGI Global. doi:10.4018/978-1-59140-989-2.ch065

Mavromoustakos, S., Papanikolaou, K., Andreou, A. S., & Papadopoulos, G. A. (2005). Human and technological issues in the e-learning engineering process. In First International Conference on E-Business and E-learning (EBEL'05) (pp. 343-348). Amman, Jordan.

McGowan, P., & Armstrong, G. (2007). Determinants of Internet adoption by SME agri-food companies. *Journal of Small Business and Enterprise Development*, *14*(4), 620–640. doi:10.1108/14626000710832730

Meera, S. N., Jhamtani, A., & Rao, D. U. M. (2004). *Information communication technology in agricultural development: A comparative analysis of the three projects from India*, (pp. 1-14). Agren Network Paper No. 135.

Mehrotra, N. (2010). *Emerging patterns in share of small farms in production and credit implications for policy formulation*. Presented at 12th Annual Conference on Money and Finance in the Indian Economy, Indira Gandhi Institute of Development Research, Mumbai, India. Retrieved February 7, 2011, from http://www.igidr.ac.in/~money/Emerging%20Patterns%20in%20Share%20of%20Small%20Farms%20in%20Production%20and%20Credit_%20Implications%20for%20Policy%20Formulation.pdf

Michiels, I. S., & Van Crowder, L. (2001). *Discovering the "magic box": Local appropriation of information and communication technologies (ICTs)*. Sustainable Development Department (SD), Food and Agriculture Organisation of the United Nations SD Dimensions Special. Retrieved from www.fao.org/sd/

Miller, S., & Berry, L. (1998). Brand salience versus brand image: Two theories of advertising effectiveness. *Journal of Advertising Research*, 79.

Millward Brown, I. M. S. (2008). *Consumer ICT survey: Quarter 2, 2008, A review of findings*. Millward Brown IMS. Retrieved from http://www.odtr.ie/_fileupload/publications/ComReg0849.pdf

Millward Brown, I. M. S. (2010). *Consumer ICT survey: Quarter 4, 2009, A review of findings*. Millward Brown IMS. Retrieved from http://www.odtr.ie/_fileupload/publications/ComReg1022.pdf

Millward Brown, I. M. S. (2010). *Consumer ICT survey: Quarter 2, 2010, A review of findings*. Millward Brown IMS. Retrieved from http://www.odtr.ie/_fileupload/publications/ComReg_1062r.pdf

Millward Brown, I. M. S. (2010). *SME and corporate ICT research; First half, 2010*. Millward Brown IMS. Retrieved from http://www.odtr.ie/_fileupload/publications/ComReg1033.pdf

Ming, Z., Xin, L., & Zhijun, Z. (2008). *Research on mobile supply chain management based ubiquitous network*. 4th International Conference on Mobile Commerce and Service, May 25-28, Kunming, Chaina.

Ministry of Agriculture, Indian government report, New Delhi. (2008). *Indian agricultural scenario*.

Minton, M. C. (2000). *Is your organization ready for e-Learning? Seven key questions that need to be answered*. Retrieved from http://www.comproj.com/Minton.htm

Misra, D. C., & Vijayaditya, N. (2006). Development informatics - Reaching the rural India role of NIC. *Information Technology in Developing Countries, 16*(1). Retrieved February 9, 2011, from http://www.iimahd.ernet.in/egov/Feb2006 /article2.Html

Mohan, C. P. (2007). *Overview of agriculture and rural sector*. RB channel, Knowledge Bank, College of Agriculture Banking, Reserve Bank of India, Pune, India. Retrieved February 6, 2011, from http://www.cab.org.in/Lists/KnowledgeBank/Attachments/4 /OverviewAgriRurSector_2_17_2007.pdf

Mohanty, K. (2007). Power of information in the Indian rural set up – An analysis. *International Journal of Rural Studies*, *14*(1).

Moni, M. (2004). *Impact of economic reforms on Indian agriculture sector: Application of geomantic technology to reduce marginalization and vulnerability of small farmers in India*. Retrieved February 9, 2011, from http://www.gisdevelopment.net/application/agriculture/production/agric0003pf.htm

Montgomery, K. (2000). Youth and digital media: A policy research agenda. *The Journal of Adolescent Health*, *27*, 61–68. doi:10.1016/S1054-139X(00)00130-0

Morat, J. (1997). Weiterbildung für Fahrer von Krannvollernter in Deutschland. *Forsttechnische Informationen*, *3*, 38–40.

Morford, S., & Hollstedt, C. (2007). *Revisiting a forest extension strategy for British Columbia: A survey of natural resource practitioners and information providers*. Victoria, Canada: B.C. Ministry of Forest Range, Resources.

Morrow, L., Kelly, T., & Kirley, T. (2003). *ICT – Its potential as a channel for enhanced extension services*. Paper presented at the AIAEE conference, University College Dublin, June 2003.

Moss, C. B., Schmitz, T. G., Kagan, A., & Schmitz, A. (2003). Institutional economics and the emergence of e-commerce in agribusiness. *Journal of Agribusiness*, *21*(1), 83–101.

Moss, C., Schmitz, T., Kagan, A., & Schmitz, A. (2003). Institutional economics and the emergence of e-commerce in agribusiness. *Journal of Agribusiness*, *21*(1), 83–101.

Mueller, E. (2000). Emergent e-commerce in agriculture. *Agriculture Issues Center Brief, 14*.

Mueller, R. (2001). E-commerce and entrepreneurship in agricultural markets. *American Journal of Agricultural Economics*, *83*(5), 1243–1249. doi:10.1111/0002-9092.00274

Mukhebi, A. (n.d.). *The challenges and opportunities of a young commodity exchange in emerging market economy: The experience of Kenya Agricultural Commodity Exchange*. Retrieved from http://www.google.co.ke/#hl=en&source=hp&biw=1245&bih=486&q=The+challenges+and+opportunities+of+a+young+commodity+exchange+in+emerging+market+economy&btnG=Google+Search&rlz=1R2SKPB_enKE362&aq=f&aqi=&aql=&oq=&gs_rfai=&fp=ceb31e2ddf411e47

Munyua, H. (2007). (Submitted to). ICTs and small-scale agriculture in Africa: A scoping study. [International Development Research Center.]. *Draft Report, 1*.

Muto, M., & Tekashi, Y. (2009). *The impact of mobile phone coverage expansion on market participation: Panel data evidence from Uganda*. World Development Report, 2009.

Mutula, M. S., & Brakel, P. V. (2006). E-readiness of SMEs in the ICT sector in Botswana, with respect to information access. *The Electronic Library*, *24*, 402–407. doi:10.1108/02640470610671240

Mutula, S., & Brakel, P. (2006). E-readiness of SMEs in the ICT sector in Botswana with respect to information access. *The Electronic Library*, *24*(3), 402–417. doi:10.1108/02640470610671240

Namani, M. (2009). *The role of information technology in small and medium sized enterprises in Kosova*. Fulbright Academy Conference, Skobje, Republic of Macedonia

Narayanan, A., & Chakraverty, G. (2009). Running ICT telecenters in rural India. *Information Technology in Developing Countries, 19*(1). Retrieved on February 9, 2011, from http://www.iimahd.ernet.in/egov/ifip/feb2009/arvindd-narayanan.htm

Nath, V. (2003). *Digital governance models: Moving towards good governance in developing countries*. United Nations Development Project Report.

National Education Research and Development. (1996). *Agricultural science for secondary schools* (p. 219). Ibadan, Nigeria: University Press Plc.

NCC. (2010). *Subscriber data at a glance (year 2009 – October 2010)*. Retrieved from http://www.ncc.gov.ng

New Connections. (2002). *A strategy to realise the potential of the Information Society*. Second Government Action Plan, Government of Ireland. Retrieved from http://www.taoiseach.ie/attached_files/Pdf%20files/NewConnectionsMarch2002.pdf

New, S. (2010). Get the best from your people: The transparent supply chain. *Harvard Business Review*, *88*(10), 76–80.

Niederhauser, N., Oberthur, T., Kattning, S., & Cock, J. (2008). Information and its management for differentiation of agricultural products: The example of specialty coffee. *Computers and Electronics in Agriculture*, *61*(2), 241–253. doi:10.1016/j.compag.2007.12.001

Niranjan, S., Dhandapani, A., Sharma, O. P., Garg, D. K., & Verma, P. V. (2007). Pest management information system decision support tool for pest management in Basmati rice. *ICAR News: A Science and Technology Newsletter, 13*(2), 15-16.

Norris, P. (2001). *Digital divide. Civic engagement, information poverty, and the internet worldwide*. Cambridge, UK: Cambridge University Press. doi:10.1017/CBO9781139164887

NPT. (2000). *National policy on communication*. Ministry of Communication. Retrieved from http://www.ncc.gov.ng

Ntaliani, M., Costopoulou, C., & Karetsos, S. (2008). Mobile government: A challenge for agriculture. *Government Information Quarterly*, *25*(4), 699–716. doi:10.1016/j.giq.2007.04.010doi:10.1016/j.giq.2007.04.010

Nugent, J. D. (1982). Videotape recording – A practical tool for training forest workers. In *Seminar on the Management of Forest Worker Training,* Vol. 2, (pp. 21-28).

O'Donnell, S., McQuillan, H., & Malina, A. (2003). *E-inclusion: Expanding the Information Society in Ireland*. Report to the Information Society Commission, Dublin.

Obayelu, A., & Ibadan, I. (2006). Analysis of information communication technology (ICT) for gender empowerment and sustainable poverty alleviation in Nigeria. *International Journal of Education and Development using ICT, 2*(3), 45-69.

Odekunle, T. O., Orimoogunje, I. O. O., & Ayanlade, A. (2007). Application of GIS to assess rainfall variability impacts on crop yield in Guinean Savanna part of Nigeria. *African Journal of Biotechnology, 6*(18), 2100-2113. ISSN 1684–5315.

OECD. (1995). *Developing environmental capacity. A framework for donor involvement*. Paris, France: OECD.

Ogawara, S., Chen, J., & Zhang, Q. (2003). Internet grocery business in Japan: Current business models and future trends. *Industrial Management (Des Plaines), 103*(9), 727–735.

Ogur, S. O. (2003). *Animal production in tropical zone* (pp. 76–82). Premier Publications.

Okello, J. J., Al-Hassan, R., & Okello, R. M. (2010). A framework for analyzing the role of ICT on agricultural commercialization and household food security. *International Journal of ICT and Research Development, 1*, 38–50. doi:10.4018/jictrda.2010010103

Okello, J. J., Ofwona-Adera, E., Mbatia, O. L. E., & Okello, R. M. (2010). Using ICT to integrate smallholder farmers into agriculture value chain: The case of DrumNet Project in Kenya. *International Journal of ICT Research and Development in Africa, 19*(1), 23–37. doi:10.4018/jictrda.2010010102

Okello, J. J., Okello, R. M., & Ofwona-Adera, E. (2009). Awareness and use of mobile phones by smallholder farmers in Kenya. In Maumbe, B. M. (Ed.), *E-agriculture and e-government for global policy development: Implications and future directions*. Hershey, PA: IGI Global. doi:10.4018/978-1-60566-820-8.ch001

Okello, J. J., Okello, R. M., & Ofwona-Adera, E. (2010). Awareness and the use of mobile phones for market linkage by smallholder farmers in Kenya. In Maumbe, B. M. (Ed.), *E-agriculture and e-government for global policy development: Implications and future directions*. Hershey, PA: IGI Global. doi:10.4018/978-1-60566-820-8.ch001

Okyere, K. A., Davis, K., & Aredo, D. (2008). *Advancing agriculture in developing countries through knowledge and innovation–Synopsis of an international conference*. ISBN: 10-digit: 0-89629-780-2 13-digit: 9788-0-89629-780-7

Oladele, O. I., & James, O. (2008). Perceived effect of privatization of extension services among researchers, extension agents and farmers in Oyo State, Nigeria. *Agricultural Journal, 3*(5), 409–413.

Olson, K., & Boehlje, M. (2010). Theme overview: Fundamental forces affecting agribusiness industries. *Choices (New York, N.Y.), 25*(4).

Omamo, S. W. (1998). Farm-to-market transaction costs and specialization in smallscale agriculture: Explorations with non-separable household model. *The Journal of Development Studies, 35*, 152–163. doi:10.1080/00220389808422568

Otolo, P. U. (2006). *Agricultural information needs of rural women farmers in Nigeria: A case study of oil-producing communities in Delta State – Nigeria* (pp. 11–12).

Otolo, P. U. (2008). The impact of information and communication in extension services to rural farmers in Niger Delta. *The Information Technologist, 5*(2), 37–38.

Otuka, A., & Sugawara, K. (2003). A labor management application using hand-held computers. *Agricultural Information Research, 12*, 95–104.

Otuka, A., & Yakamawa, T. (2003). An application for insect field survey using a hand-held computer. *Agricultural Information Research, 12*, 113–124.

Oyelaran, O. B. (1996). *NISER survey of business condition, experience and expectations in the manufacturing sector*. Nigeria Institute of Social and Economic Research (NISER) publication.

Pandey, A., Chowdary, V. M., & Mal, B. C. (2007). Identification of critical erosion prone areas in the small agricultural watershed using USLE, GIS and remote sensing. *Water Resources Management, 21*, 729–746. doi:10.1007/s11269-006-9061-z

Pandey, V. K., Panda, S. N., Pandey, A., & Sudhakar, S. (2009). Evaluation of effective management plan for an agricultural watershed using AVSWAT model, remote sensing and GIS. *Environmental Geology*, *56*, 993–1008. doi:doi:10.1007/s00254-008-1201-8

Park, R. D. (1997). *Animal production and health* (pp. 38–46). Enugu, Nigeria: Moris Publishers.

Paul, P. (2003, January 06). The Mandi is now on the Net. *Deccan Herald*. Retrieved from http://www.itcpotal.com

Paz, J., Batchelor, W., & Pedersen, P. (2004). WebGro: A web-based soybean management decision support system. *Agronomy Journal*, *96*, 1771–1779. doi:10.2134/agronj2004.1771doi:10.2134/agronj2004.1771

Phillps, A. W. (1958). The relationship between unemployment and the rate of change of money wages in the United Kingdom 1861-1957. *Economica*, *25*(100), 283–299.

Planas, J. (2001). *Continuing training in Europe: A problem in common*. CEDEFOP Workshop on Continuing Training Policies in the European Context. Retrieved from studyvisits.cedefop.eu.int/assets/.../lifelonglearning_initialcontinuing_en.pdf

Popoola, L. (2006). Socio-economic potentials of the Nigerian rural landscape and ways of utilising them toward rapid economic and industrial revolution. *Proceeding of The fifteenth Annual Congress of the Nigerian Rural Sociological Association* (pp. 1–8).

Porter, L. R. (1997). *Creating the virtual classroom: Distance learning with the internet*. New York, NY: Wiley.

Poschen, P. (1997). Forestry and employment – Much more than meets the eye. In *Proceedings of the XI World Forestry Congress*. Retrieved from www.fao.org/montes/foda/wforcong/PUBLI/V5/T34E/Default.htm

Poschen, P. (1993). Forestry, a safe and healthy profession? *Unasylva*, *172*(44), 3–12.

Poulton, C., Kydd, J., & Doward, A. (2006). Overcoming market constraints on pro-poor agricultural growth in sub-Saharan Africa. *Development Policy Review*, *24*, 243–277. doi:10.1111/j.1467-7679.2006.00324.x

Prahlad, C. K. (2004). *The EID Parry story. The fortune at the bottom of the pyramid*. Retrieved February 6, 2011, from http://www.eidparry.com/casestudy.asp

Project Management Institute (PMI). (2008). *A guide to the Project Management Body of Knowledge (PMBOK Guide)* (4th ed.). Newtown Square, PA: Project Management Institute.

Punch. (2009). LCCI move to curb youth unemployment. *The Punch Newspaper*.

Radian Consultants. (2006). *Cape Gateway evaluation: Report for the Centre for E-Innovation Provincial Government of the Western Cape*. Cape Town, South Africa: Author.

Raghbendra, J. (2005). *Land fragmentation and its implications for productivity: Evidence from Southern India*. ASARC Working Paper 2005/01.

Rahul, D. (2005). *The impact of Indian e-government initiatives: Issues of poverty and vulnerability reduction and conflict*. Bangalore, India: Indian Institute of Management.

Rao, B., & Minakakis, L. (2003). Evolution of mobile location based service. *Communications of the ACM*, *46*(12), 61–65. doi:10.1145/953460.953490

Rao, N. H. (2007). A framework for implementing information and communication technologies in agricultural development in India. *Technological Forecasting and Social Change*, *74*, 491–518. doi:10.1016/j.techfore.2006.02.002

Rathenberg, E., & Mielk, A. (1986). *Wie entwickelt man Medien für die Berufliche Bildung?* Mannheim, Germany: Deutsche Stiftung für Entwicklung.

Reddy, K., & Ankaiah, R. (2005). A framework of information technology-based agriculture information dissemination system to improve crop productivity. *Current Science*, *88*(12), 1905–1913.

Reding, V. (2001). *The role of the European Community in creating the knowledge and Information society*. Speech at the Zentrum für Europäischen Integrationsforschung in the University of Bonn on 7.3.2001. Retrieved from http://europa.eu/rapid/pressReleasesAction.do?reference=SPEECH/01/108&format=PDF&aged=1&language=EN&guiLanguage=en

Relyea, H. C. (2004). Homeland security and information sharing: Federal policy consideration. *Government Information Quarterly*, *21*, 420–438. doi:10.1016/j.giq.2004.08.007

Reynolds, K. M., Borges, J. M., Vacik, H., & Lexer, M. J. (2005). ICT in forest management and conservation. In Hetemäki, L., & Nilsson, S. (Eds.), *Information technology and the forest sector - IUFRO World Series* (*Vol. 18*, pp. 150–172). Vienna, Austria: IUFRO.

Reynolds, K. M., & Schmoldt, D. A. (2006). Computer aided decision making. In Shao, G., & Reynolds, K. M. (Eds.), *Computer applications in sustainable forest management including perspectives on collaboration and integration* (pp. 143–169). Springer.

Rifkin, S. B. (1999). Capacity-building: An approach to people-centred development. *The International Journal of Health Planning and Management*, *14*(4), 336–338.

Robinson, S. M. (1998). Microfinance: The paradigm shift from delivery to sustainable financial intermediation. In Eicher, C. K., & Staatz, J. M. (Eds.), *International agricultural development* (3rd ed., pp. 390–415).

Ross, B., & Waksman, G. (2004). *Internet use in agriculture: Progress driven by demand, not by supply.* Retrieved from http://www.acta-informatique.fr/telechar/NewAg/progress_driven_by_demand.pdf

Samuelson, P. (2008). *Thought about the Phillips curve* [Electronic Version]. Retrieved from http://www.docstoc.com/docs/52714771/Thoughts-About-The-Phillips-Curve-Paul-A-Samuelson-3

Sanganeria, S. (2003, February 13). ITC e-Choupal in wheat buying spree. *Business Standard*, p. 12.

Sanusi, J. O. (2003). *Special remarks on information and communication technology for the social transformation of Nigeria.* Lecture delivered at the Dinner Investor Forum organized by Leapfrog Venture Partners Ltd. The National Summit on SMIE, Federal Palace Hotel, February 15, 2003.

Schaefer, G. (2008). Tracking and tracing-A challenge for system organization and IT. *Journal of Information Technology in Agriculture*, *3*(1), 19–25.

Schuster, E. G., Leefers, L. A., & Thompson, J. E. (1993). *A guide to computer-based analytical tools for implementing national forest plans*. Moscow, ID: U.S. Department of Agriculture, Forest Service, Intermountain Forest and Range Experiment Station.

Sehai, E. (2006). *Knowledge management in Ethiopian agriculture.* Paper presented on14th Annual Conference Ethiopian Society of Animal Production, September 7th, 2006 (International Livestock Research Institute – IPMS Project). Retrieved from mahider.ilri.org/handle/10568/614

Sern, G. (2001). The broadband divide: A widening chasm in higher education? *EDUCAUSE Review, Jan/Feb 2001* (pp. 60-61). Retrieved December from www.educause.edu/ir/library/pdf/erm0117.pdf

Shao, G., & Reynolds, K. M. (2006). Introduction. In Shao, G., & Reynolds, K. M. (Eds.), *Computer applications in sustainable forest management including perspectives on collaboration and integration* (pp. 3–16). Springer.

Sharma, A., & Yadav, A. (2003). *What works: AKASHGANGA'S IT tools for Indian Dairy industry –Using IT to increase efficiency in rural dairy co-operatives*. Retrieved February 17, 2011, from http://www.nextbillion.net/archive/files/AKASHGANGA.pdf

Sharma, D. (2005 June 11). The business of hunger: ICT and millennium development goals. *Motion Magazine.*

Sharma, D. (2007). ICT and rural livelihoods: Whose livelihoods are we talking about? *Information Technology in Developing Countries, 17*(2). Retrieved February 9, 2011, from http://www.iimahd.ernet.in/egov/ifip/aug2007/devinder-sharma.htm

Sharma, O. P., Dhandapani, A., & Singh, N. (2004). Computer based decision support system for integrated pest management. In A. Singh, H. R. Sardana, & N. Sabir (Eds.), *Validated IPM technologies for selected crops* (pp. 181–190). New Delhi, India: National Centre for Integrated Pest Management.

Shastri, P. (2001, May 1). Digitizing land records key to reform. *Hindustan Times*. Retrieved from http://www.hindustantimes.com/News/business/Digitising-land-records-key-to-reform/Article1-219609.aspx

Shelke, R. D. (2009). Role of information technology on agriculture and its scope in India. In Ahmed, R. (Ed.), *Globalization and economic development –Part – 1* (pp. 207–222). New Delhi, India: Mittal Publications.

Shenoy, M. (2002). Transforming agribusiness the e-way. *Business India*, *634*, 102–104.

Shumer, R. (2001). *A new, old vision of learning, working, and living: Vocational education in the 21st century. Journal of Vocational Education Research, 25(2)*. Virginia Polytechnic and State University.

Shunmugam, V., & Bahl, S. (2009). ICTs as enablers of process change in agriculture. *Information Technology in Developing Countries, 19*(1). Retrieved February 9, 2011, from http://www.iimahd.ernet.in/egov/ifip/feb2009/shunmugam.htm

Silva, H., & Ratnadiwakara, D. (2010). ICT policy for agriculture based on a transaction cost approach: Some from Sri Lanka. *International Journal of ICT Research and Development in Africa, 1*(1), 51–64. doi:10.4018/jictrda.2010010104

Singh, A. (2005). *Case study: Community based information system, India EID Parry Indiagriline*. Retrieved February 6, 2011, from http://www.fao.org/rdd/doc/EID Parry Indiagriline.pdf

Singh, A. P. (2005). Lokvani: An e-effort to empower citizens. *Newsletter of the IFIP Working Group, 15*(2). Retrieved February 9, 2011, from http://www.iimahd.ernet.in/egov/ifip/nov2005/article1.htm

Singh, A., Sharma, O. P., Dhandapani, A., Singh, N., Nagarajan, S., Satya, K., & Mehta, N. (2006). Pesticide advisor (Version 2005 1.0): An expert system for judicious use of pesticides for management of pests. *ICAR News: A Science and Technology Newsletter, 12*(4), 17-19.

Singh, S. (2008). Time for agriculture to metamorphose in agribusiness. *GBPUAT Newsletter, 6*(3), 4-5. Retrieved from http://ssrn.com/abstract=1526051

Singh, S. (2010). *Potential of information technology as a tool for transforming Indian agriculture*. Retrieved from http://ssrn.com/abstract=1548882

Singh, S. (2004). Leveraging ICT for agricultural development: A case study of e–Choupal of ITC paradigm. *Journal of IMT, 3*(1).

Smith, P. (2008, February 23). No Broadband in 75% of rural homes in Ireland. A national survey of Irish Farmers' Association 945 branches conducted in 2007. *Irish Farmers Journal*, 8.

South Africa Government Gazette. (2000). *Promotion of access to Information Act*. Pretoria, South Africa: Office of the Presidency.

South Africa. (2006). *White paper on transforming public service delivery*. Retrieved from www.polity.org.za.html/govdocs/white_papers/transform.html

Sultan, F., & Rohm, A. (2004). The evolving role of the Internet in marketing strategy: An exploratory study. *Journal of Interactive Marketing, 18*(12), 6–19. doi:10.1002/dir.20003

Swaminathan, M. S. (2009). Agricultural renewal and agrarian prosperity. In Ahmed, R. (Ed.), *Globalisation and economic development –Part – 1* (pp. 3–20). New Delhi, India: Mittal Publications.

Swaminathan, M. S. (2009). Can science and technology feed the world in 2025? In Ahmed, R. (Ed.), *Globalization and economic development –Part – 1* (pp. 21–36). New Delhi, India: Mittal Publications.

Tenhunen, S. (2008). Mobile technology in the village: ICT, culture, and social logistic in India. *Journal of the Anthropological Institute, 14*(3), 514–534.

Tewari, D. (2005). A field trip to rural tele-centers in Gujarat, information technology in developing countries. *A Newsletter of the IFIP Working Group, 15*(2). Retrieved February 9, 2011, from http://www.iimahd.ernet.in/egov/ifip/nov2005/article3.htm

The Punch. (2006a, July 7). Obasanjo inaugurates computer for all Nigerians initiative. *Punch Newspaper, Business Today*, p. 33.

The Punch. (2006b, September 25). New initiatives to drive Nigeria's PC penetration. *Punch Newspaper, Computer and IT Education*, p. 48

The Star. (2006). *Cyber 'kingpin' nabbed*. The Star. Retrieved from www.star.co.za

Thompson, S., & Sonka, S. (1979). Potential effect of information technologies on the economic performance of agricultural and food markets. *American Journal of Agricultural Economics, 79*(2), 657–662. doi:10.2307/1244167

Tienne, D. (2002). Addressing the global digital divide and its impact on educational opportunity. *Educational Media International*, *39*(3/4), 211–222.

Tollens, E. (2006). *Market information systems in sub-Saharan Africa: Challenges and opportunities*. Poster paper prepared for presentation at the International Association of Agricultural Economists Conference, Gold Coast, Australia.

Tompson, S., & Kunda, E. (2000). *E-commerce and agricultural markets*. OFOR Paper Number 00-03.

Treinen, S. (2004). *New ICTs for research, extension linkages: Opportunities and challenges*.Paper presented at the FAO – GTZ workshop on Effective Communication between Agricultural Research, Extension and Farmers, Laimburg, Italy.

Trienekens, J. H. (2011). Agriculture value chains in developing countries A framework for analysis. *International Food and Agribusiness Management Review*, *14*(2), 51–82.

Trinque, B. M. (1994). *Dynamism and enervation contrasts in India's Agrarian economy*. Discussion paper series No. 94-18, November 1994 at Department of Economics, Southern Illinois University, Carbondale, IL. Retrieved February 7, 2011, from http://129.3.20.41/eps/dev/papers/9505/9505001.pdf

Tsai, Y., & Yeh, J. (2010). Perceived risk of information security and privacy in online shopping: A study of environmentally sustainable products. *African Journal of Business Management*, *4*(18), 4057–4066.

Tsioras, P. A. (2004). *Analysis and formation of modern forest workers' training systems for the Greek Forestry*. Unpublished doctoral dissertation, Aristotle University of Thessaloniki, Greece.

Tsioras, P. A., & Efthymiou, P. N. (2007). Need for forest workers' training in Greek forestry. [In Greek with abstract in English]. *Geotechnical Scientific Issues*, *18*(2), 46–56.

Turban, E., Lee, J., King, D., & Chung, H. M. (1999). *Electronic commerce: A managerial perspective* (p. 520). Upper Saddle River, NJ: Prentice Hall.

U.S. Department of Commerce. (2004). *A nation online: Entering the Broadband age*. Retrieved from http://www.ntia.doc.gov/reports/anol/NationOnlineBroadband04.htm#_Toc78020939

UNDP. (2004). *E-governance- UNDP practice area: Democratic governance synthesis of lessons learned*. Human development report 2001, No.15, April 2004.

Unegbu, M. C. (1999). *Impact assessment of information: Agricultural technology on the performance of the Nigerian sector* (pp. 72–74). A project submitted to the African Regional Centre for Information Science, University of Ibadan, Ibadan.

UNEP. (2001). *Climate change 2001: Impacts, adaptation and vulnerability*. Summary for policy makers and technical summary of the Working Group II Report. UNEP Policy Series. Retrieved from fire.pppl.gov/ipcc_SWG2_2001.pdf

UNESCO & International Labour Organization. (2002). *Revised recommendation concerning technical and vocational education. Paris, France: UNESCO*. Geneva, Switzerland: ILO.

UNESCO. (2003). *Newsletter, 1*(106).

UNESCO. (2005). *E-governance toolkit for developing countries*. New Delhi, India: UNESCO Publishers.

United Nations. (2008). *Trends in sustainable development: Agriculture, rural development, land, desertification and drought*. New York, NY: Department of Economic and Social Affairs of the United Nations. Retrieved from http://www.un.org/esa/sustdev/publications/trends2008/fullreport.pdf

Unni, R., & Harmon, R. (2007). Perceived effectiveness of push vs. pull mobile location-based advertising. *Journal of Interactive Advertising*, *7*(2), 28–40.

Upadhyay, D. K. (2009). ePanchayats ICT for rural local government in India. *Information Technology in Developing Countries, 19*(3). Retrieved February 9, 2011, from http://www.iimahd.ernet.in/egov/ifip/oct2009/upadhyay.htm

Upton, R. M., & Fuller, V. A. (2004). *The ITC e-Choupal initiative*. Boston, MA: Harvard Business School Publishing.

USAID. (2003). *Future directions in agriculture and information and communication technologies (ICT)*. USAID/ Economic Growth, Agriculture, and Trade/Agriculture and Food Security.

Van der Vorest, J., Van Dongen, S., Nouguier, S., & Hilhorst, R. (2002). E-business initiatives in food supply chains: Definition and typology of electronics business models. *The Journal of Logistics: Research and Applications, 5*(2), 119–138. doi:10.1080/13675560210148641

Van Dijk, J., & Hacker, K. (2003). The digital divide as a complex, dynamic phenomenon. *The Information Society, 19*(4), 315–326. doi:10.1080/01972240309487

Vander Vorst, J., Dongen, S., Nouguier, S., & Hilhorst, R. (2002). E-business initiatives in food supply chains: Definition and typology of electronic business models. *International Journal of Logistics: Research and Applications, 5*(2), 119–138.

Vanguard. (2009). Unemployed youth transformed to become entrepreneur. *The Vanguard Newspaper*.

Vanguard. (2010). Government to invest in Nigerias' ICT environment. *The Vanguard Newspaper*.

Varshney, U. (2003). Location management for mobile commerce applications in wireless Internet environment. *Transactions on Internet Technology, 3*(3), 236–255. doi:10.1145/857166.857169

Varshney, U. (2003). Mobile and wireless information systems: Applications, networks, and research problems. *Communications of the Association for Information Systems, 12*(11), 155–165.

Venniyoor, S., & Sah, G. (2007). Spreading the ICT revolution in rural India: Experiences and examples. *Information Technology in Developing Countries, 17*(2). Retrieved February 9, 2011, from http://www.iimahd.ernet.in/egov/ifip/ aug2007/sajan-venniyoor.htm

Vlosky, R. P., Fontenot, R., & Blalock, L. (2000). Extranets: Impacts on business practices and relationships. *Journal of Business and Industrial Marketing, 15*(6), 438–457. doi:10.1108/08858620010349510

Vorst van der. J., Beulens, M., & Van Beek, P. (2005). Innovation in logistics and ICT in food supply chain networks (pp. 245-292). Wagengen, The Netherlands: Wagengen Academic Publisher.

Vosloo, S. (2006). *Multilingualism in a South African e-government portal.* Cape Town, South Africa: Center for e-Innovation. (Unpublished document)

Wagner, T. (1997). Softwaregestützte Simulationstechnologie. *Forsttechnische Informationen, 7-8*, 96–99.

Walsham, G. (2010). ICT for broader development of India: An analysis of literature. *The Electronic Journal on Information Systems in Developing Countries, 41*(4), 1–20.

Wang, N., Zhang, N., & Wang, M. (2006). Wireless sensors in agriculture and food industry- Recent development and future perspective. *Computers and Electronics in Agriculture, 50*, 1–14. doi:10.1016/j.compag.2005.09.003

Wikipedia. (2011). *Unstructured supplementary service data*. Retrieved from http://en.wikipedia.org/wiki/Unstructured_Supplementary_Service_Data

Wild- Eck, S., Zimmermann, W., & Schmithüsen, F. (2006). Extension for private forest owners: Insights from a representative opinion poll in Switzerland. *Small-Scale Forest Economics. Management and Policy, 5*(2), 161–174.

William, S., & Ziff-Levine, W. (1990). Evaluating tourism campaigns: Conversion vs. advertising tracking studies. *Journal of Travel Research, 28*(3), 51–55. doi:10.1177/004728759002800312

Williamson, O. E. (1985). *The economic institutions of capitalism: Firms, markets, relational contracting*. New York, NY: The Free Press. doi:10.1007/978-3-8349-9320-5_6

Wilson, D. N. (2001). Reform of TVET for the changing world of work. *Prospects: Quarterly Review of Comparative Education, 31*(1), 21–37. doi:10.1007/BF03220047

Wims, P. (2007). Analysis of adoption and use of ICTs among Irish farm families. *Journal of Extension Systems, 23*(1), 14–28.

Windrum P., & de Berranger, P. (2003). *Factors affecting the adoption of intranets and extranets by SMEs: A UK study.* Maastricht Economic Research Institute on Innovation and Technology, report no 023s.

Wolter, D. (2008). *Ghana: Agriculture is becoming business*. Paris, France: OECD. Retrieved from www.oecd.org/dev/publications/businessfordevelopment

Wong, S. K., Nguyen, T. T., Chang, E., & Jayaratna, N. (2003). Usability metrics for e-learning. *Lecture Notes in Computer Science, 2889*, 235–252. doi:10.1007/978-3-540-39962-9_34

Woodside, A. G., & Ronkainen, I. A. (1982). Travel advertising: Newspapers versus magazines. *Journal of Advertising Research, 22*(3).

Woodside, A. G., & Soni, P. K. (1990). Performance analysis of advertising in competing media vehicles. *Journal of Advertising Research, 30*(1).

Woodside, A. G., & Soni, P. K. (1991). Direct-response advertising information: Profiling heavy, light and non users. *Journal of Advertising Research, 31*(6).

World Bank. (2003). *Higher education development for Ethiopia: Putting the vision.* World Bank Sector Study, Addis Ababa, Ethiopia. Retrieved from http://www.siteresources.worldbank.org/.../Ethiopia_Higher_Education_ESW.pdf

World Bank. (2007). *World development report 2008: Agriculture for development.* Washington, DC: World Bank. Retrieved from http://siteresources.worldbank.org/INTWDR2008/Resources/WDR_00_book.pdf

World Bank. (2008). *Innovative uses of information technology to link farmers to markets in India and West Africa* (p. 121). World Development Report, World Bank.

World Development Report. (2008). *Innovative uses of information technology to link farmers to markets in India and West Africa* (p. 121). Washington, DC: The World Bank.

Yin, R. (1989). *Case study design: Design and methods*. Thousand Oaks, CA: Sage Publication.

Zarini, M., Varis, T., & Mar, N. Y. (2009). Information and communication technologies (ICTs) in vocational education and training. In Maclean, R., & Wilson, D. (Eds.), *International handbook of education for the changing world of work*. Springer.

Zavaglia, T. T., Pierozzi, I., Rodrigues, N. P., & DeCastro, A. (2011). Knowledge management and communication in Brazilian agricultural research: An integrated procedural approach. *International Journal of Information Management, 31*(2), 121–127. doi:10.1016/j.ijinfomgt.2010.07.001

Zeinab, K. (2006). Trust, privacy and security in electronic business: The case of GCC countries. *Information Management & Computer Security, 14*(3), 270–283. doi:10.1108/09685220610670413

Zeinab, K. (2008). *Final report: ICT and small scale agriculture in Africa: A scoping study*. The International Development Research Centre.

Zeinab, K. (2011, February 25). Mobile applications help farmers profit. *Africa Report.*

Zhao, G., Shao, G., Reynolds, K., Wimberly, M. C., Warner, T., & Moser, J. M. … Li, F. (2005). Digital forestry: A white paper. *Journal of Forestry, 103*, 47-50.

ZhiGong. P., Yu, L., Di, X., & Lei, W. (2009). Application of RS and GIS technique for water-saving potential estimation of regional crops. *Transactions of the Chinese Society of Agricultural Engineering, 25*(7), 8-12. ISSN 1002-6819

Zirkle, C. (2003). Distance education and career and technical education: A review of the research literature. *Journal of Vocational Education Research, 28*(2), 161–181. doi:10.5328/JVER28.2.161

About the Contributors

Blessing Mukabeta Maumbe graduated with a PhD in Agricultural Economics with specializations in agribusiness management and marketing from Michigan State University (MSU) in 2001. Blessing has received fellowship awards from the Rockefeller Foundation, W.K. Kellogg Foundation, MSU Thoman Foundation, and the Southern Africa Centre for Cooperation in Agricultural Research/Germany Agency for Technical Cooperation (SACCAR/GTZ). He has taught undergraduate and graduate courses in agribusiness management at various universities in Zimbabwe, South Africa and the United States over the past 18 years. He was the founding Lecturer and Coordinator of Agribusiness Management studies at Africa University (AU) in Zimbabwe. After that, he was appointed Senior Land Bank Lecturer at the University of Fort Hare (UFH) in South Africa. He later served as an Associate Professor in Business and Community Informatics at Cape Peninsula University of Technology (CPUT), in Cape Town, South Africa. Blessing is currently working as an Assistant Professor of Agribusiness Management at West Virginia University, USA. Prior to this position, he was an Associate Professor responsible for the agribusiness management program at Eastern Kentucky University (EKU).

Charalampos Z. Patrikakis was born Greece, in 1970. He received his Dipl.-Ing. and his Ph.D from the Electrical Engineering and Computer Science Department of NTUA, Greece. He is an Assistant Professor at the Dept of Electronics of the TEI of Piraeus. He has participated in more than 27 National, European, and International programs (ESPRIT, RACE, ACTS, IST, ICT, eContent, eContentPlus), in 12 of which he has been involved as technical coordinator or principal researcher. He has published 21 chapters in books, 27 papers in international journals, 53 papers in international conferences, 6 reports, and has 2 contributions in national legislation. He is a member of the editorial committee of 11 international journals, 30 international conferences, and has acted as editor in the publication of 2 special issues of international journals, 3 conference proceedings volumens and coedited one book. His research interests include: mobile network applications and services, mobile and pervasive computing, broadband network communications, IP data communications, P2P and ad hoc networks, personal networks, and multimedia streaming. He is a senior member of IEEE, a member of the Greek Computer Society, a certified trainer by the National Accreditation Centre of Vocational Training Structures and Accompanying Support Services and a member of the Technical Chamber of Greece.

* * *

Muhammad Shakil Ahmad is PhD scholar at the Department of Management and HRD in Universiti Technologi Malaysia, Malaysia. He is also Lecturer at Department of Management Science in COMSATS Institute of Information Technology, Pakistan.

Sylvester Ozioma Anie is a staff of Delta State Polytechnic, Ozoro, Delta State of Nigeria. He has served as Polytechnic Librarian and Dean of Students Affairs. He is a member Nigerian Library Association. His qualifications include BSc (Ed) Library Science; MSc Library and Information Science; and presently, a PhD student in Delta State University, Abraka, Delta State of Nigeria. His research interest is on the impact ICT on life and the environment. The best award/ Commendation was received from Emerald Literati Network (Emerald Group Publishing Limited) which has qualified the author to be a member of Emerald Literati Network. His recent publication is entitled "The Social Benefits of ICT Policies in Nigeria" published by Library Philosophy and Practice in March, 2011.

Isaac Kwadwo Asare is currently working as a Project Manager of the AGRA-funded Ghana Agro Dealer Development Project (GADD) project implemented by the International Fertilizer Development Center (IFDC). Mr Asare holds a BSc. Degree in Agricultural Technology from University for Development Studies, Tamale, Ghana, a Postgraduate Diploma and Master of Philosophy in Sociology from the University of Ghana. His research and development interests include: participatory program planning, implementation, monitoring, and evaluation of ICT applications in agriculture and rural development. He has participated in short courses in Participatory Planning, Monitoring and Evaluation – Managing and Learning for Impact, conducted by the Wageningen International, the Netherlands and in Project Management, conducted by the Management for Development Foundation (MDF) in Accra, Ghana. In the five years of professional experience, he has two key publications to his credit published in the *Urban Agriculture Magazine* and the *Journal of the Ghana Science Association*.

Ayanlade Ayansina is a Lecturer and Researcher in Department of Geography, Obafemi Awolowo University, Ife, Nigeria. His research interest is in applied GIS and remote sensing, environmental remote sensing and agro-climate modelling and monitoring.

Theodora Betsidou, was born in Alexandroupolis Greece in 1984 where she still lives. She holds a Bachelor's degree in Engineering from the Department of *Environmental Engineering* of the Democritus University of Thrace (DUTh) and a MSc in "*Sustainable Management of the Environment and Natural Resources*" (specialized in "*Management of Mountainous Water Resources with Intelligent Systems and GIS*") from the Department of *Forestry and Management of the Environment and Natural Resources* (DUTh). Her postgraduate thesis was entitled "*Soft Computing Modeling of the Wild Fire Risk Indices: The Risk Profile of Peloponnesus Region in Greece*." She is currently teaching of Environmental Education in a school that is specialized for adult learning. She has attended several International Conferences about fractures-wastewater utilities. Most recently, she attended the 16th European Conference of Fracture in Alexandroupolis (July 3rd 7th 2006) the 13th International Conference on Experiment Mechanics in Alexandroupolis (July 1st 6th 2007) and the 2nd International Conference on Water Economics, Statistics, Finance and Asset Management of Small and Medium Wastewater Utilities in Alexandroupolis (3rd to 4th of July 2009).

Someshwar Bhagat (Senior Scientist, Plant Pathology, NCIPM, New Delhi) has awarded his Ph.D. degree in Plant Pathology from Bidhan Chandra Krishi Viswavidyalaya, Mohanpur, West Bengal, India. He has started his professional career by joining ARS in March, 2003 as Scientist (Plant Pathology) in Indian Council of Agricultural Research, New Delhi. He has more than nine years of professional experiences in plant disease management with special emphasis on biocontrol of plant diseases and Integrated Disease Management. He has more than fifty (50) publications in his credit. He is awardee of Jawaharlal Nehru Outstanding Post Graduate Studies in Plant Protection for his Outstanding Ph. D. Research work.

Archana Bhardwaj (Research Associate, NFSM Project) awarded MCA from IGNOU, Nagpur, Maharashtra. She has started her professional career as Research Associate in the year 2001. She has professional experience of 9 years working in ICT based applications for farmers & technical assistants.

Siegfried Kofi Debrah has 26 years experience in International Agricultural Development. A Ph.D from the University of Kentucky, Dr. Debrah holds a Project Management Professional (PMP) credential from the Project Management Institute (PMI). He doubles as IFDC's Representative in Ghana and Project Manager of the Millennium Challenge Compact (MCC)-funded agricultural project in the Northern Ghana. He also manages the Alliance for a Green Revolution (AGRA)-funded "Linking Farmers to Markets" project. His research and development interest includes soil fertility improvement, agro-dealer associations building, farmers' organizations capacity building, market development and ICT-based market information, and decision support systems. Dr. Debrah was previously Chief of Party of the USAID-West Africa funded Market Information Systems for Traders Organizations in West Africa (MISTOWA) which made innovative use of ICT to link agricultural traders in the West African sub-region. In 2006, Dr Debrah was awarded the maiden IFDC Board Chairman's Outstanding Professional Staff Award, the highest honor given to a professional staff for outstanding service and innovative contributions to international agriculture. He has published over 30 articles in refereed journals and workshop proceedings.

Abrahm Getachew is a Monitoring and Evaluation Officer, ILRI, IPMS Project. Abrahm has over ten years of professional experience and holds B.Sc in Agricultural Economics from Alemaya University (Ethiopia) and M.A in Regional and Local Development Studies (RLDS) and Development Evaluation and Management from Addis Ababa University (Ethiopia) and University of Antwerp (Belgium) respectively. He has worked with UNDP, Ethiopian Agricultural Research Organization and St. Mary's University College before joining ILRI.

Dirk Hoekstra is IPMS Project Manager for ILRI. He is an agricultural economist, with IR degree from Wageningen Agriculture University (The Netherlands). Dirk has worked for ICRAF and coordinated the East and Central African Highlands Agro-forestry Research Network and also managed a Dutch funded project in Ethiopia before joining ILRI.

Inuwa Ibrahim is a graduate of the Federal University of Technology in the School of Pure and Applied Sciences (SPAS) Department of Mathematics and Computer Science, P M B 2076, Yola Adamawa State, Nigeria. He is currently serving in ABTI American University Yola, Adamawa State, Nigeria.

Lazaros S. Iliadis is Associate Professor of Forest Informatics in the Department of Forestry and Management of the Environment in the Democritus University of Thrace. He has published 30 research papers in scientific Journals, 60 in Proceedings of International Conferences and he is author of one book "Intelligent Systems and Applications in Risk Estimation." He is Guest Editor in the journals: *Information Sciences, Neurocomputing, Neural Networks, Computers and Electronics in Agriculture,*

Engineering Intelligent Systems and member of the Editors' board of the *Universal Computer Science* Journal. He has been Program Chair in the following Conferences: EANN 2007, AIAI 2009, AIAI 2010, ICANN 2010, EANN/AIAI 2011 and Organizing Chair in HAICTA 2008. He has been Keynote Speaker in EANN 2009 London and Tutorial Presenter in AIAI 2010 Cyprus. He was a Visiting Professor at the University of London Birkbeck College, in University of East London, and in University of Western Makedonia Greece, in University of Makedonia, Greece. He is member of the Technical Committee 12, Working Group 12.5. of the IFIP, member of the board of ENNS (European Neural Network Society) and ex-president of the HAICTA.

Kamlesh Kumar (Senior Research Fellow, NFSM Project) awarded MCA from IGNOU, New Delhi and has one and half years professional experiences in computer application in e-pest surveillance based decision supporting system in pulses.

Okanlade Adesokan Lawal-Adebowale is a Lecturer in the Department of Agricultural Extension and Rural Development, Federal University of Agriculture, Abeokuta, Ogun State, Nigeria. He had his PhD in Agricultural Communication in June 2009, from the same University. His PhD thesis focused on ICT usage in agricultural research-extension-farmer linkage system in southwest Nigerian. Ever since then, the author's research work has been on Organisational Communication, Communication Media/ ICT usage in Agricultural and Extension Issues. Arising from his research activities, the author had 14 publications on ICT in Agriculture to his credit in both local and international journals, and Learned Society/ Association's proceedings and a chapter in book. He is a member of Learned Associations such as Agricultural Extension Society of Nigeria (AESON), Agricultural Society of Nigeria (ASN), Farm Management Association of Nigeria (FAMAN), the Nigerian Rural Sociological Association (NRSA), and African Society for Information and Communication Technology (ASICT).

N. Mathiyalagan is an Associate Professor at the Post Graduate and Research Department of Communication, PSG College of Arts and Science, Coimbatore. He has an academic experience of over 25 years and his specific area of interest includes e-governance implementation, developmental communication, new media activities and prime developmental projects undertaken by the government machinery. He has been guiding research programs leading to M.Phil and Ph.D in media studies for the past 25 years.

Neelam Mehta (Technical Officer, NCIPM, New Delhi) is awarded PGDCA and has 13 years of professional experience. She is actively involved in ICT based decision support system in various crops and implementation of IPM in various crops.

Julius Juma Okello is a Lecturer and the Coordinator of Agribusiness Management Program in the Department of Agricultural Economics, University of Nairobi, Kenya. He earned his PhD in Agricultural Economics from Michigan State University with specializations in international agricultural development, agricultural marketing and resource and environmental economics. His current research focuses on the role of ICT in linking smallholder farmers to high value domestic and international markets, food safety and risk, sustainable management of natural resources. His other research activities are on safety of peri-urban fresh vegetables, sustainable water management and sustainable management of community forests. His latest publications include "From circle of poison to circle of virtue: Pesticides, export standards and Kenya's green bean farmers," *Journal of Agricultural Economics (2010); "*Control of invasive Liriomyza leaf miner species and compliance with international food safety standards by small scale snow pea farmers in Kenya," *Crop Protection (2010); "*Using ICT to integrate smallholder farmers into agricultural value chains: The case of DrumNet project in Kenya" *International Journal of ICT Research and Development in Africa (2010).* Dr. Okello is currently coordinating an IDRC funded Africa-wide project on effectiveness of ICT on promoting smallholder farmer commercialization and market linkage. He has in the past worked for various international organizations including the World Bank, Food and Agriculture Organization (FAO) and the Oxfam (America), International Livestock Research Institute, and International Food Policy Research Institute as a consultant.

M. Olaniran Olawole is a Lecturer in the Department of Geography, Obafemi Awolowo University, Ile-Ife, Nigeria. He holds Master's degrees in Geographic Information Systems and Geography from University of Ibadan and Obafemi Awolowo University Ile-Ife, Nigeria, respectively. His research interest is in applied GIS and remote sensing, transportation, and regional development.

Bolarin Olufemi is a Lecturer in the Department of Agricultural Extension and Rural Development, Faculty of Agriculture, University Of Ilorin, Nigeria. His research interest is in application of ICT in Agricultural Extension.

Nathaniel David Oye received his M.Tech OR (Operations Research) degree from the Federal University of Technology Yola, Adamawa State Nigeria in 2002. He is a Lecturer in the department of Mathematics and Computer Science in the same University for the past 15yrs. At the moment he is a PhD student in the Department of Information Systems in the Faculty of Computer Science and Information Systems at the Univeristi Teknologi Malaysia.

Senthil Priya is a Research Scholar from the Department of Communication, PSG College of Arts and Science, Coimbatore, India and she has completed her MS (Communication) from Manipal Institute of Communication, Manipal and MPhil from PSG College for Arts and Science, Coimbatore. Her main area of interest is new media and communication. She has published two book chapters from her MPhil dissertation that analyzed the implementation level of two e-governance projects in revenue administration at district and taluk levels within Coimbatore District, Tamilnadu state. Her major research interests are e-governance implementation, new media and developmental communication. She is currently pursuing PhD research on implementation of E-Agriculture in Coimbatore taluks.

Ermias Sehai is a Knowledge Management Specialist, ILRI, IPMS project. Ermias has over 18 years of professional experience and holds B.Sc in Mathematics from Cumberland College (US) and M.Sc in Management Science (Specializing in Project Management) from the George Washington University (USA). He has worked with Ernst and Yong USA as senior consultant and with Price Water House Coopers as principal consultant before joining ILRI.

Ayanlade Oluwatyin 'Seun holds B.Sc. in Computer Science and Engineering and PGD in Geo-informatics. Her research interest is information and ICT management, database development, and management.

O. P. Sharma (Principal Scientist, Plant Pathology), NCIPM, New Delhi) has awarded his Ph.D. degree in Plant Pathology from Jiwaji University Gwalior, Madhya Pradesh, India. He joined Agricultural Research Service (ARS) in December, 1984 as Scientist (Plant Pathology) in Indian Council of Agricultural Research (ICAR), New Delhi. He has professional experiences of 28 yrs working in plant disease management, especially Integrated Pest Management (IPM), and he has published more than 100 publications. For the last two decades, he has been working for implementation of IPM in pulses, rice and cotton crops, with special emphasis on ICT based decision support system in collaboration with large number of farmers across the country. He is awardees' of ICAR Outstanding Team Research work.

Abebe Shiferaw is a Research & Development Officer, International Livestock Research Institute (ILRI), Improving productivity and Market Success (IPMS) of Ethiopian Farmers' project. Abebe is Ph.D student at University of Bern and hold B.Sc in soil and water conservation (Asmara University) and M.Sc degree in Soil Science (Zhejiang University, P.R. China). He has over 15 years of work experience. He worked for Asmara University as Graduate Assistant, Alemaya University as Assistant Lecturer and Mekelle University as Senior Lecturer. Prior to joining ILRI, he worked as a program development coordinator and Senior Agriculture Officer for World Vision Ethiopia.

Niranjan Singh (Scientist, Senior Scale, NCIPM, New Delhi) has been awarded post graduate degree in Statistics from Aligarh Muslim University, Aligarh, Uttar Pradesh, India. He started his professional career as scientist (Computer Application in Agriculture) in the year 1998 and got posted in National Centre for Integrated Pest Management, New Delhi. He has been working since 1998 in the field of development of decision support systems for pest management in the mentioned Institute and has published more than 20 publications in his credit. For the last three years, he has been involved in implementation of IT based pest surveillance and IPM in pulse crops in collaboration with various stakeholders such as state department, State Agricultural Universities and large number of farmers across the country.

Saurabh Singh is a Business Professional registered with AIMA; Professor in Domain of Business Administration at G. B. Pant University of Agriculture & Technology, Pantnagar, India. Many titles have been conferred on him, such as Accredited Management Teacher by AIMA, India; Professional Member by ISM, Arizona, USA; Fellow of the Society by ISM, India. He has earned his MBA with Certificate of Merit from Barkatullah University; Bhopal, India preceded by his B. Sc. from University of Allahabad, India. He started his career with Corporate Sector and stayed with it for six years before entering in world of academics, research, and consultancy in 2003 and still continues to work for the same cause. He has supervised the Master's Research/ Project/ Thesis work of more than fifty students. He has already authored six monographs published in Great Britain and USA and significant number of research papers in scholarly journals which achieved top ten slots in many reputed B-Schools internationally, besides having authored few popular articles. He has contributed towards development of new courses, study materials, and played key role in development and successful launch of one Bachelor and two Masters' Degree Programs in the University.

Petros A. Tsioras has acquired an MSc (equiv) Degree in Forestry and Natural Environment from the Aristotle University of Thessaloniki (1999) and a Doctorate from the same University on the subject "Analysis and formation of modern forest workers' training systems for Greek forestry" (2004). He is currently (Oct 2010 – Apr 2011) conducting postdoctoral research on ergonomics in forest operations at the University of Natural Resources and Life Sciences (BOKU – Vienna, Austria). Since 2000, he has been working in many national and international research programmes (e.g. DeSurvey - EU Contract number: IP-003950, PINDOSLIFE – NAT/GR/000291). He has been working as a Research Assistant at the Laboratory of Forest Utilization – Aristotle University of Thessaloniki since 2006. His main research interests are forest workers' training, ergonomics, and wood harvesting systems. Some of his papers have been published in peer-reviewed journals (e.g. Forest Ecology and Management, iforest), and he has also reviewed papers for other international journals.

S. Vennila (Principal Scientist, Agricultural Entomology, NCIPM, New Delhi) was awarded her Ph.D. degree from University of Agricultural Sciences, Bengaluru. She has started her professional career as Scientist (Agricultural Entomology) in Indian Council of Agricultural Research, New Delhi and she has professional experiences of more than 20 yrs in insect pest management specially Integrated Pest Management (IPM). Recipient of 14 awards during B.Sc.(Ag.), gold medals during M.Sc.(Ag.) and Ph.D, she has published more than 100 publications in her credit. She developed genetic, biological, forewarning, curative and educational tools of cotton IPM. She is Honoured Dr. Panjabrao Deshmukh Woman Agricultural Scientist Award. She is currently engaged as team leader for electronic networking and database management through surveillance and advisories for pest management and research into pest dynamics *vis a vis* climate change and pest forewarning.

Tizita A. Wasihun is currently a PhD student and Graduate Research Assistant at West Virginia University in the Department of Natural Resource and Environment Economics. She completed her Master's degree in Agricultural Resource Economics in the same University. Before coming to the United States, she earned her Bachelor's degree in Development and worked for almost four years in a research center and local and international non-government organizations on different community development projects in Ethiopia. Her research interest mainly is on agribusiness, regional and developmental economics.

Pádraig Wims has been Lecturer in Rural Development at University College Dublin, Ireland since 1987. His main academic interest is rural development, and his main research interests include the contribution of ICT to agricultural extension and rural development, the application of ICT to rural business and rural communities, and E-business in agribusinesses. Prior to joining University College Dublin, Dr. Wims worked in the Information Section of the Irish public agricultural extension service, on the development of a videotex service for Irish farmers.

Index

F

G

I

K

L

M

U

V

W

www.ingramcontent.com/pod-product-compliance
Ingram Content Group UK Ltd.
Pitfield, Milton Keynes, MK11 3LW, UK
UKHW050951290926
15470UKWH00005B/1752

9 781466 626553